RUNNING THE RAILS

RUNNING THE RAILS

Capital and Labor in the
Philadelphia Transit Industry

James Wolfinger

CORNELL UNIVERSITY PRESS   ITHACA AND LONDON

First published 2016 by Cornell University Press
Printed in the United States of America

Library of Congress Cataloging-in-Publication Data

Names: Wolfinger, James, author.
Title: Running the rails : capital and labor in the Philadelphia transit industry / James Wolfinger.
Description: Ithaca : Cornell University Press, 2016. | Includes bibliographical references and index.
Identifiers: LCCN 2015045186 | ISBN 9781501702402 (cloth : alk. paper)
Subjects: LCSH: Local transit—Pennsylvania—Philadelphia—History. | Industrial relations—Pennsylvania—Philadelphia—History.
Classification: LCC HE4491.P52 W65 2016 | DDC 331.88/11388420974811—dc23
LC record available at http://lccn.loc.gov/2015045186

Cornell University Press strives to use environmentally responsible suppliers and materials to the fullest extent possible in the publishing of its books. Such materials include vegetable-based, low-VOC inks and acid-free papers that are recycled, totally chlorine-free, or partly composed of nonwood fibers. For further information, visit our website at www.cornellpress.cornell.edu.

Cloth printing   10 9 8 7 6 5 4 3 2 1

For Amy and Elizabeth

Contents

Acknowledgments

It is a truism to say that a book, even one with a single author on the cover, could never have been written without the support of many. In my case, the gratitude is vast but the space is limited. So I begin with the many archivists and librarians who over the years offered marvelous support but also a caution: "This book," they repeatedly said, "cannot be written." The reasons were many: the sources had never been assembled, they had been destroyed, they were too diffuse, with pockets of primary materials scattered among many libraries and archives. This was, it turned out, a classic case of each archivist and librarian knowing one small piece, his or her piece, of the puzzle, but not knowing there was a bigger picture. I am happy, as this book makes its way into the marketplace of ideas, to prove them wrong, but I am even happier to publicly acknowledge that I could not have done it without them. My first and greatest debt in this category is to Matthew Lyons at the Historical Society of Pennsylvania, who granted me unparalleled access to the Harold E. Cox Papers even before they were cataloged. Staff at Temple University's Urban Archives and the Philadelphia City Archives were also a tremendous help, as were those at the Pennsylvania State Archives, the Hagley Museum and Library, and the Transport Workers Union Archives. The DePaul University Library helped me greatly with its holdings and interlibrary loan requests, and I could not have completed this project without the assistance of librarians at Temple University, the University of Pennsylvania, and the Transportation Library at Northwestern University.

Within the academy, audiences at many conferences and seminars gave me invaluable feedback. I am particularly indebted to Erik Gellman, who helped me expand my vision of this project at the Newberry Library's Labor History Seminar, and to David Witwer, who invited me at a critical stage to present my research at the Pennsylvania Labor History Workshop. I learned much from giving papers at these meetings as well as at conferences held by the Organization of American Historians, the Urban History Association, the Business History Association, the Pennsylvania History Association, the Labor and Working Class History Association, and the Chicago History Museum. Carlos Galviz and his associates at the University of London also took an early interest in my work, inviting me to present at the "Going Underground" conference that celebrated the 150th anniversary of the London mass transit system and by publishing chapters in two anthologies, *Going Underground* and *Undergrounds*.

Two organizations provided me with crucial support to complete this book, and I thank both heartily. The American Philosophical Society, one of Philadelphia's great scholarly institutions, funded a significant part of my research through a Franklin Research Grant; without its backing I could not have completed this book. DePaul University, my academic home for twelve years now, supported my work with university and college grants and also release time as a Fellow at the Humanities Center. Presenting to my faculty colleagues at the center as well as in a history department seminar enlivened my prose and sharpened my arguments. I thank all who participated in my talks for their critical engagement. DePaul also provided funding for two research assistants, Emily Busse and Elisa Caref. Thanks to both of them, especially Eli, who was a marvelous researcher and intelligent assistant whose ideas and politics challenged my thinking and ultimately helped shape this book.

I also must thank my editors and anonymous reviewers at Cornell University Press. The readers' reports were critical, detailed, and penetrating. They made this a significantly better book. My editor, Michael McGandy, has had a light touch with this manuscript, always supporting the project as a whole but pushing me to expand my arguments when the need arose. When I first worked with him on another project, I knew Cornell was where I wanted to place this book. My interaction with Michael has confirmed my high expectations. Bethany Wasik came to the press as my manuscript made its way through the editorial process. She has been diligent and attentive, and I appreciate her work to make this book the best it can be. It has truly been a large team that brought this book to fruition, and the fault for any problems that my academic audiences, my reviewers, and my editors missed lies with me.

Finally, I thank my family. Writing a book takes much time, energy, and mental space. Amy and Elizabeth, with lives of their own to lead, always understood that, or at least tried to. Knowing I had them and their support did not make the writing easier, but it did make the time away from the project a welcome respite.

RUNNING THE RAILS

FIGURE 1. Map of Philadelphia, created by Karen L. Wysocki for *Philadelphia Stories: A Photographic History, 1920–1960,* by Frederic M. Miller, Morris J. Vogel, and Allen F. Davis. Used by permission of Temple University Press. © 1988 by Temple University. All rights reserved.

CAPITAL AND THE SHIFTING NATURE OF SOCIAL CONTROL

O, ye traction magnates . . . will you still continue to fatten your dividends and your purses at the expense of human life, human suffering, and in spite of an aroused public opinion.

—*Catholic Times* of Philadelphia, 1895

When the *Catholic Times* posed its question—more a pointed challenge than a true query—to Philadelphia's traction magnates in 1895, it tapped into a pervasive feeling of discontent in the city. That discontent stemmed from Philadelphians' mixed, sometimes tortured, relationship with their transportation system. Boosters believed transportation integrated, and integrated smoothly into, a growing city, fostering commerce while helping develop space for housing and industry. As early as 1859 men such as Alexander Easton, writing in his *Practical Treatise on Street or Horse-Power Railways*, captured the view of transportation as a nearly unadulterated good with an idyllic description of the horsecars employed in the early days of the system: "There is no crowd, for the little cars glide along rapidly and frequently, accommodating every body; at a slight signal the bell rings, the horses stop, the passenger is comfortably seated, no rain drops in from the roof, the conductor is always ready to take the fare when offered, and the echo, 'great improvement, this,' is constantly repeated." Two generations later, after electric trolleys supplanted horsecars, the journalist Christopher Morley was equally charmed. Describing Chestnut Street, he wrote of "the light sliding swish of the trolley poles along the wire, accompanied by the deep rocking rumble of the car . . . the clear mellow clang of the trolley gongs, the musical trill of fast wagon wheels running along the trolley rails, and the rattle of hoofs on the cobbled strip between the metals." Such sounds gave Chestnut Street "a music of its own, [a] genial human symphony [that] could never be mistaken for that of any other highway." Observing the Chestnut trolleys from a balcony, Morley

1

experienced a sense of tranquillity born of the familiarity of streetcars in early twentieth-century Philadelphia.[1]

Yet at the same time, many Philadelphians came to understand the problems the transit system caused in their city. Horsecars and especially the electric-powered trolleys that replaced them brought danger and violence to children and heedless adults in Philadelphia's tightly packed working-class neighborhoods. Corruption ran rampant, leading citizens to claim that "Traction owns the town . . . corrupting the municipality and controlling legislation for their selfish purposes." And, most importantly, private ownership of this public service engendered worker exploitation and class conflict that periodically threw much of the city into turmoil. During the era of private ownership, major transit strikes shook Philadelphia in 1895, 1909, 1910, 1944, 1949, and 1963. The city witnessed many smaller strikes, authorized and wildcat, as well. In the history of Philadelphia's transportation system, key themes emerge across generations, including the competing visions of transportation; the violence it brought; the recurring charges and reality of corruption, venality, and graft; the problems posed by private ownership of a public service; and the incessant conflict between labor and management. These themes became common threads in this book for understanding the history of the system from the late nineteenth century to the post–World War II period. Yet, of them all, the central importance of labor relations, class relations, stood out clearly.[2]

This book explores capital's quest to control labor over nearly a century. Beginning in the 1880s and ending in the 1960s, it tells a tale of workers and management in Philadelphia's public transportation system. This system, I argue, offers an ideal venue for exploring the changing nature of workplace relations as the company called at various times the Philadelphia Rapid Transit Company (PRT), the Philadelphia Transportation Company (PTC), and finally the Southeastern Pennsylvania Transportation Authority (SEPTA) helped shape and also took advantage of broader social currents that set the contours of how management and its workers could negotiate with each other and at times engage in open, violent conflict.

To tell this story, I first and foremost offer a labor history of the men and a few women who toiled on the system. *Running the Rails* details the work they performed and how it changed over time. It examines the unions that these working people built and analyzes their accomplishments and failures. And it explores the ways transit workers negotiated, tangled with, and sometimes fought management. In some ways, then, this book at first appears to be an old-school labor history focused on the job, the workplace, and the union.

This is true, but only to an extent, for *Running the Rails* also draws on insights advanced by a recent generation of historians of capitalism, labor, and the working class who have explored myriad issues central to our understanding of late nineteenth- and twentieth-century United States history. In doing so, this book takes seriously recent calls by historians of the working class to investigate the history of capitalism in ways that "more fully [engage] the questions of class power and exploitation central to the older history of labor." In particular, *Running the Rails* takes up questions of how capital used its power to control workers' lives on the job, how the "labor question" shaped Americans' understanding of workers' and managers' prerogatives and constraints in the workplace, how the meaning of class and the nature of class relations changed from the late nineteenth to the mid-twentieth century, how the development of urban space and the growth of the city were outcomes of the contest between capital and labor in the transit industry, how the financial condition of the transportation industry shaped the system's development and the relationship between management and labor, how the politics surrounding the provision of public services shifted from one generation to the next, how technological advancement impacted workers and urban residents and they in turn reacted to it.[3]

This is a wide-ranging set of questions, and for a number of reasons Philadelphia's transportation system provides an exemplary laboratory for examining them. For one, the system consolidated in the late 1800s and remained in private hands for some eighty years. In fact, Philadelphia had one of the earliest public transportation systems in the United States and was the last large privately owned transportation system in the country, going public decades after those in New York City, Boston, and Chicago. Despite the size and long history of Philadelphia's system, unlike transit companies in comparable cities on the East Coast and in the Midwest it has been the subject of only limited sustained historical analysis. *Running the Rails* does not so much seek simply to fill this gap as to use the untold history of Philadelphia's transportation system to highlight the role it played in American history, especially the nation's labor history.[4]

The long span of time covered by this book helps to highlight the way labor relations shifted in different contexts set by economic, political, and social developments. Regardless of the period, there was a constant of the transit company agilely, if not always subtly, using the socially acceptable tools at its disposal to attempt to control its workforce. Early in the story, from 1880 to 1910, PRT management readily employed strikebreakers and raw violence to suppress worker organizing and rebuff demands for a safer, better-paying workplace. That violence became intolerable after the traumatic clashes that rocked the city in the general strike of 1910, and in the 1910s and 1920s the PRT developed nationally famous versions of welfare capitalism and company unionism that lasted until the Great Depression. When welfare capitalism became untenable in the financially challenging 1930s and the federal government declared company unions illegal with the 1935 National Labor Relations Act, the PRT again shifted its effort to control its workers, this time playing on and exacerbating racial biases in its white workforce to aid and abet a World War II–era "hate strike" that pitted some white workers against black and threatened to undermine the employees' recently elected strong union, the Transport Workers Union (TWU) of the Congress of Industrial Organizations (CIO). After World War II, as race relations at the transit company calmed, management shifted yet again to deploy two arguments: the first portrayed unions as the "other," which helped pit many Philadelphians against the TWU, and the second focused on the supposed imperatives of capitalism that emphasized maximizing profits over the social and economic needs of the city of Philadelphia, its transit workers and riders. Both allowed management to challenge a militant TWU for supremacy on the buses and subways and, just as important, in the court of public opinion.

Naked violence, welfare capitalism, race-baiting, smearing unions as outsiders, and ideas about the requisites of capitalism, all were tools to which management turned to control its workforce. Some were starker, more obvious. Some were wielded to greater effect. All were manifestations of and gave further shape to their times. *Running the Rails* thus develops an argument, often implicit, about the value of a local study that encompasses a long period of time for analyzing capital's evolving stance on labor relations. Such an approach highlights the methods of management as a tool kit, a set of strategies, that had to be wielded according to behaviors acceptable in the particular time and place. In some ways then, those strategies transcended any individual corporate official and instead reflected the historical moment. But in other ways, they matched the personalities and talents of different managers who often knew and recruited each other to the transit company. The company persisted, albeit with different names for eighty years, but focusing on "the company," this book demonstrates, can obscure the people who implemented the policies and were connected to each other across that span of time. They may have used different strategies, but all—and this point was seldom stated, often completely unspoken—sought to achieve a core goal, the control of the company's workers, which of course served a related purpose of enhancing corporate profits.

In addition to the transportation system's long life as a private company, the nature of the transit business also provides a revealing look at labor relations. Recent studies of how companies deal with their employees—Jefferson Cowie's *Capital Moves* in particular—emphasize capital's mobility, especially in a globalized world. Historically, when workers organized they demanded higher wages and better working conditions, and in response companies moved to the U.S. South and then to other countries. But a transit company, needless to say, cannot

move. The PTC could not threaten to pull up its tracks and head to South Carolina, Mexico, or China. Instead, management had to tap into other methods of
worker control. How this company that could not move or even threaten to do
so shifted its tactics over the decades is a central concern of this book because it
highlights the multifaceted nature of capital's efforts to control its workforce.[5]

The nature of the transit industry also helps sharpen this book's analysis of
class relations, because transportation work made the entire city of Philadelphia
the site of labor conflict and contestation. This was, to coin a phrase, the city as
shop floor. Philadelphia had one of the largest transit systems in the nation, and
its subway and elevated tracks, streetcar lines, and bus routes snaked their way
into every neighborhood. Strikes and other forms of class conflict at a steel plant
or a textile mill were localized events, generally reported in the press but not
played out in most people's communities. But with public transit, residents from
Kensington to South Philadelphia to Manayunk knew the workers personally and
used the perpetually overburdened system every day. To working-class Philadelphians, especially in the pre–World War II period, the transit system offered an
obvious manifestation of class relations in industries and neighborhoods across
the city. The transit system and the class relations that played out upon it were
central to people's lives and common fodder for public debate.

By employing a labor history approach, this book deepens our knowledge
not only of the history of the American working class, but also of the public
transportation industry. Most studies of public transit in the United States have
focused on two areas: technological advances that made urban systems faster
and more efficient, or the impact public transportation had on the growth and
development of cities. These works have illuminated how transportation shaped
American cities, but, with the notable exception of the works of Josh Freeman
and Scott Molloy, these histories have not deeply analyzed the experience of the
workers who made the systems run or what those workers and their companies tell us about labor relations and the generations-spanning politics of public
transportation.[6]

Although this book chiefly focuses on work, workers, and labor's relationship
with management, it cannot tell this history of public transit without incorporat-
ing other topics more fully explored by previous historians, especially technology,
urban growth, and finances. This book does not primarily tell a general history of
the Philadelphia transportation system, but it does examine how all three of these
topics were connected to each other and helped fundamentally pattern the lives
of transit employees. Philadelphia's transit system was one of the earliest adopt-
ers of new transportation technology throughout the late nineteenth and twentieth centuries. In the early years, workers on the system operated stagecoaches
and omnibuses, then shifted to horsecars as transit companies laid down rails on

many of Philadelphia's streets. This was well-known technology, not all that different from the wagons and buggies that people had been using for centuries, and local transit entrepreneurs knew it was insufficient for moving tens of thousands of people a day in a metropolis the size of Philadelphia. They thus experimented with other motive technologies, such as cable and steam, in the late nineteenth century, but neither proved effective or efficient. It was the shift to electric power in the 1890s that expanded the reach of the transit system, as faster trolleys with the ability to carry far larger passenger loads replaced horsecars. A decade later, with demand for transportation growing steadily, the PRT built one of the first subways in North America, the Market Street line. The Broad Street line came two decades after that, and other smaller lines and spurs were constructed over the years into the post–World War II period. As Philadelphia's population grew and public transit remained in high demand in the first half of the twentieth century, the PRT experimented with other modes of transportation: motor buses, taxis, interurban routes, and even the city's first air service. Only the first of these became a staple of public transportation in Philadelphia as the system sped its shift from trolleys to buses after World War II, a pattern followed in most American cities. By the 1960s, Philadelphia's transit company had settled most of the questions about technology that had marked the first decades of the industry as a period of growth and experimentation.

This technological development was intimately tied to Philadelphia's geographic expansion. Horse-powered vehicles generally traveled at four to six miles per hour, which limited most people to living about two miles from Center City or where they worked. Electric-powered trolleys essentially doubled that speed, which meant people and goods could flow through the city more quickly and efficiently and Philadelphians could live twice as far from their place of work. The Market Street and Broad Street subways multiplied this effect, as the underground trains did not have to obey the same speed limits set for trolleys or deal with congestion caused by pedestrians and other vehicles. The two main subway lines especially brought land to the north and west of City Hall into the marketplace for developers to create neighborhoods and industrial centers. Trolley lines connected Philadelphians in these new communities to the subways or brought them directly to Center City, and the land between the Delaware and Schuylkill Rivers (and beyond) filled in within three generations. Without the transit system, or alternatively, as this book suggestively points out, if it had been built into a more robust network, the city of Philadelphia would have looked far different.

Building transportation lines is expensive, however, and finances were always key in the history of Philadelphia's system. Because privately owned companies provided Philadelphia's public transportation during the period covered by *Running the Rails*, two major conflicts emerged in the nineteenth century and recurred in different forms until the 1960s. The first concerned the vexed connection between transportation as a public necessity and a private commodity. Philadelphia's political and business leaders, as Sam Bass Warner Jr. pointed out in his classic book *The Private City*, had a long, deep, and problematic commitment to privately held city services and privately controlled urban development. To these leaders, it was beyond dispute that transportation should be run as a private industry. The profit motive, they argued, would ensure the best possible service. Yet to many Philadelphians, especially workers and Socialists (not always the same people) and, later, liberal politicians, particularly after World War II, transit was obviously a social good and one that should not generate profits just for a few financiers, company managers, and investors. After all, the transportation company used public streets, helped the city grow geographically, and increased financial activity for the benefit of everyone. The transit system thus provided specific content for a larger, decades-long debate about the nature of social services and whether individuals or the broader society should "own" such systems.[7]

The second conflict was driven by class tensions, as management's pursuit of profits in a privately held industry repeatedly led to class conflict on the system. With managers doggedly seeking to keep wages low and hours long so they could maximize profits, Philadelphia's transit system became a leading example of how workers and employers grappled with the "labor question"—who will work for whom and under what conditions—over eight decades. Historians have long understood that the labor question stood at the center of social conflict in the late nineteenth century. As Rosanne Currarino has recently argued, the labor question was fundamentally about the survival of democracy, not just political rights but economic opportunity and some semblance of security, in America. But to Currarino and other scholars, that question faded in the early 1900s as the central issue became full participation in social life rather than labor and the fruits of one's toil. *Running the Rails*, by taking a long view of the transit industry in Philadelphia, suggests that the "labor question" is not an artifact of the late nineteenth century, but instead remained central to workers' lives and urban politics, albeit with different syntax in different periods, through the 1960s when this study ends. Regardless of the technology they utilized to move people around the city, transit workers on horsecars in the 1880s, trolleys in the 1920s, and buses in the 1950s confronted notably similar issues in the workplace and in their relations with management. The right to unionize, the expression of grievances, the demand for better wages, the campaign for better hours and working conditions, and many other struggles played out in similar ways for eighty years. To be sure, transit workers' campaigns for a more satisfactory answer to the labor question were not timeless, ahistorical battles. They were always shaped by the broader context of the particular era, which played a crucial role in what management and workers could do and what they could not. But regardless of the class violence, the racism, the demonizing of organized labor that American society found acceptable, this book demonstrates how the labor question remained central to workers' lives.[8]

Finally, by exploring a significant component of the city's working class and the transit system that they ran, *Running the Rails* deepens our understanding of Philadelphia's history. For many years, Philadelphia was one of the most understudied U.S. cities, especially its twentieth-century history. A far greater literature explored the city's experience in the colonial, Revolutionary, and early American periods, but for historians time and the country seemed to move on from Philadelphia by the Progressive Era. Some of that undoubtedly had to do with the growth of New York City as a center for finance and immigration, the expansion of the federal government in Washington, D.C., the emergence and subsequent decline of the nation's great mass-manufacturing cities, and the movement south and west of the country's population. Yet Philadelphia, despite its reputation for stodginess and the decline of its economy, persisted. And a new generation of historians, notably led by Matthew Countryman and Guian McKee, have rediscovered the city and produced fine studies, particularly focusing on black and working-class Philadelphians. This book adds to the growing discussion about Philadelphia's central place in twentieth-century American history, especially its labor history. Events on Philadelphia's transportation system, as *Running the Rails* demonstrates, often highlighted, and sometimes led, trends shaping the larger course of American history.[9]

In the end, *Running the Rails* historicizes the work and workplace relations of public service workers who labored in transit. Written in the second decade of the twenty-first century, at a time when these workers and their unions come under frequent attack for being a "special interest" or a hindrance to the smooth functioning of society, this book offers readers a different, historically grounded, way of thinking about these workers' experience. Working in public transit is a difficult job now, as it was a century ago. The benefits and decent wages these workers secured came as a result of fighting for decades against their exploitation. Their advances were hard won and well deserved, and readers of this book will, I hope, gain a deeper appreciation for the struggle transit workers have waged and the victories, albeit often limited, they have achieved.

To make the case about how work and class relations on Philadelphia's transit system developed over time and in the process illuminated broader issues in American history, *Running the Rails* follows a chronological format. Chapter 1 sets the context for the book by examining Philadelphia's pre-twentieth-century urban development and the need for adequate public transportation as the city's population and commerce grew. Starting in the mid-nineteenth century, entrepreneurs experimented with different transportation technologies—stagecoaches, omnibuses, horsecars, and finally trolleys—in an effort to meet the city's needs and make a profit in the transportation marketplace. Over time, certain entrepreneurs prospered, Peter Widener and his associates in particular, and they used their political ties and financial power to consolidate the city's transportation system. In doing so, they fueled Philadelphia's growth but also engendered animosity from working-class communities and their elected officials who found private control of a vital public service to be anathema. Modes of transportation and methods of financial control implemented in the late nineteenth century set a pattern of system development for decades to come. They also sowed the seeds of conflict about that development over the same decades.

Chapter 2 focuses on Philadelphia's transit workers: their daily work, their relations with the system's owners, their early efforts at unionization. Transit workers faced harsh conditions on the job and for their efforts received low pay and little respect. To improve their lot, they turned to organized labor, first with the Knights of Labor and then the Amalgamated Association of Street and Electric Railway Employees—"the Amalgamated," a member of the American Federation of Labor (AFL). Doing so put them at considerable risk because of the tense, often violent, nature of labor relations at the time. Overall, Philadelphia's transit workers found strong support among working-class residents of the city, especially during their strike in 1895, but they lived in a difficult era marked by widespread class conflict, state repression, and organized corporate power embodied most conspicuously by the National Association of Manufacturers (NAM). The labor question was front and center for the city's transportation workers, but despite their efforts it was clear by the turn of the century that they had little power to obtain the more favorable pay and working conditions that they sought.

Chapter 3 situates transit workers' experience and the great strikes of 1909 and 1910 in broader social currents of the turn-of-the-century period. The construction of the Market Street Subway nearly bankrupted the PRT and led the company and city to sign the 1907 contract that gave the city increased power over its transit system. This contract serves as a lens for this chapter's exploration of the debates over the provision of public services by private enterprise, including cries of or claims for socialism. Chapter 3 also explores debates about the "open shop" drive that employers across the nation, including at Philadelphia's transit company, used to try to keep unions at bay. In organizing Philadelphia's transit workers, the Amalgamated countered the transit company's staunch antiunionism with language focused on workers' rights, equality, and fairness. Philadelphia's transit workers found broad support in the city's working-class communities when they struck in 1909 for more pay, better conditions, and the right to unionize. That strike resulted in a limited victory for the transit workers and set the stage for a far greater conflict a year later. The 1910 clash on the transit lines culminated in a general strike of some 140,000 people that left twenty-nine dead and a city in turmoil. Conflicts in Philadelphia's transit industry showed the violence brewing in Progressive Era labor relations and convinced city leaders that they could not let that conflict loose on the city again.

Chapter 4 details the age of Thomas Mitten, one of the nation's most famous transportation company managers and a key figure in the development of company unions and welfare capitalism. This chapter primarily focuses on how Mitten tapped into broader intellectual currents of the 1910s and 1920s, when corporations more frequently sought subtler means of worker control, to develop the Mitten Plan. A combination of a company union, welfare capitalism, and an employee stock purchase program, the Mitten Plan, coming in the wake of the 1910 general strike, at first captured the loyalty of much of the workforce and the imagination of pundits, academics, and government officials. Some thought Mitten had finally answered the labor question, had solved the crisis of the age. By the late 1920s, however, critics began to question whether the Mitten Plan really helped workers or merely used different means to keep them under the control of management. Nonetheless, Mitten's plan held sway for two decades, foundering only when the Depression made welfare capitalism too expensive to the capitalists, the value of PRT stock cratered, and some workers began to demand more independent representation. In addition to his nationally famous attempt to solve the labor question, Mitten demonstrated entrepreneurial skills that expanded the PRT from a trolley and subway concern into a full-service transportation company that used buses, interurban lines, and even airplanes. His vision brought new technologies to Philadelphia's transit system and made the PRT as prosperous as it had ever been. Mitten's commitment to private enterprise, however, restricted the system's growth, as he devoted scarce capital to stockholder dividends and limited the extension of subways and elevateds. His tenure, as prosperous as it generally was, raised troubling questions for city officials about whether a privately held company could do what was best for the entire city.

Chapter 5 starts with an examination of the hard times of the 1930s that highlights the powerful impact finances had on labor relations in the transit industry. Like transportation companies across the country, the PRT faced an annual drumbeat of falling ridership, declining income, and darkening prospects. The company's employees lost jobs and took pay cuts for several years. Their dissatisfaction with the PRT, coupled with the gathering strength of the CIO, led them to abandon their company union and organize in the TWU. Management, pressed to the wall by its financial situation, knew how much the TWU would cost them at the bargaining table, and turned to racist techniques that were sharpened by racial animosity within the workforce fueled, although not started, by World War II–era demographic changes. The transit company had for decades run a racist shop that limited African American workers to menial jobs, but a combination of large-scale black in-migration, a demand for jobs often linked to wartime rhetoric of equality and democracy, and begrudging federal support created a moment where black workers would not be denied. They demanded equal access to driving jobs, and the federal Fair Employment Practices Committee (FEPC) backed their claims. PTC management (the company changed names when it emerged from bankruptcy in 1940), sensing an opportunity to break the TWU on the shoals of racial animosity, played on white racist sentiment to spur one of the largest hate strikes in American history, when many white workers walked off the job just weeks after D-day rather than accept black drivers. Those white workers, and the company behind them, ultimately lost the strike, but it highlighted a racial divide that management attempted to use in its quest to control its workforce.

Chapter 6 explores workers' experience at the PTC in the context of a larger postwar conflict between capital and labor. After World War II, the United States witnessed a wave of strikes that challenged capital's prerogatives, including the right to run an open shop, the right to deploy financial and personnel resources as it saw fit, and the right to make planning decisions without workers' input. At the PTC, the TWU pushed management to increase pay and improve working conditions, but also challenged the notion that management should have the sole right to hire and fire workers, determine routes, or even set fares. Some of these conflicts revolved around another change in transportation technology, as the PTC sought to replace a number of trolleys with buses and in the process eliminate jobs by instituting one-man operation. Workers' challenges to management's prerogatives pitted the two classes against each other in ways as powerful but not as violent as two generations earlier and led to a series of strikes. The battles played out in the press as much as on the transit lines, with management painting the workers as greedy, overly powerful union members, and, in a hint at the Cold War–era language of anticommunism, outsiders bent on, or at least indifferent to, destroying the PTC and the city of Philadelphia. Workers responded that they had a right to organize, should receive pay that kept pace with inflation, and deserved to work a forty-hour week like most Americans. More, they laid claim to defending the interest of the public. They were the ones who did the work to provide a necessary service and deserved fair compensation—income that came back to working-class communities, rather than going into the deep pockets of management and large shareholders. Labor conflict on the PTC in this period became a window into larger arguments about social services and citizens' "rights," especially as hundreds of thousands of Philadelphians suburbanized and the PTC's finances weakened.

Chapter 7 examines the policies implemented by National City Lines (NCL), the debates about the PTC going public, and the process that created today's SEPTA. NCL, which gained notoriety for its supposed role in a conspiracy to take over the major transit systems of the United States and convert trolley routes to diesel bus lines, thus benefiting Firestone, Standard Oil, and General Motors, figures prominently in this story, but not because of any conspiracy. NCL began buying stock in the PTC in the mid-1950s and took control of the company in 1955. NCL did convert many remaining trolley lines to buses, but for the purposes of this book, the company matters because of the way it employed its notion about the imperatives of capitalism to browbeat the TWU, repeatedly cut the workforce, defer maintenance, and eliminate transit routes, all in an effort to improve company finances, which raised stock prices and increased dividends. NCL's policies created much consternation for PTC employees, Philadelphia residents, and Democratic political leaders, such that by 1963, liberal politicians and their supporters realized that private control of this vital public service was no longer tolerable. That year they set in motion plans to purchase PTC and transform it into a publicly run regional system, SEPTA. NCL drove a hard bargain and dragged out the negotiations for five years. By the time Philadelphia took control of PTC, deferred maintenance and layoffs had gutted the system. In today's world, where political conservatives routinely claim government is incapable of providing services, this chapter highlights how private interests extracted what value they could from an industry and then turned it over to public management after it had been stripped of its parts and was no longer profitable. Although *Running the Rails* does not pursue this history into the era of public ownership at length, an epilogue makes clear that SEPTA's problems were not a function of bad management or inept politicians so much as a reflection of larger economic and demographic trends, as well as the way capital's imperative to maximize profits often came at the expense of the public, transit workers, and the transportation system itself.

In tracing this history of capital's quest to control its workforce over four generations, this book ultimately shows the lasting and multifaceted nature of capitalist power. Raw violence was useful and condoned at times. So too were welfare capitalism, company unions, racist arguments, and the demonization of organized labor. The times changed, the strategies changed, and so did the people. But in the end, capital's quest to control its workforce and run the rails endured.

BEGINNINGS

"Transit problems," wrote the boosters of Frankford, a Philadelphia neighborhood incorporated into the city in 1854, "are indigenous to the [city's] northeast." Horses strained on the hilly terrain before the advent of steam and electric locomotion. Hard winters and frequent rains turned dirt roads into mud holes that jostled passengers and broke axles. And streams and rivers carved up the countryside, making quick passage impossible. William Penn himself wrote to the Provincial Council in 1700 asking when they were going to "build a bridge over the Pennypack and the Poquessing, so that he could come to town in comfort from his up-river home," Pennsbury Manor.[1]

One did not have to be William Penn to want improved transportation. Philadelphia and its environs had little more than "ancient dirt roads" such as Darby Road and Old York Road four decades into the eighteenth century, and city "streets" were little better. Travelers complained that they were little more than hard, rutted paths in dry times and muddy bogs when wet. A grand jury in 1738 informed the public that Philadelphia's "streets were impassable," which sparked a push to pave thoroughfares such as Front Street, High Street (now Market Street), and Sassafras Street (Race Street). Turnpikes built in the 1790s and 1800s alleviated some of the problem, and the Lancaster Pike was widely lauded as the United States's first paved road. But roads such as the Lancaster and the Cheltenham & Willow Grove Turnpike only connected Philadelphia to outlying towns and did little to improve transit within the city. Ferries too, helped move Philadelphia's residents, visitors, and their goods, carrying the first passengers across the Delaware River in 1688 at the order of the County Court of Gloucester, New Jersey, and operating continuously until the Delaware River Bridge (now the Benjamin Franklin Bridge) finally rendered the service unprofitable in 1952.[2]

Despite these sporadic advances, it was clear by the early 1800s that the city needed a better system to facilitate the movement of people and goods. William Penn may have found his journey uncomfortable, but for most residents, the lack of transportation created graver problems. The city, laid out along the west bank of the Delaware River, was putatively two square miles, but in reality most Philadelphians lived and conducted business in a narrow strip of land along the river. As late as 1860, there were few businesses west of Sixth Street, and farms occupied most of the surrounding land. For the city to expand and the economy to grow, Philadelphians knew they had to develop a better transportation system that would carry them west toward the Schuylkill River and farther north along the banks of the Delaware.[3]

To do so, they had to grapple with two key issues. Technology was the first. Philadelphia, like most large cities in North America and Western Europe, started with stagecoaches and omnibuses in the mid-nineteenth century, then experimented with steam locomotion and cable cars, and finally shifted to electric power at the turn of the century. Decisions about technology, especially the adoption of electricity, set the course for transportation in Philadelphia until the rise of the motor bus twenty-five years later, and the bus was not fully adopted for another twenty-five years after that. A second issue was the financial structure and control of the system as it developed. Once technology advanced beyond stagecoaches and omnibuses, laying rails and obtaining equipment required large outlays of capital and benefited from strong political connections. Entrepreneurs would not finance the system unless they knew they would turn a profit, but as soon as they claimed public streets for private purposes, a generations-long debate, albeit at times sporadic, began about who would own, and thus control and profit from, a public service. The development of transportation technology and growth of financial power and manipulation at the transportation company engendered sharp criticisms about the transit system's impact on the safety of Philadelphia's streets and the rights of the people versus the financiers. Decisions made, policies set, and contestations begun in the late nineteenth century helped determine the contours of Philadelphia's transit industry for generations to come.[4]

Technology

Philadelphia's public transportation began with a newspaper advertisement in December 1831 that represented the first privately owned entrepreneurial effort to provide routine transit for the city. "James Boxall," the ad read, "having been requested by several gentlemen to run an hourly stagecoach for the accommodation of the inhabitants of Chestnut Street, to and from the lower part of the city, begs to inform the citizens generally that he has provided a superior new coach, harness and good horses for that purpose. Comfort, warmth and neatness have in every respect been carefully studied." Boxall's Accommodation, as it was known, served the most built-up part of the city in the early nineteenth century, running along Chestnut Street from what is now Sixteenth Street to the Merchant's Coffee House at Second Street. All told, the ride took half an hour each way, and Boxall was able to turn a profit at ten cents a ride.[5]

The size of Philadelphia's population in the early nineteenth century and the city's steady, if uneven, geographic growth ensured a transit company had the opportunity to succeed. The city proper—Penn's original two square miles—held approximately eighty-one thousand people and the outlying districts some ninety thousand more. "Getting there" was quite the challenge, whether by horse, wagon, or foot, and Boxall showed there were profits to be made from a substantial number of Philadelphians willing to pay ten cents a trip. Other stagecoach lines soon followed, especially running north along the Delaware River to Frankford. The coaches were not terribly comfortable, but they provided, in the words of one commentator, "satisfactory transportation facilities to the rapidly growing city."[6]

As is the case with almost any new market, other entrepreneurs saw the possibility of profits and moved in. Omnibuses—essentially large stagecoaches—represented the next wave of transit and replaced most of the stagecoaches that were better suited to long trips rather than local start-and-stop traffic. The first omnibus line ran from the Navy Yard in South Philadelphia to Kensington along Second and Beach Streets. Within a few years, omnibuses dominated Philadelphia transit, radiating out from the Merchant's Exchange on Dock Street to carry passengers on nearly every major city street. By 1848 eighteen transportation lines employed over six hundred people to run 138 omnibuses. The buses were known for their "handsome teams of from four to six horses . . . [and interiors] luxuriously upholstered, with velvet cushions." The omnibuses may have been luxurious, but the cushions could not smooth out the ride: the "ponderous vehicles," wrote Frederic Speirs in his 1897 analysis of Philadelphia's street railway system, were constantly "rattling over the cobblestones and floundering through the mud-holes of Philadelphia's main thoroughfares." The public, he continued, paid handsomely for "the privilege of jolting over the cobbles." Omnibuses, problematic as they were, provided the great bulk of Philadelphia's public transit for some twenty-five years.[7]

To improve the experience and attract more customers, transportation entrepreneurs knew they had to smooth and thus speed the ride. For people in the nineteenth century, that meant one answer: rails. The means of propulsion was not necessarily a settled matter, although horses—"living machines," as the historians Clay McShane and Joel Tarr have called them—were the most obvious power source. Horse-drawn railcars got their start in New York City with the New York and Harlem Company in 1832, and New Orleans, Chicago, Boston, and Philadelphia followed over the next twenty years. Horsecars with their metal wheels on metal rails were more efficient than omnibuses or stagecoaches, but they still traveled only four to six miles per hour. Nonetheless, they were faster than walking and helped cities grow to an average of two miles radius from their business districts. Philadelphia took part in this national movement when the Frankford and Southwark Philadelphia City Passenger Railroad Company established the city's first line in 1858 on Fifth and Sixth Streets, running from the northeast down to Southwark. The line was an immediate success, and its fifteen cars carried ten thousand riders on the first day. Other companies scrambled to secure routes, and by the end of the year seven passenger railways held contracts covering some fifty miles and transporting forty-six thousand passengers daily. Tremendous growth continued: by the end of the Civil War, twenty companies operated lines over 129 miles of track and transported forty-five million people per year; by the 1890s the city had fifty-three street railway routes that traversed 436 miles and carried 224 million riders annually. These lines, with charters granted on an individual, ad hoc basis, ran mostly on north–south streets that mirrored the geographic distribution of the population and put in place a haphazard collection of routes rather than a coordinated system that purposefully served the public's needs. As the city's population and geographic size grew and transportation became increasingly important, the vestiges of these early routes made it difficult to rationalize Philadelphia's transit system.[8]

Whatever the system's inefficiencies, government officials and business leaders wasted no time trumpeting the burgeoning network. To them, street railways promised to reverse the city's decades-long decline that saw it falling ever further behind New York City as a vital center of trade and industry. Mayor Alexander Henry told the city council in 1859 that in his view "no public improvement . . . has ever promised more general benefit to the community." Strickland Kneass, the city's chief engineer and surveyor, agreed. A British delegation reporting to Parliament in 1866 asked officials from several North American cities what they thought of their street railways, to help determine if London should consider developing similar systems. Kneass wrote that Philadelphia's lines were a vast improvement over omnibuses and helped greatly with the flow of traffic. "All, rich and poor, enjoy the great convenience," he reported. "Our ladies attend to their shopping and visiting by the cars; the merchant and professional man reach their place of business in comfort without fatigue; while the mechanic and labourer, residing at great distances from their work in the suburbs, enjoy the advantages of low rents and free air for their families. The capitalist has felt the advantage of the Railroad routes in having property . . . brought into the market at remunerative prices. . . . I will add, in conclusion, that they are here considered a great public convenience, and a most agreeable improvement." Alexander Easton, in his *Practical Treatise on Street or Horse-Power Railways*, was downright frothy, calling street railways the "improvement of the age." They were so successful that they "excite[d] the surprise of their most sanguine projectors, and the admiration of the community at large." The rails were so smooth, the horses so obedient, that the cars left "the streets nearly as noiseless as when covered with snow."[9]

These were boosters and they of course overstated the case. Horsecars improved upon omnibuses, to be sure, but they had many drawbacks. The riding experience was far from refined. "The seats were hard and uncomfortable,"

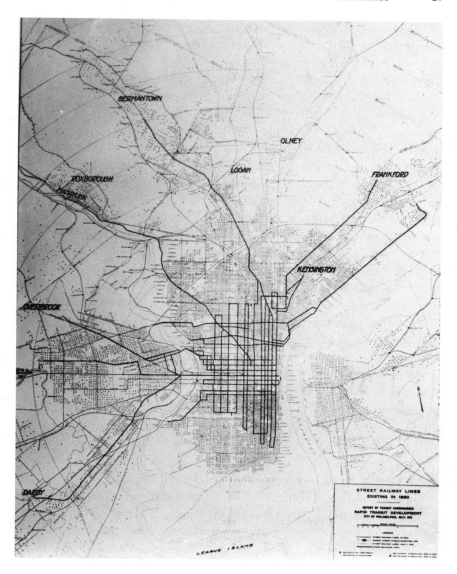

FIGURE 2.   Street railway lines existing ca. 1880. Philadelphia's nineteenth-century horsecar lines focused on moving residents in the most built-up areas rather than developing a planned, coordinated system. Photo courtesy of PhillyHistory.org, a project of the Philadelphia Department of Records.

reporter Edmund Stirling noted in the *Evening Ledger*. "Cars were lighted at night by coal-oil lamps which did little more than to make darkness visible. And, worst of all, the floors in winter were filled knee-deep with salt marsh straw, which with the accumulations of filth . . . became foul and offensive beyond description."

Horsecars could weigh 1.5 tons without passengers, and that, understandably, slowed travel to the rate of a brisk walk and wore out horses quickly. On busier routes with more passengers and starts and stops, companies often had to use extra teams, which reduced traffic speed and added expense. The historian Charles Cheape found that horses and their care consumed 40 percent of capital investment and 73 percent of gross revenues for transit companies. Overall, horses lasted only an average of four years before they had to be replaced, and by that point locals often referred to them as "walking boneyards." Horses also produced copious amounts of waste: ten pounds of manure a day and enough urine to soak the streets and make them so slick that the animals had a hard time gaining their footing. The waste offended urban residents' senses—especially those who lived near stables where transit companies stored manure for sale—but the fact that the feces contained tetanus, a fatal disease at the time, made matters far worse. So too did epizootic aphthae (foot-and-mouth disease) and equine influenza that periodically decimated horse populations, particularly in 1872. All this meant that Philadelphia's transportation companies and riders knew that horses did not provide the motive power necessary to meet the city's long-term needs.[10]

Area transit companies in the second half of the nineteenth century thus experimented with other types of power to move people along the rails. Steam locomotives provided one possible alternative, and Philadelphia was an early center of railroading. Starting in the 1830s, railroad companies laid tracks on Broad Street and along Market from the Schuylkill River to Third Street, where they then turned toward Dock Street. Lines built by the Pennsylvania Railroad, Reading Railroad, and other companies soon encircled and cut through most sections of the city. But steam railroads were not designed for the starts and stops necessary for intra-urban transport, and the dirty and dangerous locomotives sparked popular resistance in many neighborhoods. Residents of working-class Kensington rioted in 1840 against an attempt to construct tracks on their neighborhood streets, and posters warned of the railroad "Outrage" that endangered residents and robbed Philadelphians of their rights. "Mothers look out for your children!" read one poster. "Artisans, mechanics, citizens! When you leave your family in health must you be hurried home to mourn a DREADFUL CASUALTY! Philadelphians, your RIGHTS are being invaded regardless of your interests or the LIVES OF YOUR LITTLE ONES." In his study of the Great Strike of 1877, historian David Stowell demonstrated that such reactions were commonplace in the urban North—Buffalo, Albany, Syracuse, etc.—and fired one of America's great national uprisings. In Philadelphia, such conflicts as that in Kensington prefigured greater clashes to come, as transportation lines became more pervasive and electricity made transit vehicles bigger, heavier, faster, and more dangerous. In this context, Philadelphia's municipal government prohibited steam locomotives within city boundaries, forcing the railroads to disconnect engines from their trains and use horses to pull cars through the city.[11]

Although the city banned heavy locomotives, some companies, such as Frankford & Southwark, did try to use steam power in the form of "dummies." So-called because they were meant to look like a horsecar but were propelled by steam, the dummies ran along Kensington Avenue from Frankford to Philadelphia and on routes in the city's western suburbs from 1863 to 1893, when they were supplanted by electric trolleys. The dummies provided adequate service: for a ten-cent fare they made the trip from Frankford to Philadelphia in thirty minutes and seated up to thirty-two riders, a number that grew if the engine pulled a trailer. Steam showed enough potential that across the United States, cities built 524 miles of track for this mode of transit, but it ultimately had too many drawbacks. Steam engines were expensive to purchase and maintain. With trailers attached, they struggled to generate the power necessary to ascend even the gradual slopes of the Philadelphia area. And the great noise, smoke, and flying sparks that they produced frightened people and animals in congested neighborhoods. As one critic, T. C. Martin, wrote in his 1890 article "Social Side of the Electric Railway": "Steam motors are completely out of favor for use within the city limits. Their glorious record of half a century in long-distance travel does not deceive anyone traveling in a city as to the insuperable defects and nuisances of noise, smell, smoke, dust, steam escape, oil drippings, etc., which may more readily be tolerated, remotely in the open country." Given these problems, steam dummies never challenged horse power as a realistic solution to urban transportation needs (there were only ever eight or so vehicles in service in Philadelphia), yet some Philadelphians still wistfully recalled them decades after they had passed from the scene.[12]

Philadelphia, like many cities across the United States, also experimented with cable cars. All told, some thirty U.S. cities built cable car lines that hauled 400 million passengers on approximately three hundred miles of streets by 1894. Those numbers amounted to 5 percent of the total horsecar mileage and 20 percent of all horsecar passengers. The technology, which utilized a large stationary steam engine that drove a cable fashioned into a loop buried in a conduit under the street, had been under development since the early nineteenth century when it was used in mines and in scattered tunnels in London. Cable cars were more powerful than horses and particularly suited to hilly terrain such as San Francisco, where wire-cable manufacturer Andrew Halladie constructed the first modern system on Clay Street in 1873 to develop housing on Nob Hill. Within a decade, San Francisco had eleven miles of cable routes, and the technology had spread to Chicago (which developed the most extensive cable car system in the world), New York City, Philadelphia, and beyond. In Philadelphia, the Philadelphia Traction Company in 1883 installed the first cable car line on Columbia

Avenue from Twenty-Third Street to Fairmount Park, and the Union Passenger Railway followed two years later with a line that ran on Market Street from Forty-First Street and Haverford Avenue to the ferries at Front Street and Delaware Avenue. These proved successful enough that a third line opened along Seventh and Ninth Streets, connecting with Columbia Avenue. Cable cars had notable advantages: they were faster than most other forms of transportation at the time (twelve miles an hour compared to the horsecars' four to six miles an hour) and could haul many people on rugged terrain. But the technology proved expensive to build ($111,000 per mile, compared to $12,660 for horsecars in 1888), subject to breaking down, and prone to tie-ups, since cars could move only by having a gripman apply a grip to the cable; there was no way to move at variable speeds or pass slower cars. Worse, a cable break invariably shut down the entire line.[13]

By the late 1800s, then, anyone interested in urban transit knew horses were commonplace but problematic, and steam and cable represented technological advancement but had too many drawbacks to supplant the familiar animals. An inventor named Frank Sprague offered the way forward. Like so many inventors, Sprague was not the first to conceive of the new technology, but he improved it enough to make it usable on a mass scale. Since the 1830s, inventors and engineers in Europe and the United States, including Werner von Siemens and Thomas Edison, had been experimenting with electric-powered railways that used a generator and third rail or batteries to power motors onboard cars. Sprague, an engineer who graduated from the U.S. Naval Academy in 1878, worked with Edison at Menlo Park for a year in 1883 and then started his own company. The Sprague Electric Railway and Motor Company developed and sold a widely acclaimed electric motor that brought in enough funds to allow Sprague to focus on building a streetcar motor that would make electric-powered transportation practical and eventually ubiquitous. Sprague solved several problems that had bedeviled earlier inventors: mounting the motor so rough roads did not jostle it too violently; transferring energy from the motor to the axles; connecting the streetcar to its overhead power source via a trolley pole; and, most important, developing the multiple-unit control that let several cars, each with its own motor, operate in a single train. Sprague's innovations caught the eye of the Union Passenger Railway Company in Richmond, Virginia, where he signed a contract in 1887 to build twelve miles of track and operate thirty cars at a time. Richmond's hilly terrain—the track's grade was 8–10 percent in some places—challenged Sprague's new trolleys, but they made the circuit for the first time in February 1888, and word of Sprague's success spread soon after. That June, Henry Whitney, president of the West End Street Railway in Boston, came to Richmond to observe the electric trolley system and determine whether his company would switch to cable or electric power. Sprague knew he had one chance to convince the skeptical

Whitney, and one night he placed twenty-two cars in a row, started them all, and sent them down the line in quick succession. Impressed by the demonstration, Whitney returned to Boston and converted his company from horse to electric power. The first large city in the United States had adopted electricity and trolleys as the means for providing public transportation.[14]

The technology, touted as a cleaner, faster, more efficient means of transportation, spread quickly across the United States. Before Sprague, the United States had approximately ten electric railway companies using all manner of motors and power sources operating on a total of sixty miles of track. After Sprague's Richmond success, Chicago, Philadelphia, and other larger cities adopted electricity within a decade. New York City lagged many American cities on this modernization program, not retiring its last horsecar line until 1917. One 1895 study showed that by that date, already three hundred of the country's five hundred cities with population between five thousand and fifteen thousand people, and eighty-one of ninety-one cities with fifteen thousand to twenty-five thousand inhabitants, had electric railway service. By the turn of the century, track mileage had surged from eight thousand miles in 1890 to more than twenty-two thousand miles, and electric streetcars operated on 98 percent of that track. And the number of riders more than doubled, from 2.02 billion in 1890 to 4.77 billion in 1902.[15]

Philadelphia's experience mirrored and bolstered these national trends. The Philadelphia Traction Company began the electrification process with its Catharine and Bainbridge Street route in December 1892. The rest of the city's lines followed in short order, and they were all electrified by 1895, save for four outlying lines that made the switch over the next five years. Philadelphia had 400 miles of street railway tracks in 1896, making it the fifth-largest system in the country behind Chicago (659 miles), Boston (550), New York City (427), and Brooklyn (405). That mileage grew to 545 miles in 1904 as better transportation helped the city expand and the city's expansion in turn fired the need for a larger transit network. By that year, Philadelphia had one of the most extensive electrified railway systems in the United States. Ridership jumped too, growing from approximately 110 million people in 1890 to almost 300 million in 1900. At that point, most of the outlines of a transit system (with the notable exception of the subways) that shaped Philadelphia until World War II—its routes, its technology, its ridership—had emerged.[16]

Financial Control

Although Philadelphia's transit system began with competition among a welter of horsecar companies, the economic logic of coordination and then consolidation soon took hold. The consolidation of the industry of course did not take place in a vacuum; politics played a critical role. Legislation shaped how the system could develop and the profits investors could make. Philadelphia's residents and their elected representatives made demands on what transit companies owed them and their communities in exchange for operating on city streets. These claims at times extended into broader arguments about who should own a public service, whom the profits should benefit. And in a city that Lincoln Steffens famously tagged as "corrupt and contented," stock manipulations and backroom political deals played a significant part as well. The politics of public transportation—distribution of resources, financial control, legislation—started at the inception of the industry and grew along with it.[17]

Although Philadelphia's earliest horsecar industry could best be described as what the historian J. H. M. Andrews called "passenger railway insanity," that condition faded quickly as railway entrepreneurs recognized that unfettered capitalism created too much competition and doubt for their taste. Between 1857 and 1874 thirty-nine passenger railway companies received charters to operate on Philadelphia's streets, with the Pennsylvania General Assembly defending this practice by arguing, "The interests of the public demand that no corporation should have the monopoly of carrying passengers over the streets of a city between points which require the advantages of competition." The enthusiasm for street railway development paused only in 1874 because a new state constitution took away the General Assembly's right to issue special charters, and a general railway incorporation law was not enacted until 1889. Despite the separate charters, ten street railway companies in 1859 (the year after the first horsecar made its way through Philadelphia) came together to form the Board of Presidents of City Passenger Railway Companies under the leadership of James Verree, president of the Second and Third Streets Passenger Railway Company. This board, which grew in size over the course of its thirty-six years of operation, took as its chief mission regulating fares, limiting competition, and protecting against "blackmailing politicians." It was effective in its work, pushing through fare increases in the 1860s that raised the price of a ride from five cents to six and then seven. Philadelphians, Frederic Speirs wrote, "protested vigorously against the illegal attempt to stifle that healthful competition which the General Assembly intended to establish [and even denounced the increase] as conspiracy punishable under the common law." But the Board of Presidents paid scant attention, and investors in those companies watched their average returns climb to 10 percent per year. That number grew even higher, to 13.6 percent, by 1892.[18]

Those were attractive numbers, but certainly in line with returns across the United States in an overheated industry. In his treatise *Street Railway Investments*, analyst Edward Higgins informed investors that street railways had at times failed to live up to expectations, but that the industry's "future is bright [and] its intrinsic earning power is certain to compare most favorably with that in other fields of industrial enterprise." With profits rising and no peak in sight, investors flocked to the street railways. The value of roads and equipment across the United States jumped from $389 million in 1890 to $2.2 billion in 1902. That number did not reach its peak until it hit $5.1 billion in 1917. Operating revenues stood at $91 million in 1890, $248 million in 1902, and hit their apex of $925 million in 1922. Investors pumped so much money into transit entrepreneurs' hands that in 1896 street railways were capitalized at $95,000 per mile of track, while steam railroads had a capitalization of only $46,000 per mile. And the steam railroads had to build bridges and fences, develop terminals, and construct stations with their funds.[19]

With so much capital pouring into the system, sharp entrepreneurs in charge of larger companies bought out smaller, weaker ones. In the process, they made themselves immensely wealthy. Historians have recounted this process—"merger mania," in the historian Brian Cudahy's terms—in Chicago with Charles Yerkes, in New York City with Thomas Fortune Ryan, in Boston with Henry Melville Whitney, and elsewhere. In Philadelphia, three men who were at times referred to as the city's "transit czars," William Kemble, Peter Widener, and William Elkins, dominated the field. Kemble was born in New Jersey in 1828 and was the oldest of the three. After running small businesses, he served as a federal revenue stamp agent during the Civil War and afterward held the office of Pennsylvania state treasurer. With this background, he became one of the leaders of Philadelphia's political machine and worked with several Philadelphia politicians to establish the Union Passenger Railway Company, of which he became secretary. Widener, born to a Philadelphia brick maker in 1834, made his start in a butcher shop, which helped connect him to local politicians. By his forties, Widener had developed a political base in the Twentieth Ward, served as the city treasurer, and established close ties to Harrisburg. Elkins, born in West Virginia in 1832, made his money in Pennsylvania's nascent oil industry and worked as Widener's official bondsman in the treasurer's office. Together, they understood that one of the next great industries in the United States was street railways, and they took controlling interest in Chicago, New York City, and elsewhere. The contemporary journalist Burton Hendrick accurately labeled Widener "the best-known street-railway man in the United States" and estimated his fortune at some $50 million.[20]

To consolidate Philadelphia's transit companies, Kemble, Widener, and Elkins manipulated the law through their influence in Harrisburg and Philadelphia's political machine. They first got the state legislature to pass a law in 1883 that gave companies the right to install mechanical or electrical traction on current horsecar lines. That same year, they chartered the Philadelphia Traction Company and signed 999-year contracts that leased many of the city's lines, starting with Kemble's Union Passenger Railway Company, the Ridge Avenue Company, the Grays Ferry Company, and the Thirteenth and Fifteenth Streets Company. Philadelphia Traction controlled the industry for approximately a decade, until two other consolidated companies, Electric Traction Company and Peoples Traction Company, emerged to challenge it in 1893. Philadelphia Traction was the strongest of the companies, controlling 203 miles of track; Electric Traction had 130, and Peoples Traction had 73. At that point, the Widener group secured further legislation in Harrisburg that allowed the three companies to consolidate under their leadership, and Union Traction Company was born in 1895. Six years later, the Widener group, who lost Kemble upon his death but added Widener's son George as well as Thomas Dolan, who was president of the United Gas Improvement Company, established the Philadelphia Rapid Transit Company (PRT) to complete the system's consolidation. In 1901, then, Widener and his partners created the sole public transportation company that moved Philadelphians across the city from the turn of the century until it went bankrupt during the Great Depression. In doing so, they found their share of support in Philadelphia, but their political and financial manipulations made the entire system unstable and ripe for public criticism.[21]

In consolidating, electrifying, and expanding Philadelphia's transportation system, Widener and his compatriots joined and helped foster a national wave of urban growth. Across the country, from Boston and New York City to Chicago and Saint Louis, urban boosters understood that horsecars and then especially streetcars fostered the expansion of cities. To these boosters, this was an obviously beneficial development: transit made more land available for housing, urban populations could thus increase, population growth created more economic opportunity, tax revenues expanded commensurately, and workers could live farther from their smoky factories. Everybody won. Of course this was as much rationale as fact. Cities, to be sure, grew in part because of transit systems, but historians have noted how residents of Chicago, Detroit, Los Angeles, and elsewhere repeatedly found that public transportation did not live up to boosters' billing. In Milwaukee, to cite one example, real estate developers promised trolley service but reneged once people purchased outlying suburban houses and thus had little recourse.[22]

Philadelphia's business and political class strongly backed Widener and his consolidation and development of the system. The city's population grew from 847,000 in 1880 to 1.05 million in 1890 to 1.3 million in 1900. It became a truism to city officials that "universal cheap transportation [was] the most important factor in the growth of cities and in the every day life of their inhabitants." Retailers, land speculators, middle-class residents, and prominent businessmen all backed the extension of transit lines. One pamphlet explained to readers that Philadelphia had no choice but to build out its transportation system. Transit, the authors argued, increased "the value of property and additional revenues from taxes to the city." It improved the city's roads and decreased time spent traveling. A chief problem, they added, was that "the enterprise of the house builders has outrun the enterprise of the railways, and is year by year forcing upon them the problem of better transit." The authors concluded by exhorting patriotic citizens to back the cause. If Philadelphia, other pamphlets asserted, would follow the lead of Widener and his fellow builders of the transit lines—men of "unselfish public spirit," as one account of the PRT's leadership put it—then the city would one day become "even more populous than the city of London." Any other course would see Philadelphia wither, falling ever further behind New York City. In this heated atmosphere, some boosters even went beyond electric trolleys and began making the first push for elevateds and subways, although those were not built until after the turn of the century.[23]

There was honesty in back of this boosterism, an earnest desire to make Philadelphia larger, more prosperous, competitive with New York City and London, but political corruption and financial abuse too often tarnished such desires. Years before Steffens told his readers about Philadelphia's corruption and contentment, the city in the last third of the nineteenth century had already witnessed Republicans marshal corporate power at the gas system (the so-called "Gas Ring") and use it to make "King" James McManes the first citywide political boss. The Gas Ring used its political power to allow McManes and his associates to leave financial obligations to the city unpaid and expand their influence over other companies such as the Philadelphia Traction Company. In the transit industry, graft took many forms, although the most glaring example was "watered stock." Across the United States, the Federal Electric Railway Commission found "an amazing story of financial manipulation," "a sordid background of ruthless exploitation of a great public-serving industry to make financial killings for manipulating insiders." Many cities—Chicago being a prime example—participated in this organized theft, but Philadelphia perfected the art. In brief, watered stock (a term apparently coined by cattle dealer Daniel Drew, who fed his cattle salt and then had them drink water before being weighed for sale, which artificially raised their weight and thus their value) referred to the practice of overvaluing the stock of a company so that investors paid in far more than the stock representing the company's real assets was worth. One study found that Philadelphia transit company stocks had brought in $18 million but that the companies had assets worth only $12.3 million, meaning nearly a third of the stock price was water. This extra money was then used to pay dividends to men such as Widener who had bought or been given shares at much lower prices. In addition to taking dividends, they also sold their stock at the artificially high prices and then rebought it when the bottom fell out of the market, as repeatedly happened.[24]

Consolidation also provided an opportunity for great profits that may not have been illegal, but certainly added nothing to the value of the system and in fact undermined its capacity to provide optimum service. Union Traction consolidated the other transit companies in 1895 by buying out their leases with the promise of guaranteed dividends to their stockholders. The owners of the Philadelphia Traction Company sold their 999-year lease (a period of time common in Pennsylvania, whereas 25–50 years was typical of the rest of the country) for a guaranteed dividend of 8 percent; Electric Traction Company owners received 4 percent guaranteed; Hestonville system owners received guarantees of 4 percent for common shares, 6 percent for preferred shares; and the list went on. The PRT did the same thing just six years later, buying out speculators who secured franchises on streets not occupied by Union lines and promising to pay 6 percent on Union stock by 1905. In all, the PRT had to operate the lines of fifty-two underlying companies and pay 44.5 percent of gross receipts to those companies that provided no service and many of which had never laid a mile of track. The arrangement drained away capital that could have been used to extend lines, maintain equipment, and replace obsolete vehicles. Future regulators and attorneys found the PRT's finances so convoluted that it was almost impossible to determine the company's value. Charles C. McChord, a former chairman of the Interstate Commerce Commission, wrote in his 1927 investigation of the company that the "pyramiding [of] guaranteed dividends on guaranteed dividends" had created "an accumulation of fixed charges probably never paralleled in the history of street railway exploitation." The situation, he continued, irritated anyone not holding underlier company stock, but it was not illegal. That may have been technically true, but one newspaper rather accurately labeled Philadelphia's transportation magnates "traction thieves."[25]

Politicians proved a mixed bag in controlling the activities of transit companies. Some promoted and tried to enforce laws that protected the common good and made companies pay for the right to use city streets. But others looked for their main chance, a way to profit from the transit boom. Philadelphia's earliest street railway ordinance, "To Regulate Street Railways," which went into force in 1857, was notably favorable, but rather ineffective, for the city. It required, among other things, street railway companies to maintain and repair all streets occupied by their tracks; pay an annual fee of five dollars for each car intended to run in the city; hire "careful, sober and prudent agents, conductors and drivers"; and give the city the right to purchase the lines at any time. The state legislature, however, retained the right to grant railway charters, and the companies generally avoided paying taxes or making anything more than minimal street repairs. With the passage of the so-called Railway Boss Act of 1868, the state legislature dropped all pretense of home rule and took away Philadelphia's right to regulate street railroads without specific authorization from Harrisburg. The city even lost the right to control how icy streets would be salted. In 1873, Pennsylvanians, recoiling at how often the state legislature granted special privileges to narrow interests, including Philadelphia's transit companies, approved a new state constitution that revived the principle of local control and prohibited the construction of new street railways without local consent. This stopped the land grab of the city's streets, but at the same time the absence of competition allowed Widener to take control of the system over the next fifteen years. At that point, the state passed its general incorporation law of 1889 that allowed new transportation companies to enter Philadelphia and granted the right to electrify all lines. The city council now had the leverage it needed to drive a hard bargain that would last for the rest of the nineteenth century. The street railway companies, threatened by possible competition and expecting greater profits from electric-powered transportation, agreed to repave the entire length of city streets with improved materials, take down overhead trolley lines when required by ordinance, ensure those lines did not interfere with the construction of an elevated rail line, and limit fares to five cents. "The city," wrote Stirling, "was in a few years transformed from one of the worst to one of the best paved cities in the United States." All told, street railway companies paved 258 miles of city streets at a cost estimated to be between $9 million and $14 million. Philadelphia's local government thus demonstrated that it could extract concessions from private interests that benefited the city as a whole. This arrangement, which the transit companies found increasingly onerous over time, lasted until the city and its transit company signed a new agreement in 1907.[26]

Although Philadelphia's politicians could drive a hard bargain for the benefit of their city, they too often failed to do so. Widener and his group forged close connections with Matthew Quay, the state Republican leader who dominated Pennsylvania politics in the decades after the Civil War. Quay protected Widener's interests, including smoothing the process of consolidation, in return for the traction magnate's political support. Politicians in Philadelphia and Harrisburg recognized the alliance and became friends of the Philadelphia Traction Company and its successors. The city council, for example, waived the five-cent fare in 1886 and granted new rights to Philadelphia Traction subsidiaries. The two men's alliance soured only in 1899 when Widener refused to continue supporting Quay politically. Quay retaliated by passing street railway legislation that opened Philadelphia's streets to John M. Mack, a businessman who made a fortune in the paving industry and foresaw great profits if he could break the monopoly of Union Traction. In the first two weeks of June 1901 Quay and Mack forced through "ripper bills" that created thirteen new transit companies in Philadelphia, all under Mack's control. The bills, signed into law by Quay loyalists Governor William Stone and Philadelphia mayor Samuel Ashbridge, would have given Mack access to Union Traction's rails and the right to build subways and elevateds, which would have ruined Widener. They compelled Widener to consolidate again, adding more water as he created the PRT.[27]

Observers, seeing the raw power politics and manipulations that benefited the wealthy, found such moves repugnant. The National Municipal League's Clinton Rogers Woodruff wrote that these events marked an "unparalleled record of franchise-looting." "I doubt," he continued, "if ever the machinery of government has been more brazenly prostituted to private ends and profit; if ever there has been a more conscienceless betrayal of public trust. I use these words advisedly, and not overlooking the record of the Tweed regime and of the present Tammany administration of Greater New York." John Wanamaker, one of Philadelphia's great merchants, believed Mack and his political allies had plundered his city. Wanamaker, in an effort to protect the interests of all Philadelphians, offered to pay the city $2.5 million for the franchises, guaranteed he would charge only a three-cent fare during certain hours, and vowed to turn over the lines to the city anytime within ten years if he received the money he invested in return. Quay, Mack, Widener, and the city council all ignored him. Little wonder one of the state's few independent Republicans, Representative E. A. Coray, was left to plaintively cry that "logic, reason, the law, the constitution, even the truth itself, is lost." The state was overcome with "wickedness."[28]

Critics

Many Philadelphians, for a variety of reasons, agreed with Representative Coray and launched critiques of the city's transportation industry throughout the last third of the nineteenth century. Much of the criticism came from working-class Philadelphians, who were most impacted by the transit lines that sent horsecars and then electric trolleys barreling down their streets. They of course understood that public transportation, especially its consolidation and electrification, had rationalized travel in Philadelphia, paved most of the city's streets, and made it easier to live farther from their places of industrial work. But at the same time, they voiced grave concerns about the way traction magnates became wealthy at the expense of the city and its fare-paying public; about the lack of investment in the system, especially the cars; and about the harm that transit vehicles brought to their communities. These critiques at times jelled into a larger political and philosophical public debate about who should control a public good as vital to the general welfare as transportation.

Public criticism of transportation companies pervaded the United States in the decades from the Civil War until World War I. Most Americans across the country, wrote Thomas Conway Jr. of the University of Pennsylvania's Wharton School (who went on to head several transit companies and spearhead the development of the President's Conference Committee car that tried to modernize trolleys in the 1920s and 1930s) expressed "a spirit of antagonism and disbelief in the honesty and good intentions of electric railway managers." "One of the most firmly fixed beliefs in the public mind," he continued, "is that the street railway industry is one of abnormally large profits, which, moreover, increase in arithmetical ratio with little or no effort on the part of the company." Although such sentiments were largely unjustified in Conway's view, they held particular sway in Philadelphia. The city's transit companies, the reporter Stirling wrote, never had a Cornelius Vanderbilt to infamously say "the public be damned," but dating back to the introduction of horsecars, the industry faced persistent criticism for bringing "unholy personal profits to the insiders" and "a mountain of disgrace" to the city.[29]

It was consolidation, though, that particularly intensified critiques of transit financiers. Philadelphians, especially in the working class, felt the bite of five-cent fares, and nothing infuriated them more than price hikes and having to buy two tickets for one ride, as happened after Union Traction abolished free transfers in 1895. Transit was a political scheme, read one pamphlet, that guaranteed nothing but "gang supremacy and the increase of gang profits." The financiers, especially Widener, worshipped Mammon alone, read a letter to the editor, with "dollar-worship, money, being their god." Transit magnates, the press routinely reported, "milked" the city and "robbed" its people, with higher fares and exorbitant contracts for hauling street sweepings and ashes and performing other duties. So many people complained that they picked up the sobriquet "kickers," and Mayor John Reyburn even threatened theater owners with revocation of their licenses if they did not ban vaudeville songs that mocked the PRT. "I thought this was the town of our liberties," said performer Jack Norworth, although he promised to stop singing his popular ditty about the "rumpled . . . public [that] might start kicking" because his song would harm the city's theaters. Transit magnates, in the eyes of many Philadelphians, cared only for profits, not the public.[30]

Those profits, Philadelphians observed, always seemed to stay with transit investors and operators rather than finding their way into system improvements. "The original traction companies," reported investigators for the Pennsylvania State Railroad Commission, "subordinated comfort, convenience and popularity almost entirely to cost. In addition to buying the cheapest cars obtainable there seems to have been little attempt on the part of the management to determine the fine points of car design as embodied in the wishes of the public and the standards in use in other cities." Cars were known for being small and old (as late as 1911 two-thirds of the system's 3,292 cars dated from original electrification in the mid-1890s). Most of the larger cars were actually two small cars spliced together. A decade into the twentieth century, only 36 percent of Philadelphia's fleet was composed of large double-truck cars; Chicago's fleet was 75 percent, New York City 75 percent, Brooklyn 89 percent. Small cars meant miserable conditions: the German-American League called them "inhumanly and beastly overcrowded." Worse, despite the public's complaints, the cars had poor ventilation and dim lighting and still lacked adequate heat as late as 1910. Philadelphians protested that the cars caused "serious discomfort" and even "serious illness," and a rider, as one man put it, had "a right to ask relief against being compelled to endanger myself." One petition told state officials that the signers "emphatically condemn the asinine management for its willful neglect to furnish proper accommodations." Philadelphians too often experienced a jolting ride, in part because of old equipment, but also because the PRT used what civic leader John C. Bullitt called "the worst and cheapest" paving materials, turning his beloved city's streets into "a shame and a reproach." Ubiquitous political cartoons mocked the PRT and the service it provided, but perhaps the sharpest rebuke came from an elderly woman whom the historian Harold Cox interviewed in the 1960s about the early transit cars: "They rode like hell!" she snarled and stalked away.[31]

Surrendering control of the city's streets to the transit companies made the expensive fares and slipshod service all the more galling. For one protest, 7,253 households representing some thirty-six thousand people signed a petition demanding that the government defend their rights against the grasping transit magnates. Their representative, John C. Bullitt, asked the city council how it could even consider allowing Widener and his associates control of the city's thoroughfares when they had failed to pay a bill of $485,000 to pave Philadelphia's streets. Steam railroad companies had to invest their funds in leveling the land and laying track, in effect building their own roads. Why, Bullitt wondered, did the street railways not have to do the same? Perhaps, he intimated, their political connections had something to do with why the street railways got free access and nearly perpetual rights to public space. Warming to his task, Bullitt then argued that the people who lived along the streets where the transit lines ran had a moral right to control their neighborhoods and protect their families and property values. But the city council always ignored them, giving away "absolute rights, these vested rights, these rights to spread this system all over the City, without having a single safeguard." The city council, at the command of the transit magnates, he concluded, was willing to "trample upon the views and expressed wishes of every man upon the lines of the streets." Hearing these complaints and their echo in some quarters of the business community distrustful of a transit monopoly (especially the Trades League, which became the chamber of commerce), Widener agreed to improve street paving and made a few other concessions but overall had the political power and wealth to largely ignore the protests.[32]

When those protests focused more on safety, family, and community, they were not so easily ignored. By the 1890s, Philadelphia already had a forty-year history of protesting public transit's impact on residents' communities. The earliest conflicts arose in 1859 when horsecar companies wanted to operate their lines on Sundays. Clergymen and their flocks took umbrage, asserting that such transit service would "require [communities] to double your police force on the Sabbath, it will throng your mayor's court on Monday morning, it will fill your almshouses with starving children during the week, it will decrease your Sabbath schools and increase your prisons, it will open the road to vice and fill the highways with its votaries." State law (dating to 1794), they argued, prohibited "worldly employment or business whatsoever on the Lord's Day," and the transit companies would not only violate it, but clearly not care about doing so. It took eight years of wrangling in the courts and the state legislature (where one bill allowing Sunday service was referred to the "Committee on Vice and Immorality") for horsecar companies to gain the right to operate on the Sabbath. The charge that transit critics resisted any change, that they practiced an "intolerable fogyism," as one railway promoter put it, certainly had some merit. But in truth, streetcars really did remake the city's streets in new and often dangerous ways.[33]

Horsecars could be dangerous on crowded city streets, but they were nonetheless familiar. Because they moved relatively slowly and carried lighter loads, drivers could usually stop their horsecars quickly before anyone was seriously injured. The greatest danger often came when something spooked the horses and they bolted or tried to escape their harness. But overall, people knew to listen for a horse-drawn vehicle and could get out of the way if they were paying attention. Philadelphia rarely averaged more than one death a month from horsecar accidents from the start of the technology's use until electrification in the mid-1890s. That is when life for many Philadelphians took, or so they believed, an ominous turn. (Table 1 highlights the surge in deaths and injuries caused by electrification.)[34]

The mere use of electricity, especially via overhead lines, caused much conflict—"violent opposition," in Speirs's terms. Such opposition was common around the world: London, Paris, Berlin, Manhattan, Washington, D.C., all banned overhead wires. In Philadelphia, people worried that transit companies would "gridiron the very best parts of [the] City with wires and poles." Those wires were unsightly, pulling Philadelphia even further behind New York City and Chicago, where lines were mostly buried, but also dangerous. William

TABLE 1. Philadelphia street railway accidents and fatalities, 1861–1895

YEAR(S)	KILLED	INJURED
1861–1864	29	44
1865–1869	50	87
1869–1874	56	80
1874–1879	38	97
1879–1884	59	171
1891	18	210 (1891–1892)
1892	31	197 (1892–1893)
1893	35	243 (1893–1894)
1894	67	379 (1894–1895)
1895	65	
1896	66	

Source: Speirs, Street Railway System, 123.

Note: No reports filed 1885–1890; no injury data for 1896.

Marks, president of the Edison Electric Light Company of Philadelphia, told audiences that they should fear any overhead trolley lines. Touching those lines would electrocute people or animals; if they broke, they could start fires. And so many lines, added Mayor Edwin Stuart, would interfere with fire crews trying to do their jobs. Residents on streets impacted by electrification, such as Sixteenth, signed petition scrolls ten feet long, and along Spruce Street, 616 out of 618 households affixed their names. The city, they argued, had to protect their right to a safe community. To further their point, some protesters played on gendered notions of women as the weaker sex. "Is it to be wondered at," asked Bullitt, "that they should feel that dread and apprehension of these wires coming before their homes, when they have before them the declaration of the chief official of the City government that these overhead wires are a constant menace to life? I take it that all or nearly all of you, are husbands, having wives and daughters, and I ask you to bear in mind the nervous organization of women; when there is placed before them that which is a constant menace to life, it reaches down to the deepest recesses of their nature, and produces a fear and dread and apprehension which is unknown to our sex; it produces a fear and terror which makes life a burden." We must, he concluded, "save these women who are entitled to their protection from this dread and apprehension which is calculated to render their lives wretched and miserable." To be sure, court records from the era show a number of female complainants who sued transit companies because sparks from trolley lines gave them an "awful fright" or made them nervous about traveling. But Bullitt might have been chagrined to know men far outnumbered women as litigants against the PRT.[35]

Electric lines may have brought some danger into Philadelphia's neighborhoods, but it was the trolleys themselves that truly threatened the city's residents. Streets had long been the province, especially in working-class communities, for children to play and adults to conduct business, but faster, heavier electric-powered streetcars with steel wheels on steel rails made the roads more thoroughfare than common ground and thus much more dangerous. This new, more hazardous use of the streets rankled Philadelphians in the early days of the streetcar, and those feelings only intensified over the years, finally exploding, but not ending, in the strikes of 1909 and 1910. Most streetcars weighed fifteen to twenty tons and traveled an average of eight miles per hour, with the city speed limit set at ten miles per hour. It took years for braking technology to catch up with the size and speed of electric vehicles, and the PRT, like many transit companies, was reluctant to pay the cost of replacing hand brakes with safer air brakes. The city's courtrooms and newspapers were filled in the two decades after electrification with stories of men and women killed when trolleys skidded into their wagons, when cars ran them over in the street, when passengers stepped from a trolley into the path of an oncoming vehicle. In 1907, Philadelphia hit its apex with nearly 37,000 accidents resulting in 113 transit fatalities. Philadelphia's system killed one person for every 5.7 million passengers carried; New York City stood at one per 6.9 million, Brooklyn one per 7.5 million, Boston one per 10.7 million. Such accidents, coming in the worst years more than twice a week, created incensed crowds. John Smith's story offers one example among scores. An employee of a dyeing establishment at Seventeenth Street and Fairmount Avenue, Smith was riding his bicycle home from work one evening when he was struck by an ash car with such force that his badly mangled body became so wedged in the brake rigging that the driver had to call for the wrecking wagon to lift the streetcar off him. Smith's groans attracted over one hundred people, who angrily protested how long it took for the PRT to dispatch its crew and pull him from the wreckage. The company, they believed, did not care who it hurt. Smith died on the scene.[36]

Adult deaths were tragic enough; children dying inspired far greater horror. Fully embracing the yellow journalism of the day, Philadelphia's newspapers devoted hundreds of stories to what the *Record* called "the slaughter of children," "the crushing of victims under the cruel wheels." In lurid detail, the *North American* told of the death of Jeanette Schwartz, a three-year-old girl who was playing with friends on North Twenty-Second Street when she was struck by a trolley. "Her left arm was severed and her head was crushed," wrote the paper. "As in numerous other cases where children have been crushed to death by Rapid Transit Company cars, this car was not equipped with jacks." A policeman, knowing Schwartz could not afford to wait, crawled under the trolley and after several minutes managed to pull her from under the car. She died on the way to the hospital. The *Evening Bulletin* told of Frank Cooper, an eight-year-old newsboy, struck and killed at Fourth and Chestnut. Cooper, "the little chap" as the paper called him, was part of a family of twelve, and his job provided a chief source of financial support, since his father was ill and unable to work. Dinnertime had come for the boy, the reporter wrote, and "it was in his eagerness to sell his last paper that he met his death." The *Inquirer* shared with readers a photo of eight-year-old Sarah Kapelnick clutching her teddy bear. She had tried to dart across the tracks in front of a trolley at Third and Fitzwater, only to slip and lose her balance. She avoided the wheels, but her body made contact with the motor and rail, completing the circuit and, as the *North American* put it, "sending the death-dealing volts through the unfortunate child."[37]

Such horrible accidents routinely brought throngs to the scene. Sometimes, overcome with emotion, they attacked the trolleymen. At Kapelnick's accident, for example, hundreds of people rushed to the streetcar and tried to seize the driver. "Lynch him!" several people cried. Club-wielding police beat them back from the platform of the streetcar, arresting one man and sending twenty to a nearby drugstore to have their heads bandaged. But removed from the heat of the moment, most Philadelphians, especially in the working class, had sympathy for the drivers and conductors. A letter to the editor in the *North American* was typical. "Force [the PRT], by law if possible or by popular sentiment, to stop the slaughter of little children, to stop the anxiety of loving mothers who send their little ones to school with a fear in their hearts that they may be brought home to them with broken limbs or, worse still, a mass of lacerated flesh and bones," it read. "Do not arrest the motormen, who cannot stop the car in time, on account of slippery rails or defective brakes. . . . Arrest and prosecute those higher up and force them by all legitimate means to provide adequate protection for the citizens, by whose grace they occupy the streets of the city."[38]

The PRT, in most Philadelphians' estimation, had done too little to stop the carnage. Safety equipment, as the historian Clay McShane has pointed out, was a constant sore spot for transit companies and their publics, especially since those companies often delayed improvements because of their cost. In Philadelphia, as in many other cities, the chief point of contention was the rope fenders that would have been laughable if not so deadly. The fenders were essentially a rope basket attached to a pipe frame that extended twenty-nine inches out in front of the trolleys. In theory, they were supposed to scoop the unwary out of the way; just as often, they forced children under the wheels. Mass meetings called on PRT management to lower the fenders and implement better, if more costly, safety devices, such as the automatic-drop wheel guards used in New York City and Chicago. Rather than adopting available technology, though, the company stalled for time and finally ran a competition in 1911 for local amateur inventors

FIGURE 3.   Spring Garden Street Bridge, with horse-drawn wagons and trolleys using the same street. Two ways of life merged, often uneasily, for a time. Photo courtesy of PhillyHistory.org, a project of the Philadelphia Department of Records.

to develop better fenders. A Mrs. Anna Thompson won acclaim for her invention that simply wrapped the rope basket farther back toward the trolley's front wheels. Other prototypes resembled giant ice tongs or rope lassos to snatch people out of the way, brushes that would sweep people away from the car and clean the street at the same time, and cowcatchers common on steam railroads. After all the contests and public meetings, the company finally settled on the one practical solution: the widely available H. B. Lifeguard that placed a gate under the front of the car that when struck released a tray that fell into place, protecting people from the front wheels. All told, it took sixteen years for companies running electric trolleys in Philadelphia to adopt this technology and make the streets safer. It was in this same period that the PRT instituted other safety measures, such as preventing people from entering and exiting trolleys on both sides (often in the way of oncoming traffic) and stopping children from stealing rides and hanging on the fenders and sideboards. Company officials spoke of their apparently newfound commitment to safety; critics suggested that the $1 million cost per year of accidents had more to do with it.[39]

These critiques, about how the transit system's finances benefited the wealthy, about how money rarely got invested in roads and equipment, about the danger electric streetcars brought into people's communities, at times fed a larger debate about who should own such a vital industry and why. These debates were, put another way, part of a broader argument central to the Progressive Era about private economic power and its relationship to the public good. Arguments in Chicago about the transit system, in San Francisco about the water supply, and similar contestations in many other cities reflected a nation grappling with what was termed "gas and water socialism"—that is, who should own those public service industries. Streetcar systems, the historian Clifton Hood has pointed out, often played central roles in these debates because they represented an intrusive monopoly and because public control of their operation, including their routes, offered the opportunity to remake the city.[40]

Philadelphians from the earliest horsecar days understood the tensions created by a privately owned business that provided a public good. "Street railways," wrote the Philadelphian Alexander Easton in his handbook on horsecar railways, "must realize that they are not an unfettered part of the economy, free to make business decisions solely in response to market forces. Street railways will operate in close proximity to the public and political domain, and must accommodate their corporate aims and goals to the needs of the public sector." Horsecars, the *Nation* told its readers, were "not simply private property; they are wheels in our great social machine." Strickland Kneass, Philadelphia's chief engineer and surveyor, made these claims concrete in 1859, writing that having private companies build the lines but pay fees for the use of the streets would best serve Philadelphia's interests. Doing this, he argued, "will place in the power of the city the means to correct abuses, preventing imposition upon its citizens and creating what should be a public benefit. . . . The city will then know where to look for the proper repair of its highways, and not be, in a measure, dependent upon a company of individuals whose only interest will be the amount of dividends realized." He added that the city should reserve the right "to obtain the possession of the roads as soon as the financial condition of its treasury will permit."[41]

Although Kneass pointed to municipal purchase as a way of cementing control, Philadelphia leaned toward regulation as the way to solve what the contemporary analyst Emory Johnson called the "street railway problem." Thus the city put in place paving requirements, demanded a five-cent fare, and so on. With this approach, Philadelphia fell in line with most American cities. The exceptional examples of public ownership proved the rule: Brooklyn and New York City owned a one-mile cable line over the Brooklyn Bridge from 1883 to 1898; Monroe, Louisiana, started a line ca. 1907, as did Bismarck, North Dakota; and San Francisco had one starting in 1909. The construction of subways, which required more capital and entailed greater risk, led to more public involvement in transit across the United States. But overall the era of public ownership did not really grow until the end of World War I. Before that time, public transportation in the United States was almost exclusively a privately owned industry, and most cities—Chicago offering the prime example—relied on regulation as a means of exerting control over the transit companies ("regulated capitalism," in the historian Paul Barrett's terms). Pervasive regulation, however, did not mean powerful regulation: Emerson Schmidt, surveying urban transportation in the 1930s, found that early regulatory laws had lost their teeth well before his time, and Charles Cheape's work on New York City has shown how the law lacked the power to direct transit development in the public interest.[42]

With the political and financial power of Philadelphia's transit interests making regulation feeble, some residents pushed for municipal ownership, albeit to little effect at the time. Philadelphia's political and economic elites, who gave the city its justly earned pro-business reputation, helped brake any such movement. So too did Philadelphians' experience with the notoriously corrupt city-operated gas company, which was finally privatized after much public clamor in the mid-1890s. Yet the monopoly power of the PRT raised searching questions for many residents: How could the city control a vital system for the betterment of all? Were publicly owned systems, more common in Europe, actually the way forward? Earlier competition had already led to consolidation and monopoly, so how could competition even develop now that the PRT was so entrenched?[43]

These questions mostly simmered below the surface, but at times they did boil over, especially when the PRT or its forerunners changed ticket prices. This first happened in 1864 when the Board of Presidents raised fares from five cents to six and the city council responded by preparing an ordinance that would have allowed Philadelphia to buy any companies charging the higher price. Nothing came of that effort, and the companies charged the higher fare until 1887, when the city council finally negotiated the price back to five cents. Widener bided his time, and eight years later, after forming Union Traction, he had his company eliminate free transfers in order to generate more income. A firestorm ensued. Some 40 percent of public transit riders used transfers, and they were outraged by what one commentator called the "arbitrary disregard for the public interest." Union's action, wrote Speirs, fulfilled the prophecy of one opponent of street railway consolidation. "You may rest assured your street railway system is destined very soon to be an absolute monopoly," said prominent Philadelphia attorney Wayne MacVeagh. "The only question remaining is whether the monopoly shall be owned by the Traction Company or the whole body of the good people of Philadelphia." Philadelphians held mass meetings at the Academy of Music to protest the change in fares, denouncing the company's "defiant contempt of public rights," and a grand jury called the abolition of free transfers a "public evil" and asked that city attorneys challenge Union's right to exist.[44]

Although this 1895 skirmish set the tone for a generation to come, Union Traction managed to fight off the challenge for a while, in part by selling stock to many Philadelphians in hopes that owning a small piece of the company would lead them to defend Union from public attack. But the controversy would not die. Over the next decade, reformers in the Municipal League charged that the company (now the PRT) was bilking the public and demanded that the city consider purchasing the system for the common good. Businessmen who wanted better service and had concerns about a transit monopoly also wanted the city to overhaul its relationship with the the PRT. Newspapers, especially the *North American*, and trade unions also demanded action. The protests grew so bitter that company officials worried that they "discouraged potential investors and might lead to municipal recapture." To stave this off, PRT officials agreed to reinstate free transfers in 1902 and began negotiations with the city on a new contract that gave city officials greater power when it went into effect in 1907.[45]

The PRT's actions assuaged the bulk of the protesters, but not its most ardent foes. To members of the city's Socialist Party, which had its greatest strength in the textile mill district of Kensington, free transfers and the prospect of adding a few oft-corrupt politicians to the company's board were simply not enough. Transportation was a public good, not an opportunity for the well-heeled and well-connected to make a profit. Throughout the late nineteenth and early twentieth century, Socialists argued that Philadelphians did not have to accept the privately owned system. The PRT's ownership had "stolen all its wealth from the citizens of Philadelphia," the party argued, and it was now time to "Educate and Organize." Private ownership, the party continued, "invariably leads to the oppression of the workers and to struggles between employers and employed; . . . the practice of extortion upon the public for the profit of the owners; . . . the domination of state and city by private property interests." Conversely, public ownership offered "the only road open to salvation" in an era of rampant corporate power. A few politicians, such as reform mayor Rudolph Blankenburg and director of city transit A. Merritt Taylor, at least countenanced the idea. To Taylor, city ownership meant "a great income-producing municipal asset" that would provide a considerable "range of movement and enormous and valuable time saving." Most business and political leaders, however, derided any such plans as "socialistic ideas [with no conception of] the magnitude of the problem or the burdens which it involves." The Socialist Party's call for public ownership thus fit into a strain of Philadelphia politics that called for municipal ownership but made little headway in the late nineteenth and early twentieth century.[46]

By the dawn of the twentieth century, then, most of the basic contours of Philadelphia's public transportation system—its technology, its financial structure, and the critiques of both—were in place. The city had moved from stagecoaches and omnibuses in the mid-nineteenth century, to steam locomotion and cable cars, to electric power. Improved technology combined with mass consumer demand drew entrepreneurs to the industry. The free-for-all mentality that allowed thirty-nine passenger railway companies to receive charters in the mid-nineteenth century gave way by the late 1800s to the logic of cooperation and then consolidation, which allowed Widener and his partners to dominate the field. Consolidation in some ways made the system more rational and efficient, but also created irresistible opportunities for watered stock and outright graft. Many Philadelphians questioned what their transportation system revealed about their city. The transit system undeniably enabled more people and goods to move around the city more efficiently. But at the same time bigger, faster, heavier cars brought mayhem to many communities, while the direct economic benefits disproportionately went to an elite group with financial power and political connections. Surely, critics argued, transportation was a public good, so should not the costs be borne by and the benefits redound to all citizens? That was an open question at the time, and remained so for generations to come. One group that particularly asked the pointed question about who benefited from the transit system was the employees, the drivers and conductors, who worked on its lines. They were dreadfully exploited and found themselves repeatedly browbeaten and silenced by management, yet without them the horsecars and trolleys would not run, the city would not grow. Without them, there would be no transit system.

WORKING ON THE LINE

Transportation was a hard business for the drivers and conductors who ran Philadelphia's system in the last third of the nineteenth century. Workers spent grueling hours on their feet, as many as eighteen a day. Their work exposed them to the elements in Philadelphia's harsh winters. They received little pay, at best about two dollars a day, which was low even by late nineteenth-century working-class standards. Management held them in low esteem. And dealing with pedestrian and horse traffic on city streets, especially after electrification made the cars faster and heavier, wore down the nerves of even the most competent drivers.

Most Philadelphians, particularly in the working class, understood the difficulty of the job and sympathized with the transit workers. The Toynbee Society, a local organization dedicated to improving workers' lives, put it best. "Your committee cannot avoid observing that the conditions of service required of these men are such as human beings ought not to be subjected to," the society wrote in its 1895 investigation of the Union Traction Company. "A thirteen- or fourteen-hour daily service, to which Sunday forms only partial relief, with only one interval for refreshment, is too great a tax on the physical constitutions of the men. When they have eaten outside of their working hours, as they must do twice daily, there is barely time left for sleep. The nervous fatigue, too, even when all the time left from eating and working is devoted to sleep, is such that the men in these circumstances must wear out in a few years, and then practically be thrown upon the community for support, after the Traction Company has gotten from them at a profit all the service that their life was capable of giving." The only solution, Toynbee concluded—and Union Traction's workers agreed—was to form a union that would defend their rights and push for a better workplace.[1]

This chapter explores the work lives and labor organizing experiences of Philadelphia's late nineteenth-century transit workers. It begins by detailing the broader context of the period's working-class history in Philadelphia, in Pennsylvania, and the United States. Doing so highlights how difficult life was for working people in the late nineteenth century and the challenge the city's transit workers took up when they decided to join a union. After all, in the early 1900s, local machinists and printers could justifiably claim that Philadelphia's business and economic elites had made their city a "scab recruiting station," the "enemy's country" for organized labor. And yet, by the 1890s Philadelphia's transportation workers had come to believe that the best way to better their lives was to heed the call of the American Federation of Labor's Amalgamated Street and Electric Railway Employees (the Amalgamated) when it urged them to "wake up and organize and show these magnates that the time for bulldozing employes is past and gone and we must be treated as men." Joining a union, as this history shows, was no mean feat and one the workers did not take lightly, but by 1895 the Amalgamated's deft organizing and the workers' growing frustration with their obvious exploitation made them ready to take on Widener's Union Traction in the city's first system-wide strike. The Amalgamated ultimately lost, but the strike set the stage for greater labor organizing and deeper, more powerful and profound conflicts to come fifteen years later.[2]

Labor in Philadelphia, in Pennsylvania, and the United States

Philadelphia, wrote the historian Howell Harris, was a "great and grimy industrial city . . . at the dawn of the twentieth century." Combined with its seven surrounding counties, Philadelphia was one of the nation's great industrial districts, a region that could with justification call itself the "workshop of the world." The city earned this sobriquet on the strength of its smaller industries that employed the bulk of its 171,000 industrial workers in 1870, and 311,000 in 1900. The city housed, to be sure, a number of large concerns such as Baldwin Locomotive, Cramp's Shipyard, and Midvale Steel, but those were exceptions. Of all the city's thousands of employers, only Baldwin and Cramp's employed more than four thousand people at the turn of the century. Most working-class Philadelphians worked as common laborers, or in textile mills and garment factories, metal shops and sugar refineries. Few companies, only a dozen in the entire region, had over fifteen hundred employees.[3]

The structure of Philadelphia's industrial base had grave implications for labor organizing in the city. The men and their families who founded local companies such as Disston Saw and Stetson Hat still owned and operated them in this period, and they fully believed in a smaller-shop, family-owned proprietary capitalism, which meant that they, in Harris's words, "favored tradition, simplicity, and direct, personal control just as much as they believed in the absolute rights of ownership." To these men, if "unionized skilled workers attempted to assert a claim to share in decision making about wages, hours, and working conditions . . . they would be seen as trespassers in other men's houses, in many cases houses that their proprietor-managers had built at their own risk and with their own sweat." These owners had no time, nor inclination, to negotiate with "outsiders," with "agitators." To them, unions were a personal affront, one that struck at the very core of the entrepreneurial capitalist system.[4]

Although the structure of Philadelphia's industrial base gave the city some particularities, the fundamental question at stake—the "labor question"—was a national one. Rosanne Currarino, building on the work of scholars such as David Montgomery and Alan Trachtenberg, is the most recent historian to frame clearly the central issue of the Gilded Age and Progressive Era: "Could democracy survive in industrial America?" In an era when wage work, often in industrial settings, was replacing an older agrarian economic and social system (Robert Wiebe called this period a "way station between agrarian and urban America"), how could workers secure some semblance of equality in American society? The right to vote was simply no longer enough, if in fact it ever had been. Democracy and equal citizenship for all (at least all white men) appeared incongruent with a capitalist economic order and a legal system that increasingly promoted consolidation of business and concentration of wealth. Workers and their leaders in the nascent labor movement understood that they too had to consolidate, to form unions in order to stake their claim for a different American society, one that recognized that workers could not be free if their terms of employment left them under the domination of their employer. "Political liberty with[out] economic independence," AFL leader Samuel Gompers once wrote, "is illusory and deceptive." And business's constant fight against organized labor, added the Toynbee Society, was inarguably hypocritical: "Itself an organization of consolidated capital and an illustration of the advantages to be gained by such consolidation, it is ready to fight the slightest beginnings of consolidation on the part of the labor that serves it." Workers, to the era's businessmen, had a right to withhold their labor, but not to come together in unions that would advance their interests, especially with strikes.[5]

To many workers in the decades after the Civil War, the economy seemed to be closing in around them, circumscribing their lives. The independent freeholder of antebellum America was giving way to "hirelings for life," a class of people who had no choice but to work for wages. Their pay generally hovered around subsistence levels, and employers readily pointed out to the restless that the country had plenty of unemployed people ready to take their jobs. In the 1890s, the number of unemployed hovered between three and four million, and a United States Industrial Commission study showed that between 60 and 88 percent of American citizens were poor or very poor. Frequent depressions—1873–1879, 1882–1885, 1893–1897—exacerbated workers' situations, especially in large cities such as Philadelphia. In those periods, unemployment soared, and many workers faced the prospect of starvation. With no work in their communities, they often took to the road, where they picked up the name "tramps." Some Americans saw them as adventurers, but most economically secure observers thought of them, and the unemployed in general, as a "dangerous class." And it was a large class, with studies showing that 23 to 30 percent of the workforce found itself unemployed at some point every year. Common wisdom held that the unemployed were a shiftless lot, unwilling to work but ready to steal property or cause bodily harm when the opportunity arose. To combat them, leading citizens such as Vice President Alexander Cassatt of the Pennsylvania Railroad advocated vigilance committees armed with horns, flares, and billy clubs. Across the country, states acted with less threatened violence but no more compassion, passing vagrancy laws and tramp acts to control unemployed working-class people. New Jersey passed the first such law in 1876, and forty states followed its lead by 1896. Pennsylvania passed such an act in 1879, making it illegal for anyone to be "going about from place to place begging, asking or subsisting upon charity, and for the purpose of acquiring money or a living, and who shall have no fixed place of residence, or lawful occupation in the country or city in which he shall be arrested." The penalty was imprisonment for up to three years. The Tramp Act did not apply, as the law put it, to blind, deaf, or dumb people, women, children, or anyone maimed or crippled who could not perform manual labor. They were to be prosecuted under the general Vagrancy Act of 1876. In the end, many workers had to accept starvation wages when they could find a job and faced imprisonment for vagrancy when they could not. Little wonder many believed they had to organize.[6]

Although Philadelphia offered rocky soil for unions, by the late nineteenth century the city already had a significant history of organizing. A century earlier, in 1786, the city's printers went on strike for higher wages. Journeymen carpenters and shoemakers followed over the next two decades. In 1827, the Mechanics Union of Trade Associations formed as the first union of all organized workingmen in a single American city. The Mechanics Union lasted only two years, but many affiliated groups—cordwainers, smiths, leather workers, and so on—came together again in 1834 to form the General Trades Union, which led the nation's first general strike and played a key role in the successful fight for a ten-hour workday. Soon after, the Panic of 1837 decimated Philadelphia's labor movement for a generation, but the Machinists' and Blacksmiths' Union formed in 1858 and was strong enough to lead a strike at Baldwin Locomotive two years later. These early labor movements were often small and sporadic and always met hostility from the political and business class, but they showed the willingness of Philadelphia's workers to organize despite the odds.[7]

Philadelphia's labor movement grew and became more permanent after the Civil War. The Knights of Labor played a key role, with the organization starting in Philadelphia's mill district of Kensington. Uriah Stephens and six of his fellow garment cutters founded the Knights in 1869 with the explicitly egalitarian ethos that workers would advance their cause only if all came together in solidarity, black and white, men and women, skilled and unskilled. "An injury to one is the concern of all," they proclaimed. Operating at a time when they knew the state of Pennsylvania had violently repressed the radical Molly Maguires in the nearby coalfields, the Knights of Labor formally eschewed strikes and violence in favor of political remedies. At its height in 1886, the Knights, called the Quaker City Protective Association in Philadelphia, had ninety-five thousand members in Pennsylvania (seven hundred thousand in the United States), with half of them in Philadelphia, mostly in the textile industry. That number dropped by 80 percent in 1887 as the Knights, despite their proclamations, were tarred as violent in the wake of Haymarket, while at the same time workers, especially in textiles, had seen the organization back down in a confrontation in Frankford's Wingohocking mills that made them believe they could not count on the organization's firm support. Disheartened, many working-class Philadelphians dropped out of the labor movement for a time. Into the void stepped Samuel Gompers and the American Federation of Labor, which espoused a "pure and simple unionism" that largely rejected political action and social reform in favor of a focus on basic wages and working conditions. It was in the AFL that Philadelphia's transit workers eventually found their home.[8]

As workers, pressed to the wall, joined unions to fight for workplace rights, and intransigent employers opposed them at every step, violent conflicts shook America. Chicago—think McCormick, Haymarket, Pullman—was the epicenter for these conflicts, but the list of labor strife reveals a national phenomenon: the Great Railroad Strike of 1877, the Southwest Railroad Strike of 1886, the Lattimer massacre, the Homestead strike, and many more. In all, the United States had 9,173 strikes in the 1880s and 13,620 in the 1890s; 1.9 million workers walked out in the first decade, 2.8 million in the second. It became commonplace, especially after the Great Railroad Strike of 1877, for state authorities to back the rights of corporations with army troops, the national guard, and the state police. Paul Gilje, in his survey of rioting and violence in American history, did not miss the mark when he wrote that the story of labor relations in the era was "written in blood."[9]

The Great Railroad Strike of 1877 was one of the searing moments that set the tone for labor relations in Philadelphia. The strike, as David Stowell has pointed out, had deep roots in an American political economy that had created a chasm between workers and capitalists, but its immediate origins lay in the Panic of 1873, which induced railroads and other corporations to cut workers' wages. The railroads in the 1870s were the largest corporations in the United States and were in the process of knitting the nation together geographically. As such, they were widely seen as supporting, if not synonymous with, the national interest, particularly by the railroads' officials themselves. The strike, which began on the Baltimore & Ohio Railroad and quickly spread to other lines and many cities, thus seemed to strike at the very core of the American state. The confrontation was not just about wages, but about corporate intrusion in people's lives, both in the way large companies could cut people's income when they saw fit and in the way the railroads dominated the nation's urban space with little regard for people's safety or other uses of community streets. Cities from Philadelphia to Pittsburgh, Baltimore to Saint Louis, Chicago to San Francisco all experienced class violence, which left at least one hundred people dead as the strike played out over July and August 1877. All of Pennsylvania's major railroads were affected: the Pennsylvania, the Reading, the B&O. So were many of the state's largest cities, with violence shaking Philadelphia and Pittsburgh, but also Altoona, Reading, Harrisburg, and Scranton. The strike came to a conclusion only when President Rutherford B. Hayes dispatched federal troops up and down the railroad lines to squelch the conflict. In doing so, he became the second president in U.S. history to use troops in this fashion (the first since Andrew Jackson suppressed a strike on the Chesapeake and Ohio Canal forty-three years earlier), and over the next quarter century and beyond many presidents emulated him, using the force of the state to suppress worker unrest.[10]

Although other cities, such as Chicago and Pittsburgh, witnessed more violence, Philadelphia's experience, especially in Kensington, was difficult enough that the city's elites, such as Pennsylvania Railroad president Thomas Scott, demanded state protection for people and property from the working-class "mobs." Scott, a major player in Republican politics who helped settle the electoral crisis of 1877 that placed Hayes in the White House, demanded the garrisoning of U.S. troops "at prominent points, large cities and great business centres . . . [where] their movements can be combined rapidly, and they may be directed against points of danger." "Give the workingmen and strikers gun-bullet food for a few days," Scott once said to demonstrate his outlook, "and you will observe how they will take this sort of bread." Hayes was not ready to go so far with federal troops, but the state of Pennsylvania paid heed. A decade earlier, in 1865, the state legislature had passed the Coal and Iron Police Acts, which permitted railroad and coal-mining corporations to hire private police forces, often composed of army veterans hardened by the Civil War and later the Philippine-American War. Armed with revolvers, nightsticks, and Winchesters, they treated miners as "serfs" and earned a deserved reputation as "thugs" or, in the terms of the state's immigrant miners, "Russian Cossacks." In the summer of 1877, Republican governor John Hartranft significantly expanded the force to serve the security needs of the Pennsylvania, Reading, Lehigh Valley, and Central of New Jersey Railroads. This private police force in 1905 became the first state police outside of frontier organizations like the Texas Rangers. The state also turned its militia into a National Guard, which entailed installing a military justice system, tightening the chain of command, requiring drilling in field maneuvers, and replacing ceremonial dress with combat fatigues. And in Philadelphia proper, National Guard regiments followed a national trend and began fund-raising campaigns among city elites to build armories that would, in the *North American*'s terms, protect against "any possible outbreak of lawless violence." The Pennsylvania Railroad donated $5,000 to the cause, and other corporations followed suit. Soon armories that bore a suspicious resemblance to castles dotted the landscape. The architecture of the buildings made it clear that corporate and political leaders believed the nation was on the verge of class warfare, and they intended to wield the force of the state on their side. Not until World War I—forty years later—did the National Guard become the backup for the army rather than a police force for industry, and at that point new armories changed to look like community centers. For those decades in between, working-class Americans in general and Philadelphians in particular rightly saw armories as a symbol of repression.[11]

Although corporate elites turned primarily to state violence to control workers after the 1877 strike, they also employed other tools. At a time when social Darwinism and faith in the market economy reinforced each other to create a pervasive, although not monolithic, national laissez-faire ideology that portrayed working people as culpable for their own plight and the government as responsible only for setting the rules (tilted toward capital) and providing for the national defense, they had plenty of tools from which to choose. "Yellow dog" contracts compelled workers to swear not to join a union or they would lose their jobs. Sympathetic courts—and they all seemed to be sympathetic—issued injunctions, first in Massachusetts in 1888, then more powerfully in New Orleans in 1893, and from there all across the nation, to keep workers from organizing, campaigning for workplace rights, or engaging in sympathetic strikes. The historian David Montgomery highlighted the growing use of this employers' weapon: 28 injunctions were issued in the 1880s, 122 in the 1890s, 328 in the 1900s, 446 in the 1910s, and 921 in the 1920s. Employers also used the Pinkerton Agency (a national private police force) to spy on their workers and cultivated close relations with elected officials so they could call on the local police when needed.[12]

These ideas and tools helped lead to the formation of the National Association of Manufacturers (NAM). Established in 1895, NAM first served as a key advocate for foreign trade, but it soon became the bastion of the open-shop drive. NAM spokesmen contended that they did not oppose organized labor per se, but in truth their arguments left no room for unions: employers must have complete control over hiring, firing, apprenticeships, wages, and production, they said. To them, businessmen and the market they defended were virtuous, "American"; labor unions represented "mob rule" and "foreign invasion." Philadelphia business organizations, particularly the Board of Trade, composed some of NAM's biggest backers, championing the campaign, for example, to defeat workmen's compensation laws because they represented "vicious government interference which held the employer responsible for his workers' reckless indifference to danger." One of NAM's strongest affiliates, the Pennsylvania Manufacturers Association, was founded by Joseph Grundy in suburban Bucks County in 1909.[13]

Businessmen, especially those in NAM, understood the need to get involved in politics, and they cemented a relationship with the Republican Party that put their values into policy at the national level and worked exceptionally well in Philadelphia, too. Philadelphia's businessmen were nationally known for maintaining close ties with the Republican-dominated city government and ensuring that candidates with suitably laissez-faire sensibilities such as John Reyburn and J. Hampton Moore won election. Across the state, Republicans dominated elections, winning ninety-five of ninety-six races from 1893 to 1931, and in Philadelphia GOP candidates often received 80 percent of the vote. The city's corrupt Vare political machine had something to do with those numbers; but overall, NAM, Philadelphia's businessmen, the city's middle class, and the GOP accepted the primacy of business in American life. They created a bulwark for business principles that made it exceedingly difficult for labor organizations to build their membership in Philadelphia. All told, only about 10 percent of workers belonged to unions at the turn of the century. Philadelphia's traction workers, because of national and local ideology, politics, and power relationships, had a difficult task in front of them as they contemplated organizing.[14]

The Nature of Transit Work

Transit employees in the late nineteenth and early twentieth centuries composed one of the largest and fastest-growing bodies of workers in the United States. A Bureau of Labor report in 1905 described the national transit workforce as "an army of men" (only fifteen women worked as motormen and conductors) numbering some 140,000, which made it four times larger than any other country's in the world. From 1870 to 1900, the number of Americans employed in the transit industry grew by a factor of seven, from 132 people per million to 920 per million. Most of these employees were young men—companies often set an age limit of thirty-five on new hires—with little education or industrial training and thus few prospects. The industry was concentrated in the northern states between the Atlantic Ocean and the Mississippi River, and Pennsylvania—Philadelphia in particular—was a key part of it. With 15,721 transit employees, Pennsylvania had the second-largest total in the country (about half that of New York), and the great bulk of them worked in Philadelphia. About 80 percent began working in Philadelphia transit at age thirty or under, and nearly a quarter listed their initial occupation as "farmer," with a similar number simply calling themselves "laborers."[15]

Transit companies, especially in the era of growth in the late nineteenth and early twentieth centuries, employed selective policies to hire workers. Companies across the United States, including Union Traction and later the PRT, inquired about men's work background and marital status, height and weight, educational attainment, and even whether they had venereal disease. Prospective employees could "have no use for intoxicating liquors" and had to pass a physical examination more stringent than those required for life insurance policies. Doctors paid special attention "to hearing and to acuity and range of vision . . . and to blood pressure, veins, kidneys, and feet." Applicants had to pass all the physical exams, the moral and background checks, because working in transit, as a PRT training manual put it, was a privilege that required companies to make every effort "to establish the man's character, habits and reputation; for this is an occupation calling for clean, honest, earnest men." With two applicants for every position, the company could afford to be so selective.[16]

That selectivity extended to the ethnicity, race, and gender of the workforce that management hired. Betraying their qualms about the immigrant-heavy urban population, transit companies preferred "country men owing to their greater vigor, strength, honesty, and loyalty, and their willingness to work cheaper." In Philadelphia, newspaper reporters found that most transit workers came from outside the city and relished urban life, especially when coupled with the chance to continue working outdoors. Drivers came from Montana where they fought in the Indian wars, from Delaware Bay where they shucked oysters, from Bucks County where they worked on the farm. "Gosh!" a Bucks County man gushed to the press, "A fellow wants to see something of the world, don't he?" Nationwide, 80 percent or more of the workforce of most transit companies was American citizens. Many of them came from Irish and German stock, populations that had lived in the country long enough to at least begin assimilating. That was certainly true in Philadelphia, where the immigrant population, although never as large as that in Boston, Chicago, or New York City, came chiefly from Germany, Ireland, and, by the early twentieth century, Italy and "Russia" (the last were actually Eastern European Jews). The composition of the PRT's workforce tilted heavily to the Irish, with Germans and then, by the Great Depression, Italians making up a significant, if smaller, percentage. "The Irish and Germans are about as steady as any men we have," said one PRT superintendent, adding that the Irish were "daisies," and "the Germans [were] steady, too." Only a few transit companies employed African Americans, and almost all of them worked as porters, laborers, and janitors. Philadelphia, which at the turn of the century still had only a small black population of 62,613 people (4.8 percent of the city total), did have one area company hire African Americans in 1898. Foreshadowing more difficult times to come, the Philadelphia and Western Company had to suspend the experiment when its white employees walked out. Women found opportunities in Philadelphia transit just as difficult: they got an occasional job in the office or selling tickets but were essentially locked out of the industry until World War I. Once the war was over, they lost those jobs.[17]

Transit workers took pride in their jobs at first, but such feelings faded quickly as the nature of the work became apparent. In the tough economic times of the late nineteenth century, steady employment and a sharp navy-blue uniform gave men a sense of self-respect. Many also told Philadelphia reporters that as young men they liked earning their keep and seeing urban America firsthand as they worked their way through Chicago, Kansas City, Denver, and Los Angeles, as well as Philadelphia. The job, said one driver, certainly "has enough excitement to keep me awake." Nonetheless, many aspects of the job, especially long hours, made it notably onerous. Horsecar drivers worked from morning to night, as many as twenty hours straight in some cities. "Trippers"—men who drove cars on an as-needed basis during peak times—had it particularly bad, since they had to report for work but never knew if they would make a run or not. No run meant no pay for them. Some states, such as Pennsylvania, New York, Rhode Island, and California, passed laws to try to limit transit workers' hours. To legislators of the day, such legislation was intended to safeguard the public, not to protect workers from exploitation. Even so, the Rhode Island law in particular came under legal challenge for infringing on the right of contract, but the state's supreme court upheld its constitutionality. But finding a law constitutional was a far cry from enforcing it vigilantly. In Philadelphia, Speirs found the city's transit workers an especially "sadly overworked set of men" who provided a "daily service of from fifteen to eighteen hours of continuous work with very brief respite for breakfast, dinner and supper." They had no time for normal family life. "When I want to see my children," one conductor said, "I have to see them in bed. I am off in the morning before they are awake and they are asleep when I come home at night. If they want to see me in the daytime they have to wait for my car." A ten-hour day, the men claimed, was ideal, and anything more than fourteen hours was inhumane and put the public at risk, since they had to ride with exhausted drivers. As early as 1868 Philadelphia's horsecar workers held a public meeting to protest their work hours, but the transit companies refused to budge.[18]

Harsh conditions made the long hours more difficult to endure. Workers had to stand for the entirety of their shift, frequently resulting within four or five years in vascular disease, a condition colloquially known as "trolley legs," which forced drivers to wear tight silk or rubber stockings. In the heat of the summer,

Philadelphia drivers had to wear their heavy uniform coats; it wasn't until 1918 that the company finally allowed them to drive in their shirtsleeves. Winters were worse: In Philadelphia and across the country, transit companies fought for years to prevent the installation of windshields in car vestibules, despite stories of drivers literally freezing to death. Transit workers argued that a simple windshield would shelter them from the elements, helping them do a better, safer job, and also protect the car's equipment. To transit companies, though, windshields were up to management's discretion (in one Missouri case a company argued that being compelled to outfit their cars constituted "cruel or unusual punishment"; it lost), and they preferred to save the capital outlay, since the men should know that their jobs required "great physical endurance . . . and [that] little thought was given to their creature comforts." The PRT still had not enclosed all of its vestibules by 1906, when managers worried that doing so would cost the company some $16,000. No wonder Philadelphia's *Public Ledger* could write that company horses were better off, with "shorter hours, more regular feeding time and better care than the men."[19]

Such difficult conditions made an already dangerous job worse. Reports from around the United States showed that brakes, especially early hand brakes, were notoriously balky and led to many accidents. Live wires electrocuted several

FIGURE 4.   In early streetcars, open platforms in winter exposed drivers to the elements. Photo courtesy of Historical Society of Pennsylvania.

transit workers every year. And car men found any number of other ways to die: falling off running boards, mangled in collisions, crushed between cars, going over an embankment, struck by trains, caught between the car and the platform. All told, the United States witnessed thousands of accidents and the deaths of over one hundred trolleymen per year. At the PRT, management, despite knowing the dangers of the job and how the company overworked its employees, held workers liable for any monetary damages that came from harm to people or equipment. But most Philadelphians understood transit workers' circumstances and generally refused to hold them accountable when accidents occurred. "Nor should we expect vigilance and attention from employees worn out by 17 hours of incessant labor," said a coroner's jury after one fatal accident. "The constant occurrence of passenger railway accidents demands from this jury an unequivocal condemnation of the companies who compel men to do work to which the bodily and mental frame is not usually equal."[20]

Transit workers had more stresses than just long hours and bad conditions. In a country going through the throes of urbanization, city officials did not know how to handle traffic congestion. City streets in Philadelphia, New York City, and Chicago were clogged with horsecars, foot traffic, and public transit vehicles moving with little direction. By 1905, Philadelphia's busiest intersection, at Eighth Street and Market Street (a four-track intersection with two tracks going in each direction), had a reported 391 cars pass through every day between 5 and 6 p.m.; that equated to a car every nine seconds. Other intersections were little better: Fourth Street and Chestnut had a car pass every seventeen seconds. Such stress wore men out. So too did dealing with the public and the police. Transit workers lodged their greatest complaints about dealing with drunks, especially on Sundays when young men were traveling to parks in the countryside. In PRT reports, conductors routinely complained of men, and some women, stumbling against sober passengers, vomiting on the floor, and starting fights. Conductors also tried to police smoking, spitting (tobacco and just in general), and lewd behavior that offended more refined passengers. Some felt responsible for patrolling racial boundaries, such as the conductor who reported that one black man "had been in the habit of insulting lady passengers [when he] would sit next to a lady and let his hand slide down back of her and try and work his hand in opening of back of dress." Transit workers also constantly guarded against people stealing rides by jumping on bumpers and running boards, misusing transfers, and simply refusing to pay. The police made matters worse by boarding cars, flashing their badges, and demanding free rides. Conflicts escalated quickly when conductors told them they had to pay or get off the car. "[I] would like to break [your] head. . . . You cock eating bastard," one officer told a PRT man. Conductors could of course give it right back: "You stick your report up your ass you son-of-a-bitch. . . . Go to hell and fuck yourself," a driver told one policeman who threatened to file a report on the PRT worker after a verbal conflict.[21]

C. E. Calkins, a transit worker writing in the *Motorman and Conductor*, captured in poetry the stresses of the job:

The Motorman

I.
He faithfully stands at his anxious post,
As he forces his car though the city's maze
With a skill that is born of the city's host
And a nerve which he constantly displays.

II.
Oh, his head must be clear and his eye be bright
As he clangs along on his noisy way,
And his thought must be quick as a flash of light
Else he would not last from day to day.

III.
Unthought of by most of that busy throng
Which daily depends on his watchful care
With a hand on the brake, and a foot on the gong,
Oh! Motorman, may God ever guide you there.[22]

Low pay also marked the transit workers' experience. Across the country, from the 1890s to the 1920s street railway pay usually ran between 5 and 10 percent behind that of manufacturing workers. Those pay levels held true in Philadelphia, where in the period from the Civil War to the turn of the century conductors, who handled money and interacted with the public, earned $2 per day, and drivers earned $1.50 to $1.75. In 1885, the average pay of Philadelphia's horsecar drivers amounted to 12 cents an hour. Standard practice was to pay the workers at the end of each workday, which meant that the company had no obligation to bring the employees back the next day, leaving them with no job security. With wages so low, workers found it difficult to put much into savings in case of bad times. National surveys showed that over half of all transit workers saved no money in a year, and those who did put something away averaged about two dollars per week. Philadelphia newspapers, such as the *Evening Times*, found the pay scandalous and the young men running the system sympathetic. In a number of articles, editors told Philadelphians of the plight of what they called 365-Days-in-the-Year men. Aaron Scott, for example, was a young motorman with a wife and no children who needed seven days of wages just to pay for his family's "necessaries." The Scotts had nothing set

FIGURE 5.   Crowded Market Street (you can just make out City Hall in the background) shows the stresses of the job, but also how drivers had the power to bring the city to a halt if they went on strike. Photography by Harry P. Albrecht and originally published in *Philadelphia in Motion: A Nostalgic View of How Philadelphians Traveled, 1902–1940.*

aside, no money in case of illness, and no time for the theater or other leisure activities. "With all my hard work we don't seem to get ahead very fast, my wife and myself," Scott told a reporter. "We get enough to eat and to wear, but we can't put anything aside for time of trouble." Samuel Weitzenhoefer, a conductor, recounted how hard he and his family worked, but still bounced from one house to another, always searching for cheaper rents. "With all our industry, we find it hard," he said, adding that his family "simply won't be downed. . . . I work every day in the year because I want to get ahead. I don't want to be a mere machine all my life."[23]

The rules of the job added greatly to workers' discontent. Rulebooks on various systems, including Philadelphia's, ran from twenty-five to one hundred pages, covering everything from safety equipment and collecting fares to appearance on the job and the use of profane language. Management kept a close watch on its workers, handing out demerits on a regular basis and tallying them to determine whom to fire. Jokes abounded about transit workers having sticky fingers with the fares, and many systems required employees to take out $50 bonds that supposedly insured they would not steal from the company. Workers such as

FIGURE 6. Transit workers like Aaron Scott were sympathetic figures, often receiving favorable press that highlighted their low wages and difficult working conditions. Photo courtesy of Historical Society of Pennsylvania.

these, living on the margins, in effect had to pay for the privilege to go to work. In Philadelphia and other cities, transit companies also employed detectives known as "spotters"—often women, because they were less likely to arouse suspicion— who rode the cars, spying on the workers to see if they were stealing fares, drinking on the job, or gambling. Such rules, especially about drinking while on duty, no doubt needed enforcement, although the PRT could take it too far, as when it said even off-duty workers "should never . . . be even slightly under the influence of liquor" or ran investigations of rumored card and craps games on employees' private time. Many employees complained that detectives blackmailed them to line their own pockets. PRT management even turned to passengers to request that they "report infractions [to] aid us in securing discipline, efficiency and good service." While the company thought these were prudent policies needed to discipline its workforce, the employees felt that they were being treated like criminals.[24]

Some managers undoubtedly felt that the workers were indeed criminals; nearly all viewed their workers with contempt. "I have never yet met with anybody that could place the work of a horsecar driver in a favorable light," one commentator wrote in *Electrical Engineer*. "One certainly could not fairly expect a man who spends the day with his nose at the tail of a car horse to realize a very high ideal of life and duty, especially when the whole of his work is done under conditions exhausting alike to temper and physique." A committee of managers from several street railway companies added that "the management does not regard them as skilled workmen . . . that if they were all to resign on the same day, that the company could go ahead the following." "All that is required of an applicant is an ordinary physique and a public school education," employers told government officials. "They must not be too damn smart, not too well educated." No wonder the Street Railway Association thought it could tell the public that they employed "an army of careless, coarse, and incompetent men, who, through their ignorance, carelessness and incivility, have done more to bring street railway into public disfavor than all the official acts of its management and directory." These men, wrote another manager, "were very often of the character that would drink to excess and lead the host in demonstrative cussing." Philadelphia, one commentator noted, mirrored the nation because its "car men were drawn from among the roughest members of the community." Or, as correspondence between a Philadelphia mill owner and PRT management put it: transit workers "are about as dumb as they look." They need to be reminded they "are paid to act and not paid to think." To many transit managers, their employees were mere brutes.[25]

The shift to electric power did temper some of these views, but transit workers' treatment did not substantially change. Electric cars were bigger and faster and, according to a government report, "required a higher order of intelligence" to operate them. They also exacerbated the stresses of the job. "Electricity for street-car propulsion," a committee of PRT workers wrote, "has brought with it greater responsibilities, labor of a more exhausting and exacting nature, and a mental and a nervous strain." Transit companies including the PRT understood that they needed to provide more training: workers needed to understand how electric motors worked, how to care for the equipment, why keeping to schedules mattered, and so on. This training, managers trusted, would lead to better workers—C. D. Emmons from the Fort Wayne and Wabash Valley Traction Company fancifully hoped it would even inspire an esprit de corps—and perhaps show that the work "is thus being lifted out of the class of unskilled jobs." Yet wages did not advance much, conditions remained difficult, and the rules of the job just got more onerous. The shift to electricity may have made the work more "skilled," but that was small consolation to a workforce composed of men like Aaron Scott and Samuel Weitzenhoefer that could detect little change save for the form of power that they used.[26]

The difficulties of the job, the low pay, the lack of respect led to a workforce in constant flux. Nationwide approximately one-third of transit employees quit or were fired every year. Most workers saw transit jobs as temporary employment, a means to "tide over a period of idleness [and then] leave it without much hesitation." An internal survey of New York City's Metropolitan Street Railway Company in the early 1900s found that 34 percent of the employees had less than one year of experience, 59 percent less than three years, and only 29 percent had worked for the company five years or more. The PRT had a higher turnover than the national rate every year in the early 1900s, often significantly higher. In 1906, the company turned over 65 percent of its workforce; the figure was 70 percent in 1908, and a whopping 89 percent in 1910 (an admittedly anomalous year because of the general strike). All told, the PRT's average turnover rate from 1906 to 1910 was 67 percent. "Something must be wrong," wrote PRT official Daniel Pierce, "in an industry where every year about one-third of the employes give up their work or are discharged."[27]

Pierce was correct, no doubt, but what he and his fellow managers failed to understand, or refused to articulate, was that they had created and then profited from an industry that made the workplace so low paying and miserable that few employees wanted to stay. The PRT's workers, like their brethren across the country, however, understood this, and by the 1890s they were ready to act, ready to organize. They of course recognized the difficult atmosphere in which unions had to operate at the city, state, and national levels. Joining a union was no small step. Yet after years of abuse, they were ready to listen to that call of the Amalgamated's journal *Motorman and Conductor* that told them it was time to "wake up and organize and show these magnates that the time for bulldozing employees is past and gone and we must be treated as men."[28]

Organizing Philadelphia's Transit Workers

Before the Amalgamated, attempts to organize transit workers were spotty and uneven. The earliest organizing effort started in New York City in 1861. There, John Walker, a driver on the Third Avenue line with twenty years of experience, grew tired of the treatment meted out to transit workers and organized them into a "benevolent association." Walker's organization disavowed strikes at first, but as Third Avenue repeatedly cut their wages, the men walked off the job. The outcome of their strike is unclear, and the organization soon fell apart, as did many such associations, under the pressures of the Civil War.[29]

It took two decades for another organizing campaign to emerge, also in New York City. In 1883, a printer riding the cars in New York struck up a conversation with one of the drivers and persuaded him to join the Knights of Labor. He and thirteen of his fellow drivers formed the first local assembly of horse railroad employees in September 1883, Local 2878, and by December of that year they had five hundred members. Transit company repression broke the Knights' ranks several times, but by late 1885 they were firmly enough established to win a new contract from the Third Avenue Railroad and present a list of grievances to the Sixth Avenue Line. Sixth Avenue ignored the demands, but capitulated after a five-hour strike. The Knights also won strikes on the Broadway and Fourth Avenue Lines (the latter owned by Cornelius Vanderbilt). Smaller companies across New York City saw the working-class solidarity and quickly reached agreements with their employees.[30]

The Knights' success in New York City transit led to expansion in Philadelphia, Baltimore, and elsewhere, with workers especially challenging their long hours on the job. In Philadelphia some three thousand transit workers joined the Quaker City Protective Association in 1886 and through their president, Morris Weidler, presented demands calling for a twelve-hour day, standard pay of two dollars per day, and time for meals. In the negotiations, Weidler appealed to the transit companies' self-interest. Pushing workers so hard, he said, made them more inclined to steal and damage equipment. "You can get better service out of a man if you pay him fair wages than if you make him work when he ought to sleep, and pay him nothing for the sacrifice," Weidler told the *North American*. "To make a man work from sixteen to eighteen hours a day is inhuman." Weidler's considered stance resulted in his dismissal from the transit service, and coworkers who boarded in his home moved out for fear of reprisals. Nonetheless, the Knights continued their campaign. Pressed, the Board of Presidents, the organization of city transit companies established in 1859, asked for a week to consider the demands. With the workers talking strike and the public, as one commentator put it, "so thoroughly in sympathy [with] the men that further resistance would be dangerous," the companies decided to come to terms. They agreed to limit work hours as practicable, set the pay rates requested (although many employees still put in a seven-day week), promised not to discriminate against anyone who had joined the Knights, and conceded that the workers had a right to organize. But they refused to accept that they had to negotiate with a union or that the workers had any legitimate say in who should be discharged.[31]

Given the state of labor relations in the city, state, and nation, most commentators thought the workers had done well. They got concessions on hours and wages, plus at least limited recognition. Most important, there had been no violence, largely because the transit company owners knew the public was against them, and like many in their class, they feared a broader class war. "Public opinion," wrote Speirs, "was so thoroughly in sympathy with the very reasonable demands of the men that further resistance would be dangerous." "What threatened for some time to be a bitter conflict ended happily," he continued, "[with] the reduction of the barbarously excessive hours." These accomplishments, however, proved to be the apex for the Knights. Many workers in the transit and textile industries left the organization in the ensuing months as they came to believe in light of the Wingohocking walkout that the Knights would never support strikes and they had thus achieved all they could with the union.[32]

The Amalgamated offered the next step forward for Philadelphia's transit workers. Begun in Detroit in 1891 as the streetcar industry expanded, the Amalgamated faced a difficult road to stability. The union held its first convention in Indianapolis in 1892, and two years later it had eleven locals but only two thousand dollars on hand. That convention was so gloomy, because of transit industry repression and the pall cast by the Depression of 1893, that only twelve delegates attended, and many wondered if the Amalgamated would survive. Yet it did, chiefly because the union found a leader who understood, articulated, and battled for the needs of his membership.[33]

William D. Mahon, the grandson of immigrants from Northern Ireland, grew up working in the coal mines of Ohio's Hocking Valley. There he learned that hard work too often led to nothing but abject poverty, and when he secured a job on the street railways of Columbus, Ohio, in 1886 he quickly helped establish a union for the transit company's workers and served as its first president. That same year, the AFL held its founding meeting in Columbus, where Mahon met Samuel Gompers, beginning a relationship that saw the two become allies as the AFL and the Amalgamated both grew in size and importance. Mahon abhorred violence and, scarred by the confrontations of the 1894 Pullman strikes, always pushed for arbitration. Although he is now largely forgotten, people interested in the fate of labor at the time routinely lauded him for his commitment and "common sense" views. Carl Sandburg, for instance, called him "one of the figures to give meaning to the cause of labor and the mystery of democracy." Mahon no doubt appreciated such praise, but it could border on the hagiographic, and he more humbly asked that his epitaph from working people simply read, "He was our friend." Mahon served as the Amalgamated's president from 1893 until his death in 1946, and over the course of that remarkable span he built a strong union: starting with fewer than two thousand members in the early 1890s, it reached thirty thousand by 1904, and peaked at one hundred thousand in the 1920s. At one point in its heyday the Amalgamated maintained agreements with more than 350 transit companies and negotiated annual wage contracts that nearly tripled workers' pay between 1902 and 1927 (from $615 to $1,737).[34]

Of course a union does not grow simply because of its president. Over the course of his many years in office, Mahon benefited from the growth of the AFL and his friendship with Gompers, from the expansion of the streetcar industry, from transit workers' growing class consciousness, and from moments of union strength such as World War I. But Mahon's tenure was not mere serendipity. He had a notable ability to articulate transit workers' needs and then advocate for them. From the earliest days of the Amalgamated, Mahon told workers that they deserved a union that matched their "intelligence, efficiency and skill." He advanced their demands for medical and death benefits. He pushed arbitration at every opportunity. And he strove overall "to secure employment and adequate pay for our work; to reduce the hours of daily labor; and by all legal and proper means to elevate our moral, intellectual and social condition." At times, he could be scathing, writing that his union's story was one of "wrongs and mistreatment, a story of long hours and small pay, and this not in the dark ages but under the glowing sun of our nineteenth century's civilization." He could also be sarcastic, publishing "The Railway Man's Ten Commandments," which included such edicts as[35]

> Thou shalt not seek the benefit of any trade union.
> Thou shalt make all thy humble applications in vain.
> Thou shalt commit 300 rules to memory.
> Thou shalt not have regular times for thy meals.
> Thou shalt not make any complaint against these conditions, except through thy station master, and then only for the increase of his waste paper basket.
> Thou shalt not covet any superior position. Thou shalt not do anything to benefit thy wife, but thou shalt obey thine official—be his man servant, his gardener, his ass, or anything short of being his neighbor.[36]

Mahon's Amalgamated fit comfortably into the AFL's brand of "pure and simple unionism" that focused on wages and working conditions. But he also had what legal scholar Sidney Harring termed a "social democratic political agenda" that led him to challenge corporate power. In particular, Mahon advocated for municipal ownership of public transportation as an ultimate social goal. Such advocacy placed the Amalgamated within the Socialist strain of American, and Philadelphia, politics. That agenda, however, was geared to a long-term goal—transit historian Scott Molloy correctly labeled Mahon an "evolutionary socialist"—and the Amalgamated subordinated its push for public ownership to demands for adequate pay, reduced hours, sick benefits, and arbitration of disputes.[37]

The Amalgamated, then, was a slender reed—whatever Mahon's strengths—for Philadelphia's transit workers to cling to in the years after the Knights faltered. But it was their only reed. In early 1895 several workers held an open meeting to try to form a union, and the company fired many of them immediately afterward. A conductor on the Philadelphia system, Terrence Clark, then contacted the Amalgamated and asked for help. Mahon and other national leaders went to Philadelphia to assess the situation and found it "discouraging" because of the coercion the men faced. Yet the Amalgamated, knowing the size of the system and hearing the desire to unionize that the men espoused, chose to make Philadelphia the focal point of its organizing for the year. Mahon knew the men feared reprisals if they joined the union, so he spent several months quietly visiting the carbarns, telling the workers what the Amalgamated stood for, how they deserved representation that advanced their interests. Widener, who had been warning his workers for a year that he would fire anyone involved with the Amalgamated, posted notices on company bulletin boards announcing that more than one well-known conductor had been discharged for "pretending to take an interest in his work with the company [while] secretly exerting his influence and taking an active interest in the affairs of the Amalgamated." These were intimidation tactics, Mahon reassured the men, and they would win out if they just stuck together. There could be no repeat of the aftermath of 1886 if they were to win. Mahon's words surely helped, but the fact that Amalgamated organizers were so poor that they often had to live five to a room and eat for fourteen cents per meal likely detracted from his message. Despite the difficulties, Philadelphia's transit workers hung together well enough to receive a charter a few months later (Division 477) and were ready to take on the Widener interests and Union Traction.[38]

Despite a severe national economic depression, late 1895 offered a favorable moment to challenge Union Traction because the company's workers had recently organized and because Union's recent consolidation had led to the end of the free transfer system. This move sparked a town meeting at the Academy of Music where speakers denounced the "greedy corporation," hashing out "all of the old grievances against the Company, its underliers, and their excessive rentals." From that meeting, a petition emerged calling on the city to exercise its right to purchase the street railways and urging all citizens to question political candidates on their stance about transit fares. Voters, the petition read in language clear enough to hearten the city's Socialists, should "support no candidate who will not pledge himself to sustain the rights of the people." "Loud applause," wrote one observer, greeted "every expression of the speakers in favor of municipal ownership." One newspaper covering the event warned that the company had "estranged the public and embittered the city." "War is costly," wrote the *Philadelphia Press*, and "Union Traction has chosen war. . . . *When strikes come, public opinion will favor the strikers* [emphasis in original]."[39]

Sensing this unrest among the public and the workers, leaders of the Amalgamated presented demands to management in October: two dollars per day, a ten-hour workday (twelve hours was the rule since 1886, but seldom enforced), vestibules for all cars, and the right to belong to a union without fear of discharge. Management refused to recognize the Amalgamated, refused to arbitrate, refused even to meet with Mahon, and ignored the union's demands for two months. Such intransigence infuriated observers. The Toynbee Society wrote, "The company knows that organization is or may be an advantage to the men, and it is not willing that they should have this advantage." In its *Motorman and Conductor*, the Amalgamated wrote of the Liberty Bell and other great Revolutionary relics found in Philadelphia, the freedoms that sprang from the city, and contrasted that history with the plight of workingmen who had to meet "using all the caution of a thief, lest the railway companies' spotters and spies should find out the men and discharge them." "Yes," the journal concluded, "this is the condition under the shadow of that old monument of liberty." Even middle-class Philadelphians, the historian John Hepp IV found, were angry enough about Union Traction's "Trolley Grab . . . raising fares and reducing wages" to support the workers' cause.[40]

Receiving no satisfaction from the company, Union Traction's five thousand workers walked off the job on December 17, believing their solidarity, the support of the public, and the effect their strike would have on business interests at Christmastime would carry the day. Following a mass meeting at the Academy of Music on December 16, all the workers struck at 5:00 a.m., and the system came to a halt by rush hour. Ordinary Philadelphians immediately showed their support of the strikers. They walked "for conscience sake," as one clergyman put it. They donated some $10,000 to the union cause. And across the city they protested in the streets, especially in the working-class districts of Kensington, northwest Philadelphia, and along Market Street. Sometimes they verbally challenged the company; sometimes they attacked streetcars driven by nonunion crews. The worst conflict took place outside the Baldwin Locomotive Works in

North Philadelphia, where a crowd hurled bolts and bricks at a trolley driven by a strikebreaker. They broke a woman's jaw and injured two men with flying glass. Police charged the crowd and, in the words of the *New York Times*, "clubbed it vigorously." But that was just one example; across the city Philadelphians destroyed tracks, barricaded rails, and overturned trolleys. Overall, some three hundred streetcars were destroyed. Several times the police clubbed people marching on city hall, including a young William Z. Foster, future leader of the American Communist Party, who always remembered the transit strike as part of his "baptism in the class struggle." The transit workers, it should be noted, had received an admonition from Mahon to "keep cool and let there be nothing like rioting" and generally remained peaceful. Most of the violence came from police, strikebreakers, and residents of working-class neighborhoods, people whom the *New York Times*, in the class-biased rhetoric of the day, described as "knots of scowling, depraved looking men."[41]

With violence shaking the city and the Christmas shopping season in jeopardy, city elites pressured Union Traction to settle the strike. Mayor Charles Warwick, despite repeated police violence, vowed not to use his force to break the strike but just to "keep order" and rebuffed offers of help from Pennsylvania governor Daniel Hastings. Union Traction officials apparently felt pressure from men such as John Wanamaker, who did not want to inflame the situation because he did not want to see further damage done to his city or his department store's profits (the strike nonetheless cost merchants some $2 million). A group consisting of Wanamaker, major Union Traction shareholder Thomas Dolan, Catholic archbishop Patrick Ryan, and members of the Christian League persuaded both sides to come to the bargaining table after seven days. There Union Traction and its employees reached an agreement that the company would consider issues of wages and hours, allow its workers to join a union, and rehire the fired workers if it had enough runs. But Union maintained that it did not have to recognize any labor organization or submit to arbitration.[42]

Such meager concessions sowed discord in the Amalgamated. Mahon and the Amalgamated leadership, believing it was the best they could do, backed the deal and spun it as a victory, claiming the strike was the biggest event of the year for the organization. The Amalgamated claimed it opened the door to organizing on the East Coast and trumpeted correspondence from transit workers in Baltimore and Kansas City that said Philadelphia set the pattern for their cities. More, the union claimed that in Philadelphia, Union Traction now knew its employees had "certain rights that must be respected" and that the Amalgamated was firmly established on the property. But many workers knew better. They believed that after the risk they took, they had been "sold out," and hundreds never got their jobs back, despite the company's promise. Most regarded a new grievance board as a paper tiger, especially after Union Traction's leadership denounced the Amalgamated as an "outside association" that it would never allow to come between the company and its men. For several months after the strike's settlement, many workers grumbled and called for a second strike, especially to secure the ten-hour day. Mahon, knowing how weakened his union was by the recent strike, refused to allow another walkout so soon. He appealed for their patience, asking the workers to "show your loyalty to your officers and committee and all fair minded people of Philadelphia [to ensure that matters were] properly adjusted." If the company betrayed them, he concluded, "I again assure you that our association will stand by you to the last." Behind the scenes, though, Mahon admitted to Gompers that 1895 had at best been a "successful retreat."[43]

Discontent continued to fester among the workforce. Overall, the trolleymen trusted Mahon, but they were unhappy with the return on all they had invested in the strike. Their union, which had wagered everything, had largely been crushed, and the workers were, as one commentator put it, "utterly disorganized and unwilling to reorganize." Many Philadelphians continued to believe the job was too arduous for the well-being of drivers or their riders. But with Union Traction in control, for the next few years transit workers could do little more than "call attention to the hardships of long hours in the cold, and, asserting their own helplessness, beg the assistance of the public in securing a ten hour day and the enclosure of the front platforms." The 1895 strike, then, showed how strained labor relations were in Philadelphia's transit industry, how the Amalgamated suffered from internal divisions, how many members of the public supported the workers, and how resentment and violence simmered just below the surface. It took another fifteen years, but all these issues eventually came again to the fore in one of the most violent strikes in Philadelphia's history.[44]

TIME OF TROUBLES

On March 7, 1910, two trolley cars headed down Frankford Avenue in north-east Philadelphia. It was usually a normal occurrence, but not in the midst of a brutal transit strike that had just turned into one of the largest and most violent general strikes in American history. Citizens on the street saw the cars coming and scattered. The fifty strikebreakers on board had broken out the windows and were leaning out the empty frames with a revolver in each hand. "Cursing and shooting with equal rapidity," a reporter wrote, "the men made a swift run to Allegheny Avenue. . . . Two people were shot and scores had narrow escapes." One bullet gashed a sleeping baby's head, another struck a policeman's hat, just missing the head inside. Two other men were not so lucky: they were shot by the "finks," as imported strikebreakers were called, and a third man's throat was cut mid-shave by a barber when they both jumped as a bullet smashed through the shop's plate-glass window. Enraged Philadelphians took to the streets when the strikebreakers turned toward home, stoning the cars as they passed. They did little damage, though, and four more citizens were shot in the melee. This "wild ride," as one commentator called it, only hints at the violence that swept Philadelphia, as Kensington, Frankford, Germantown, and other working-class neighborhoods saw tens of thousands of people use the general strike to confront corporate power, as symbolized by the PRT, that they saw dominating their lives. All told, twenty-nine people were killed and hundreds injured in the conflict.[1]

This chapter traces a difficult decade for Philadelphia's transportation system and its workers, and culminates with an examination of two wrenching strikes: a 1909 transit strike that resulted in the workers' victory, and the 1910 transit strike that grew into one of the great general strikes in U.S. history but ended with the Amalgamated being crushed. For the owners and managers of the PRT, the century dawned with a growing realization that their industry was likely at its apex and that they were gravely constrained by the five-cent fare. They had to continue physically building the system, including digging the city's first subway (the Market Street line); but with limited financial resources and deep reservations about working with public authorities, they were hard-pressed to construct the kind of transportation network Philadelphians needed and demanded. The cost of the Market Street line, coupled with the city's need to build a system that ensured its continued growth and vitality, led to a contract between the PRT and Philadelphia's city government—known as the 1907 agreement—that structured the future of city-company relations for decades. That contract, which gave the city greater oversight of the PRT and the right to buy the system after fifty years, sparked conflict between the two entities in the years to come as transit company management sought to maximize profits while the city pushed for greater service. The 1907 agreement, then, by the mid-twentieth century came to be a representation of, and provide an arena for, the competing needs of private enterprise and the public good, or, put more broadly, the clash between "capitalism" and "socialism" (however perverted both terms became in the public discourse).

With the construction of the system and the 1907 agreement highlighting the financial difficulties of the PRT, worker activism pushed management to the breaking point. Of course, PRT leaders were already predisposed to treat any demands as a challenge to managerial authority and the "proper" order of America's economic system. Transit workers, at the same time, were still trying to recover from the 1895 strike and harbored great discontent about their employers and workplace. By 1909–1910 they were ready to press their demands harder than they ever had before and felt little concern about the financial plight of the company when they had far greater problems of their own. In a city riven by class conflict, working-class communities knew the way that Philadelphia employers often dominated their lives. More, they knew the way the PRT in particular dominated the urban space of their neighborhoods. When the transit workers went out on strike, many were ready to support their working-class brothers. From this array of issues—the PRT's troubled finances, worker organizing and demands, managerial intransigence, the support of the city's working-class community— grew two transit strikes that culminated in one of the largest general strikes in American history. These events focused Philadelphia's, and the nation's, attention on the city's transportation system, its use of violence to control its workforce, and the labor question that it brought forcefully to the fore.

The Cost of Moving the Masses

Transit remained, in the first decade of the twentieth century, a lucrative industry, at least for some investors and managers, but boom times had passed. Few save the rosiest boosters after 1900 echoed Edward Higgins's 1895 claim that street railway investments would compare "most favorably with that in other fields of industrial enterprise." Nonetheless, a 1904 study showed that across the United States, street railways generated nearly $300 million in revenues, which amounted to 15 percent of what steam railroads earned. Net income stood at approximately $35 million, and that number climbed to some $60 million by the end of the decade. By that point, most industry leaders believed public transportation as a financial investment had reached its zenith.[2]

Ridership numbers showed transit use increasing across the Western world—in Germany and Britain, New York City and Philadelphia—from 1880 to 1910, but increasing costs threatened the underpinnings of the industry. The conversion from horse power to electricity strained every company's budget. Cars grew from eighteen to thirty to fifty feet in length and were commensurately heavier. Extra weight necessitated heavier rails, more roadwork, and greater upkeep. Material costs for items such as rails, ties, switches, and wheels skyrocketed by 87 percent between 1895 and 1910. Add in the price of miles of electric wire, motors, generators, and other equipment, and it is small wonder that analysts found stock prices declining at almost every system throughout the first decade of the 1900s. Labor costs also increased, although much more modestly, at a rate that mirrored inflation, rising by 9 percent between 1902 and 1907 (average wages went from $605 per year to $658 per year over that span). Weaker companies could not survive the new financial realities: in 1909 alone twenty-two transit companies went into receivership, and analysts widely expected the trend to continue. They proved prescient, as the number of those proceedings averaged eighteen per year until the United States entered World War I.[3]

While rising costs threatened the financial existence of many systems, transit companies remained locked into a five-cent fare that most managers found outmoded and insufficient. Five-cent fares generally began in the horsecar days half a century earlier, and omnibus lines often charged six cents in the decades before that. Across the country, the five-cent fare became a shibboleth that working people claimed as their right and politicians mandated (California, for example, set the rate by state law in 1877). "The 5-cent flat fare," geographer James Vance Jr. wrote, "was a foundation of the industry." It was "the measure of adequate transportation facilities" and opened up the city to working-class people. "For the first time, factory workers could live well away from the factories without bearing a high economic burden thereby, and the more menial office employees could suddenly think in terms of suburban living." But with inflation and longer commutes in growing twentieth-century cities, a nickel no longer covered transit companies' expenses. A rider in Chicago could travel thirty miles for five cents, in New York City eighteen miles, and in Philadelphia nineteen. At the same time, European riders paid approximately one cent per mile. The economics, wrote the Street Railway trade association, did not add up: "With the longer rides, the increasing cost of labor and materials, the increasing use of transfers, and the enlarged requirements as to service and facilities imposed by public service authorities and by public opinion, the return to the street railway industry with fixed rates of fare . . . has in general become insufficient."[4]

In an effort to stay afloat, some companies tried to implement a zone system that charged more for longer distances traveled; others wanted to charge for transfers. Nowhere did such efforts find support. New York City refused to raise its fare into the post–World War I era. Cleveland's policy so limited fares that stock analysts thought city regulation had crossed the line into confiscation. Transit companies' only chance at survival, lamented Thomas Conway Jr. in an article on decreasing financial returns in the industry, was to give a "clear and frank explanation of the matter to the public, in order that they may see the justice of the street railways' efforts to correct abuses, and may give the corrective program the support of public opinion, without which it is doomed to failure." Given most companies' well-known history of watered stock, people did not want to hear it. Fare increases, they had learned, or at least believed they had learned, simply soaked the public for the benefit of speculators. The speculators, of course, had brought this habit of mind upon themselves.[5]

Philadelphia followed and helped set this national pattern, in some ways serving as a worst-case scenario. The city's transit system carried 144 million passengers in 1890, a number that grew to 222 million in 1895, 292 million in 1900, and 390 million in 1904. The company ran a surplus in the first decade of the twentieth century, although it was seldom more than a million dollars and dipped as low as $220,000. Electrification bit deeply into profits. The cost of laying track and electric line rose from $38,500 per mile in 1890 to $86,000 in 1900. At the same time, the company still had to pay off now obsolete cable lines and continue payments to its multitude of underliers. So much money going to initial building projects and underlying companies damaged the PRT greatly. It could not afford payments to its own stockholders, deferred maintenance at all costs, and seldom bought new trolleys and other equipment. Such practices increased public criticism and made it even more difficult to encourage investment. One local businessman and Republican reformer, George Earle, who was known for rebuilding banks and railroad companies, did come on board, but even he said he did so to "serve my City [rather] than any corporation."[6]

The clamor for a subway to move traffic on an east–west axis through the city highlighted these financial challenges. As early as the 1870s and 1880s, civic boosters, the press, and ordinary Philadelphians had begun to call on the city's transit companies to build elevateds and dig subways. It became common wisdom, though, to argue that elevateds were "ephemeral structures" ill-suited to Philadelphia's narrow streets and that an underground system like London's was the only real solution. Only an underground railway, the *Philadelphia Evening Bulletin* argued, "will solve the problem of rapid transit without interfering with private rights, or obstructing and disfiguring the streets." The *Philadelphia Star* agreed. "Those who live in West Philadelphia, Kensington, Richmond, or the northern portion of the city, and who are required to spend at least two hours of valuable time each day in getting from their homes to their places of business, have a thorough appreciation of the annoyance and absolute loss sustained by them every day in the year," wrote the editors. "The time for rapid transit has come."[7]

Of course subway construction was more expensive than elevateds or streetcar lines, especially because of the snarl of wires and pipes under Market Street, and Union Traction, followed by the PRT, refused for over a decade to start digging. The public clamor for subways, combined with the PRT's mounting financial difficulties, led John Mack to enter the transportation industry and secure the rights to build a subway. His plans forced Widener to dissolve Union Traction and create the PRT in its stead. Mack's interests were absorbed into the system—with his stockholder supporters getting a robust 18 percent real return on their money—and PRT turned its attention to a proposal to build not one but five high-speed lines. These routes would have run through Germantown, Passyunk, Frankford, and along Ridge Avenue and Market Street. With the Mack threat passed and the expenses evident, the transit company killed all the proposed lines except Market Street, thus burying a plan that would have brought rapid transit to nearly every corner of Philadelphia. To pay construction costs, the PRT issued another $30 million in stock with a guaranteed return of 6 percent—a "guarantee" the company could not make good on until 1922 because of standing debts to underliers, which now included Union Traction, as well as the darkening financial picture of the transit industry.[8]

Construction proved far more difficult and expensive than expected. Digging began in April 1903 and proceeded relatively easily through earth and gravel from City Hall west to the Schuylkill. There the PRT had to bridge the river and erect an elevated line out to Upper Darby. Major problems came at City Hall and eastward, where the subway tunnel had to skirt the immense new government building and then make its way through hard rock almost to the Delaware River, where it emerged from underground to become an elevated line that served the ferries to Camden. With all of the buried gas lines, many unmapped, construction workers had to dig the entire tunnel east of City Hall with picks and shovels. Still, a gas main exploded in 1906, shattering every window within a block and severely damaging the tunnel. All told, it took crews four years to complete the tunnel west of City Hall and a fifth year to finish the line east of the building. The project cost between $18 million and $22 million, about twice the initial estimate. Observers worried that the expense would so hamstring the already notoriously parsimonious company that it would not be able to make any improvements to the system.[9]

Philadelphia's political leaders and PRT management knew this was an untenable situation for an industry vital to the city's long-term prosperity. So in 1907 the PRT approached municipal officials about putting in place a new contract that would ensure the company's fiscal health. Both sides were ready to deal: the

PRT wanted to clarify its obligations and lessen its historically onerous paving charges; the city wanted to stabilize its tax income, share in company profits, have membership on the PRT Board of Directors, and gain the right to purchase the system in the future at cost. Neither side, despite the economic problems of the transportation industry, engaged in substantive discussions about immediate public ownership. The PRT and the city both thought they got what they wanted in the agreement, at least at first. The PRT agreed to lock in its fare at five cents per ride and sell a six-strip ticket for twenty-five cents. The fare could be changed only if both parties agreed. The city received representation on the company's board, with the mayor and two citizens chosen by city council receiving seats. The city secured the right to examine the books and also got to split evenly any profits above the amount necessary to pay a 6 percent dividend to stockholders. In lieu of paving, Philadelphia received $500,000 per year for the first ten years and an additional $50,000 per year each decade afterward. Finally, the city received payments of $120,000 per year for the first decade, and an additional $60,000 per year each decade afterward, that were earmarked for a sinking fund that could be used to buy the system in 1957 or later.[10]

This contract played a key role in the company's history until the city finally purchased the system in 1968. At its heart, it tied Philadelphia's government to its transit system, and by dint of public representatives' seats on the board turned a private entity into a quasi-public one. In the short term it dampened some of the disputes over issues such as paving and snow removal. It helped stabilize the PRT's finances and headed off bankruptcy until the 1930s. It also gave city attorneys the opportunity to review the company's books and thus assure local officials that they were receiving fair compensation from PRT profits. Philadelphia's merchants and the press particularly liked the contract because it promised to rationalize the company's cash flow, which in theory would help it provide better service and bring more customers to city stores. But in the long term, the contract became deeply problematic, as it pitted company versus city needs, profit versus public interest motives, and led to intensifying corporate-municipal conflicts in later decades.[11]

Despite the short-term positives, some observers from the start recognized drawbacks in the contract from the city's perspective. City Councilman Edwin O. Lewis led this camp. To him, Philadelphia had surrendered any financial power over the PRT by doing away with the purchase option for half a century, and now had to cling to an ineffective police power that the company knew it could ignore. The city had no power to levy higher taxes because it had locked itself into a fifty-year contract. And the sinking fund, Lewis argued, would likely be misspent, and the company now had no obligation to maintain the public streets it occupied. Other critics highlighted the point that the city owned no stock in the company, and its board members were little more than glorified advisers. The 1907 contract, Lewis told the press, was "criminally one-sided," an "absolute surrender," a "stupendous blunder." Overstated perhaps, but time proved he was closer to correct than any opposing view.[12]

Return of the Amalgamated

The weakening financial condition of the transit industry, the negotiations that led to the 1907 contract that involved the city government more formally in the affairs of the PRT, and an industrial depression that same year that increased pressure on workers in all fields combined to give the Amalgamated its first opening in Philadelphia in over a decade. Crushed in 1895, it was not until 1906 that PRT workers again reached out to the Amalgamated to come in and help them organize. President Mahon sent the union's international vice president, P. J. Shea, whom Mahon sometimes described as the Amalgamated's Napoleon, to Philadelphia that year. Once there, Shea found "much sentiment for organizing" but also "a serious hopelessness which was born of the threatening attitude of the company." Shea and his compatriot Rezin Orr, the Amalgamated's treasurer, broached the subject of meeting with PRT president John Parsons, who had risen through the ranks of the Philadelphia and Chicago transit industries under the tutelage of Widener and his associates, to discuss the workers' demands—a ten-hour workday, an increase in pay to twenty-four cents per hour, and the right to arbitrate their differences with management—but were immediately rebuffed. Shea then implemented an organizing plan that took him into every carbarn in the city. There, he argued that if PRT workers joined the Amalgamated they would win better pay, improved working conditions, and the right to arbitration. Company repression forced him to hold some of the meetings in secret, but by November 1907 the Amalgamated could claim that half the workers had joined the union, and they held several mass rallies to champion their cause.[13]

As the Amalgamated gained momentum, the PRT ratcheted up its repression. General Manager Charles Kruger led the charge. Kruger, a native Philadelphian who had served as secretary and treasurer of Union Traction under Widener, was known for his hatred of unions. In an attempt to squelch the organizing, he and Parsons used company detectives to shadow union leaders and intimidate rank-and-file workers. They "invited" their employees to company-wide meetings where they could discuss the supposed dangers of the Amalgamated away from representatives of the union, who were barred at the door. One such meeting drew 850 attendees, who listened to a series of speakers tell them how happy they should be at the PRT, that they were "getting as much as they were worth."

The audience sat stone-faced, but the PRT still issued in their name not one but two public statements in which the workers avowed that they were "loyal to our employers, and not in sympathy with the present strike movement" and "perfectly satisfied with our present conditions and [willing to] use all means within our power to oppose and evade any strike whatever." The PRT, a reporter noted, had really gone to the "whip hand" now in its effort to shut down the Amalgamated. These meetings, "whip hand" or no, did not always meet Kruger's expectations. At one gathering of three hundred workers, company loyalists refused to take the stage, and instead rank-and-file motormen and conductors repudiated the "satisfaction" they supposedly got from their work and demanded that the PRT increase their pay and limit their hours. That "boomerang meeting," as the press gleefully called it, ended in haste.[14]

Kruger and Parsons next turned to forms of repression that threatened a violent clash. First, they contracted with James Farley, the nationally known "King of the Strikebreakers," to bring in eighteen hundred men, station them at suburban Willow Grove Park, and prepare them to drive streetcars in the event of a strike. Ninety percent of these strikebreakers had just participated in a nasty transit strike in San Francisco and were widely reported as a rough crowd, "brawny and ready for anything." The PRT staffed Willow Grove with an army of two hundred to feed and care for these men, as well as armed guards to keep the peace. Still, Willow Grove became known as "Hobo Park," and fights routinely broke out in the crowd of aimless young men. Most Philadelphians, especially in the working class, reading accounts of this small army of ruffians encamped on the city's outskirts undoubtedly scoffed when they heard Farley's claims that he knew "every man out there personally and a finer body of railroad men you cannot find anywhere," or read Parsons's statement that the PRT had hired these men only because the company "owed certain obligations to the citizens of Philadelphia" and was just trying to meet its civic duties. As one driver lamented, rather than paying salary, room, and board for all these strikebreakers, it would be nice if the company raised his pay instead. "Gee!" he told the press upon hearing of the strikebreakers' extensive dinner menu, "I wish I could set a table like that, but my salary is too small to think of it. It takes all I can make to buy clothes and shoes for the little ones."[15]

The PRT added to its show of force by also bringing in the city police. Old-line conservative mayor John Reyburn agreed that the company could have use of the police force, and director of public safety Henry Clay immediately mobilized his men. Clay armed his officers with riot sticks and placed them, as the *Philadelphia Record* put it, on "war footing." The Amalgamated's Orr claimed his union did not want a strike, arguing that it was Clay who was "apparently determined to force a fight first and argue the point afterward." "His order sounds almost like the dare of a hot-headed troublemaker," Orr continued. "If he really wants to 'protect the public,' he could do it more efficiently by endeavoring to bring about a peaceable settlement between the car men and their employer. These preparations for war create no good feeling." One policeman, who claimed he supported workingmen, said the force was being blatantly used as "union busters, and nothing else." "The display of force," he continued, "was made with the sole purpose of showing to the union conductors and motormen that their organization might as well disband; that the whole power of the city would be used to suppress the organization." Perhaps, members of the Amalgamated and their supporters in Philadelphia began to wonder, having the city in bed with the company was not so good for workers' interests. Given the context of the times, maybe it just gave a conservative mayor and his director of public safety an excuse to bust the union in the name of the public interest. The *North American* mocked "General Clay" as conducting a "joke" of a war against fellow Philadelphians in the name of protecting the city's profit margins. Such strong-arm tactics, wrote the *Philadelphia Press*, would have the opposite effect of that promised, likely leading in a direction that the city supposedly wanted to avoid: conflict, strike, and violence.[16]

Were it not for cooler heads in the Amalgamated—some would say an overabundance of caution—such an assessment would likely have proven accurate. Throughout the November–December campaign, though, the Amalgamated repeatedly said it did not want a strike. It merely wanted a meeting with President Parsons and Mayor Reyburn to discuss the workers' grievances. With tensions mounting, the PRT finally said it would meet with its own employees, but not "rank outsiders . . . who are identified with an organization that is manifestly inimical to our interests and those of our men. We shall simply ignore the Amalgamated." Three PRT employees took the bait and met with Parsons and Kruger on their own. Parsons refused to have a stenographer present to take minutes of the meeting and listened, rather impatiently, to the men's demands about wages, hours, and the reinstatement of forty-one employees fired for joining the Amalgamated. After receiving the list of names, Parsons looked up and informed his visitors, "Gentlemen, your interview is at an end." A few weeks later he fired the workers who had the temerity to meet with him. Kruger, meanwhile, issued an order saying that "in the event of a strike being called, any employee who declines to take his run out, when directed by his Superintendent, will be at once discharged, and never again taken back into the employ of the company." After that ignominious meeting, Shea and Orr pursued Mayor Reyburn instead. Reyburn finally agreed to a conference in mid-December, and at that meeting he seemed open to listening to the union's demands but cautioned that his first duty was to protect the public and preserve order. Reyburn did urge the PRT to meet with its workers and try to come to some agreement, despite the industry's difficult finances and the "obvious" rights of management. By this point, it was apparent to Parsons that the workers would not go on strike—Samuel Gompers had just announced that he and the AFL would stay out of the conflict because Reyburn was involved and the union's participation for some unstated reason would thus not help the workers' cause—and he began sending the strikebreakers back home. Parsons did know, however, that the outcome in 1907 was not settled and kept Farley on a $3,000-per-year retainer in case the PRT needed his services in the future. That expenditure, it should be noted, came out of the PRT's bottom line and thus the company's return on investment to the city for the express purpose of using strikebreakers against Philadelphians.[17]

Given the obvious coming resolution of the organizing campaign and strike threat, Parsons became more magnanimous. He agreed to reinstate forty-seven of the sixty employees discharged for belonging to the Amalgamated. He promised to set aside one day a month to hear worker grievances, although he made no promise to do anything about them and set up no machinery outside of his personal graces to address their needs. Parsons also refused any pay raise or alteration of hours. That was it for the workers: another organizing campaign, another defeat at the hands of the company. With management, the mayor, and the AFL opposed or indifferent to them, they voted—again—to accept dictated peace terms that gave them little for their struggles. The Amalgamated, however, this time vowed to stay in Philadelphia, set up monthly meetings of its members, and make sure that the workers' concerns would "be taken up in due time by the men at the monthly conference which Mr. Parsons has consented to arrange." Unlike in 1895, the Amalgamated would not disappear from the scene. Mayor Reyburn, meanwhile, received the praise of the press for averting the strike and the violence that would almost certainly have accompanied it, all the while safeguarding the financial interests of the city both in its need for uninterrupted public transportation and the returns the city was due as a stakeholder in the PRT. Reyburn thus set a pattern of city involvement in many strikes to come on Philadelphia's transit system.[18]

This labor agitation at the PRT fit into a broader pattern of working-class unrest that roiled the nation in the first decade of the twentieth century. The United States had 7,029 strikes between 1893 and 1898, and 15,463 between 1899 and 1904, with the number of workers involved jumping from 1.7 million to 2.6 million. Statistics are incomplete for the second half of the decade, but even allowing for a decline in strike activity because of the 1907 depression, it is apparent that worker activism gathered momentum in these years for the period David Montgomery referred to as the "strike decade" of the 1910s. The central conflicts of the period revolved around the continued campaign for an eight-hour day and workers' greater control over their workplace. These demands crashed headlong into the growing employer offensive for the open shop spearheaded by NAM, which mandated that workers could not be forced to join or financially support a union. This "militant open shop" movement, as the historian Robert Wiebe termed it, hardened in the period 1900–1915 and gained a stamp of approval from the U.S. Supreme Court in 1908 when in *Loewe v. Lawlor* it ruled that labor organizations could be prosecuted as trusts under the Sherman Act.[19]

This intensifying conflict spurred several strategies in the labor movement. The AFL, for instance, chose to focus more on formal politics as a way of countering the power of employers and their allies in the courts. More radical workers organized and developed the Industrial Workers of the World (IWW—the Wobblies), which had a robust presence in the West but also in certain economic sectors in eastern cities, such as in longshoring on the docks of Philadelphia. The IWW's assertiveness inspired some of the major strikes of the period, with the Lawrence textile strike of 1912 offering a prime case. Other strikes and class-conflict-related violence also shook the country as Los Angeles, for example, experienced a strike of its metalworkers that resulted in the bombing of the open-shop publisher of the *Los Angeles Times* in 1910.[20]

Philadelphia was one of the nation's hotbeds of class conflict in this period, witnessing strikes in industries such as elevator construction, the building trades, and textiles. A few examples highlight the city's working-class unrest across the board. The Wobblies, as the historian Peter Cole has explored, created one of the more extraordinary chapters in the IWW, with Local 8 on the riverfront simultaneously challenging employer prerogatives and the racial divide that too often crippled the working class. A plate printers strike shut down production at the Charles Elliott Company, leading Elliott to complain in overinflated rhetoric that the AFL wanted his total submission or total destruction. Your principles, he wrote to the labor organization, are "not only against the brotherhood of man, but against all the fundamental principles of a free country." "The moment [an] employee is conscious that his position does not depend upon his merits, his ambition is destroyed, talent gone, and prospects forever blighted." And the city had a massive strike in its shirtwaist industry, where thousands of women workers, called the "girl army" in the language of the era, walked off the job to protest deplorable wages and working conditions. In that strike, in which the women picked up wide support from the city's Central Labor Union (CLU), the Socialist Party, the IWW—which brought William "Big Bill" Haywood of the Western Federation of Miners to town—and trolley workers, who took up a collection to support their fellow workers, the shirtwaist workers beat long odds to win the first mass women's strike in Philadelphia's history and secure union recognition and better working conditions. Strikes at the PRT, then, fit into a broader pattern of labor conflict that disturbed Philadelphia.[21]

Although industrial conflict troubled a wide array of workplaces, many of the era's largest and most violent strikes took place on urban transit systems across the United States. These businesses were by their nature local, and thus their conflicts are less known than those in the railroad or coal industries, yet the "trolley wars," as historian Scott Molloy has accurately termed them, wracked some two hundred cities in the decades surrounding the turn of the century. Brooklyn, Chicago, Saint Louis, Cincinnati, Cleveland, Milwaukee, Houston, San Francisco, and many other locales experienced strikes that resulted in more deaths than any industry save coal. These conflicts were known for being exceptionally savage, with bombings, brawling, and murder taking place with regularity. Chicago's 1903 strike was so ferocious that the Reverend Thomas E. Sherman, the son of General William Tecumseh Sherman, called it "a war more barbaric than [that] in the Philippines," and observers likened San Francisco's conflict to the French Revolution.[22]

The key to understanding why trolley strikes were so explosive lies in the nature of the industry: transit lines ran into every neighborhood, which spread clashes throughout the city. Transit workers lived in working-class neighborhoods, often near their carbarns, and were seen as part of the community. Their neighbors knew of their widely reported low wages and appalling working conditions and had sympathy for the drivers. To working-class urban residents, in Philadelphia and elsewhere, transit companies' treatment of their employees provided an omnipresent reminder of working-class grievances about the repression of organized labor and the power of big business. More than just a reminder of class grievances, fast heavy trolleys with steel rails and electric wires also brought danger into working-class urban space. Transit companies, with too often fatal consequences, dictated how the public, community property of the streets would be used. To many working-class people, streetcars were Janus-faced, representing a modern city with free-flowing people and capital, but also death, violence, violation. Transit companies to them were dangerous intruders who had to be challenged, in a wave of urban uprisings that rattled the United States. Such clashes were, in Molloy's terms, the embodiment of "guerilla Progressivism," the working class's willingness to use violence to take more control over their neighborhoods and lives. People raised, in Henri Lefebvre's famous formulation, "a cry and a demand" for their "right to urban life," their right to some control over the city in which they lived. No transit strike of the era was greater, or more fully represented these issues, than those that shook Philadelphia in 1909 and 1910.[23]

The Transit Strikes of 1909 and 1910

President Mahon and the Amalgamated leadership knew 1909 was probably the union's last best hope to organize the PRT, and they sent their best organizer,

Clarence O. Pratt, to Philadelphia. Pratt, age forty-two at the time, was a native of Warrensville, Ohio, and had been in the Amalgamated for sixteen years. He started on the Cleveland streetcar system, became affiliated with the union there, and participated in strikes in that city, Nashville, Salt Lake City, and many other communities. Pratt was an imposing-looking man, with deep-set eyes, wavy hair, and a rugged complexion. Reporters described him as "hypnotic," "a student of human nature," and noted his rare understanding of his fellow workers, how they thought and how they felt about their companies and communities. Pratt knew how to use this connection to the working class to galvanize a workforce and their supporters in the broader population.[24]

With the Amalgamated having struggled for more than a decade at the PRT, Pratt knew the opposition he faced and that the solidarity of the workers would not be enough. So he made the strategic decision to cast the 1909 conflict in terms that explicitly tied the transit workers to the broader community and all of them to questions of people's rights that animated Progressive Era politics. Many ordinary Philadelphians already had sympathy for the workers, but the PRT's recent fare increase implemented in league with political officials (the elimination of the six-for-a-quarter strip tickets, which the press called a $2 million "fare hold-up") ignited the protests. Pratt made clear to the press that yet again the company "denied the right of the public to good service and deprived its employees of decent wages." This was "the people's fight," he continued, urging Philadelphians to attend a mass protest at the Academy of Music. Tens of thousands of people, including the Teamsters and women from the textile mills, came to the meeting. They marched in parades wearing "I Walk" signs and, in an effort to claim true allegiance to America's values, waved American flags while hundreds of children sang "The Star-Spangled Banner." Political cartoons compared Mayor John Reyburn to Benedict Arnold for his support of the company when he was supposed to be looking out for the public interest. And resolutions from the Central Labor Union condemned the PRT for levying a tax (the fare increase) on working people without their representation. It was time, said the CLU, for organized labor and outraged citizens to stop "whining like slaves under the lash" of a greedy corporation. Pratt announced he would waive demands for a pay raise if the company would reinstitute its old fares and called on the mayor to put the fare increase to a vote. Drawing on a key political issue of the day that excited the thousands of women in the audience, he then demanded women be granted the right to vote. They too faced taxation without representation, he asserted. Seeing the support coming from working-class districts, the Socialist *Evening Call* told its readers that "class lines [were] sharply defined." The outcry grew so cacophonous that the press warned the PRT that the "marching armies carrying torches and burning red fire" would not be "held up by corporate greed." Backed by popular animosity toward the PRT and support for their cause, the Amalgamated presented demands to management in May: an increase from approximately twenty-one to twenty-five cents per hour, a ten-hour workday to be completed within twelve hours, the right to buy uniforms on the open market, arbitration of differences, and union recognition. Management refused to recognize the Amalgamated and ignored its demands.[25]

Another weeklong strike followed from May 29 to June 5, 1909, and this time the transit employees were better organized and had even greater public support. When they walked out, the company immediately fired eleven hundred men who did not report for their shifts. The workers held meetings around the city, urging Philadelphians, in a recurring gendered language that linked honor, toil, and manhood, to recognize the justice of their cause as well as the connections between the transit workers and the public. Such appeals highlighting the links between community and collective action, historian Ardis Cameron has shown, often fired the activism of working-class women who understood that labor conflicts were about economic rights, but also about physical security and family and community stability. "We feel that there is a certain amount of honor which even the humblest citizen, working as motorman or conductor, must have the manhood and dignity to stand for," Pratt said. Philadelphians, he added, would back them, even if "the corporation is so strongly entrenched and so closely allied with corrupt political powers that it can ride roughshod over every constitutional right guaranteed to our American citizenship." The public widely agreed with the strikers' cause, as thousands held marches to protest the "fare holdup" and preachers denounced the "outrage against the average citizen, the wage earner and the poor man." One speaker at a rally termed the PRT and its political backers "traitors," while ex-councilman Robert Frye from a working-class district of North Philadelphia called on protesters to start a "revolution" that would sweep out entrenched corporate and political elites. Even the police, often the tool of Philadelphia's business leaders, made it clear that they would not crack the strikers' heads. The public's attitude, according to the AFL's *American Federationist*, showed a widespread belief that "the public as well as the employees have rights which [the company] would do well to respect."[26]

Violence again gripped the city, especially in Kensington and working-class districts of North and West Philadelphia, where transit lines crisscrossed the communities, dominating the urban space. Strikers stayed out of the confrontations as people greased the tracks to cause accidents, dynamited a car, and shattered the windows of many others. Near Kensington, a marine laid an American flag across the tracks and dared a nonunion driver to run over it. "Your name is mud," he shouted as the driver rumbled over the flag. A mob screamed "Kill the scab!" and a boy pulled the trolley's pole from the wire, stopping it. The driver

jumped from the car and sprinted to a carbarn, with the mob at his heels. Minutes later, the neighborhood swarmed with men who had just gotten off work
in the textile mills; the car was stoned, and police had to put in a riot call. James
Jones, an out-of-work lumberman from New York City who came to Philadelphia on the promise of three dollars a day and was apparently initially unaware
that he was there to break a strike, received a severe beating in the melee. He
wandered the streets all night until the police took him to Frankford Hospital,
where he died from his injuries. It took four days before his heartbroken wife
made it to the city to claim his body. So the violence raged across the city. Internal
PRT memos referred to Kensington and other working-class neighborhoods as
"enemy" territory where residents needed to be "disciplined" by the authorities
and their communities "amputated . . . [like] a diseased member of the body"
from the larger city. Reporters, clearly understanding the long-standing nature
of the conflict, wrote that "the temper of the people was more violent than in
the strike of 1895" and noted that violence flared whenever PRT trolleys rolled
through working-class communities.[27]

Nothing incensed working-class Philadelphians more than the strikebreakers
and the corporate disdain that they represented. Before the strike began, the PRT
replaced an ailing Farley (who died of tuberculosis in 1913) with Pearl Bergoff.
Bergoff, who was also known as "King of the Strikebreakers" (the industry had
room for more than one king in that age), claimed he could muster four hundred thousand men as strikebreakers and had an armory with eleven hundred
rifles. In Philadelphia, he placed spies among the workers and supplied several
thousand men from Boston and New York City to run the streetcars if necessary.
Newspapers widely denounced these men as "scum," "tools of the corporations
used to depress wages or prevent justice to employes." "No wonder," the *Scranton
Times* continued, "the strike-breakers are despised by labor as well as by all communities inflicted with their presence." "Despised" was the right term: in West
Philadelphia, women formed an "amazonian mob" that stripped and beat one
strikebreaker, and policemen repeatedly broke up mobs around carbarns. Elsewhere strikebreakers could not even leave the circus tents where they bivouacked.
It took hundreds of mounted police to get a thousand strikebreakers on a train
out of town at Reading Station, and along the route thousands of Philadelphians
waited with "every conceivable sort of missile, from brick-bats, to hatchets and
sledge-hammers."[28]

With the city teetering on the edge of chaos and the CLU threatening a general
strike, Republican politicians forced the PRT to settle. In this moment, just days
before the Republican primary, the combination of public pressure, violence, and
labor tension forced a split between political and business leaders. State Senator
James McNichol, who made his money in construction and controlled several

North Philadelphia wards, and the Vare family, who ran the political machine that dominated much of the rest of Philadelphia, remained sympathetic to the PRT's business interests, but they worried that reform candidates might use the chaos to upset their machine. Pragmatic politics forced McNichol to favor Philadelphia's working class, and after meeting privately with Pratt, he publicly blamed the company for not treating its men with respect and warned the PRT that if it did not settle he would have all policemen removed from the street-cars. Facing an enraged public, a united workforce, and political opposition, PRT management knew it could not win, and on the day before the primary it agreed to reinstate all workers who had gone on strike, increase pay to twenty-two cents per hour, institute a grievance procedure, and limit workdays to ten hours. The company still refused, however, to negotiate with the Amalgamated. About 30 percent of PRT employees, especially those in Kensington and other inflamed neighborhoods, believed they could get more and voted against the agreement, but the support of Pratt and other Amalgamated officers guaranteed passage.[29]

The Amalgamated exalted at first, asserting that it had won a "glorious bat-tle" that showed the close connection between transit employees and the rest of Philadelphia's workers. "Our fight," Pratt said, "strengthen[ed] the people's cause," not just because working people won, but because the company would have to add three hundred cars to its fleet to meet the new contract. The employees got better pay, the public better service. As Socialists and unionists argued, when workers did better, everyone did better. The result may not have been the Amalgamated's ultimate goal of municipal ownership, but that was never really on the table. After the victory, the Amalgamated's Philadelphia division surged to six thousand members. Union officers claimed their success spurred work-ers in other industries to organize because Philadelphians now understood that the union button represented "manhood, good fellowship, best workman, and integrity." John Mitchell of the United Mineworkers attended a victory party, where he lauded the streetcar workers for "giv[ing] courage and strength to the men of other trades," but raised a cautionary note. "A fight for rights," he said, "is neither lost nor won by a single defeat or victory." In saying this, he tapped into an undercurrent of uneasiness. Management was seething about the loss and slow to rehire fired workers. Within a few weeks, Pratt wondered if the company would live up to its promises; within a year, he saw the company's dealings as a tremendous double cross.[30]

In the summer of 1909, just weeks after the strike concluded, the PRT began plotting to smash the union once and for all. John Parsons stepped down as PRT president after the strike, replaced by Kruger, who had begun as a banker in Philadelphia and had "completely gained the confidence of the city's financial interests." He believed corporations across the nation were under attack from

Socialists and was determined to squelch labor unrest. Pratt took to calling him "Czar Kruger," and his hard-line politics, his bedrock belief in management's prerogatives and refusal to countenance workers' grievances, played a key role in the expansion of the conflict into a general strike. Under Kruger's leadership, the PRT rehired Bergoff to plant his men as spies in the carbarns and compile a list of some five hundred union leaders and company critics. The PRT also secretly established a company union, the Keystone Carmen, to offer an alternative to the Amalgamated that played on workers' loyalty to their home state and the dubious feelings some had about "outside" labor leaders. Keystone members, who had to vow that they were "unalterably opposed to strikes," received better runs, and their representatives had access to Kruger. Amalgamated workers complained about their unfair treatment compared to Keystone men, but to no avail. Able to promise better work conditions, the company union enrolled some one thousand employees by February 1910 and even offered to create a band to further the glory of the PRT. Kruger, who was trying to keep his support of Keystone quiet and knew the city had exploded in violence less than a year before, declined, saying he could not approve an organization called "the Philadelphia Rapid Transit Military Band." With Bergoff's men and the Keystone union in place, the company felt it had firm enough footing to provoke a confrontation and fired 173 Amalgamated members.[31]

The mass firing triggered another PRT strike and subsequently the largest general strike in Philadelphia's history and one of the largest in U.S. history. Convinced that management was spoiling for a fight—the *Public Ledger* claimed the company's motto was "In time of peace prepare for war"—the Amalgamated demanded full recognition, insurance coverage, a better pension plan, and reinstatement of the fired workers or it would strike. In meetings between the company and the union, Kruger told the workers: "We are prepared to wipe the Car Men's Union out of existence . . . and want to." As tensions grew, the PRT had Bergoff bring five thousand strikebreakers to Philadelphia, and Mayor Reyburn gave them free rein to do as they pleased. The Amalgamated struck on February 19, and strikebreakers were soon running trolleys with police protection or in packs to protect themselves from angry mobs.[32]

These men, almost all from out of town (many from New York City and Baltimore) and with little experience driving trolleys, served as the PRT's shock troops, bringing mayhem, overt class conflict, to Philadelphia's neighborhoods. Philadelphians, who had expressed their concerns for a generation about trolleys violating their urban space, saw strikebreakers plow through crowds, running people over and shooting them on the sidewalks. Two girls and a woman were hit by stray bullets in one incident, two women were shot in another, and the stories kept repeating. In retaliation, crowds attacked the streetcars, pulling trolley poles off the wires and beating the crew, or worse. In Kensington, one driver who had one hand on the controller and another on his revolver was dragged from his car and, as a policeman put it, "killed cold." Some policemen tried to protect the trolleys, but many sided with the crowds, who explicitly described the conflict in class terms. "This fight [is] between Capital and Labor," one officer told a non-union driver, while another pounded his nightstick on a seat and said anyone not in the Amalgamated was a "dirty scab." The Socialist Party proclaimed it "a strike of class against class with the lines sharply drawn." The capitalists, the *Evening Call* continued, had declared "unholy war on organized labor," and Philadelphia's workers were "soldiers of the common good." Across Philadelphia, any streetcar rolling down the tracks brought shouts of "Scab!" "Kill the scabs!" Mobs piled stones on the tracks to derail trolleys and smash them, and grand juries frequently refused to indict anyone involved. In all, some twenty-nine people were killed in conflicts in working-class areas from Kensington to Center City to South Philadelphia, and approximately one thousand streetcars were destroyed. Wherever the streetcar lines went, especially into working-class neighborhoods, violence followed.[33]

The violence between Philadelphians and the strikebreakers was both a symptom and a cause of the antipathy that had been growing between the PRT and the citizenry for a generation. Reporters noted that traction companies were generally disliked around the country, but that Philadelphia's was exceptionally unpopular. The PRT's fare increase and disregard for urban communities angered people, but so did the company's treatment of its workers. Local ministers called the PRT "the most brutal street car company in America," one that looked out only for its bottom line and refused to "bend their ways to suit the people who use the cars." The only fair solution to the strike, clergymen argued, was arbitration that would meet the needs of the workers and the public. Kruger refused, claiming that the company's right to hire and fire whom it pleased was inalienable. Mass marches on city hall and meetings at the Labor Lyceum brought anywhere from five thousand to twenty thousand people into the streets. Many of the supporters were women who worried about both the violence the PRT did to their families and the class solidarity the transit workers needed. Chief among these women was Viola Brown, who led the "girl army" that had recently won the shirtwaist strike. At one meeting Brown displayed the city's working-class solidarity, proclaiming that the PRT might have money and power but "the people are with you." "Remember this," she continued: "be loyal and I'll be loyal to you. In me you have a friend who will stick to you as long as she has a drop of blood." The storm of cheers, a reporter wrote, drowned out the rest of the meeting.[34]

With the city, especially its working-class sections, on fire, government officials mobilized the local police, the Fencibles (an independent military organization), and ultimately the state police. The situation was "volcanic," a "seething battle-ground," in the words of one commentator, and local forces, many of which had little interest in protecting the PRT in the first place, were soon overwhelmed. In Kensington, the fire department turned its hoses on a crowd, while elsewhere the police, swinging their clubs, charged into groups of people, hauling out as many as five hundred for arrest. Still the *New York Times* described the city as under "mob rule" and more violent than 1909. The Fencibles offered no help. Mill women in Kensington gave the transit workers the kind of assistance Brown had promised, sticking lemons on the young men's bayonets, cutting the brass buttons off their coats, and chasing them into carbarns with what *Collier's* called "a severe loss of prestige." The state police, whose chief duty was to put down labor disturbances, finally moved into the neighborhoods with the most turmoil, arresting two hundred people and squelching the worst of the unrest.[35]

As the company and its political backers tried to repress the strikers and those who supported them, the CLU implemented a multipronged plan to launch the

FIGURE 7. State police in the 1910 general strike: *Philadelphia Evening Bulletin*. The state police, often called the Cossacks by Philadelphia workers, brought the force of the state to bear in suppressing the 1910 strike. Special Collections Research Center, Temple University Libraries, Philadelphia.

general strike. People were understandably leery of taking such a momentous step—Pratt repeatedly counseled patience—and the CLU called it a last resort. But to CLU president John Murphy, the use of state violence altered the equation. He vowed to lead one hundred thousand men out on strike in retaliation for the mayhem in Kensington and said workers would fight back if necessary, pushing the city into "a carnival of riot and bloodshed that will startle the entire country." Murphy then called together 140 unions in the CLU, including the building trades, machinists, and textile workers, and formed a ten-member strike committee that voted unanimously to stage the general strike. The CLU issued a proclamation that drew on the Declaration of Independence, claiming that "the history of the Rapid Transit Company is a history of repeated injuries and usurpations, all having in direct object the establishment of absolute tyranny over the Carmen." In less florid language, the CLU argued that the PRT was responsible for the conflict because it refused to negotiate with its workers or accept arbitration and that as a semipublic corporation it had an obligation to further the public good. "The capitalists and a small group of self-seeking politicians," the CLU concluded, "hope to crush all organized labor."[36]

With the stage set, the CLU held several mass rallies around the city to continue explaining why a general strike was vital to labor's interests and invited in organizers for many unions. The *Evening Call* found the city "flooded with union organizers" who signed up twenty-five thousand people and encouraged them to back the transit workers' cause. Across the city, thousands of painters, plumbers, waiters, bookbinders, and others from nearly every conceivable occupation joined unions, knocked on doors, and campaigned in the streets to support the general strike. As they did so, they began formulating their own demands, not just supporting the trolleymen but calling for the right to organize their shops, public ownership of the transit system, and the development of a Socialist political party. Soon the city's largest groups of workers—building trades, textiles, and shipyard workers—walked out en masse, and some 140,000 people were on strike in early March. The Pennsylvania Federation of Labor even discussed expanding the general strike to the entire state. The CLU's power alone could not shut down the city, but combined with the resources of other unions and working-class Philadelphians' support, its plan worked and created what one student of labor relations called "one of the most effective [strikes] the nation had ever seen." Effective, and one of the few general strikes in U.S. history, with New Orleans in 1892, Seattle in 1919, San Francisco in 1934, and Oakland in 1946 being the other examples.[37]

Such a strike, so rare and so momentous, demanded the nation's attention. For workers, it was a moment when class exploitation was laid bare and they had an opportunity to rise in opposition. To the Amalgamated's Mahon, the strike was about the future of his union but also about an "industrial fight [in which] the workingman finds capital and corporate interests allied against him." Or as one Philadelphia worker put it, "This fight is not confined to that one union. It is an attack on organized labor, and threatens the life of every union in the country." Others believed this was the beginning of a nationwide conflict between capital and labor. German trade unionists called on all workers to support the "struggle of the working people of Philadelphia against organized capital," to make a "mighty movement of the poor and oppressed." E. E. Greenwalt, the head of the Pennsylvania State Federation of Labor, argued that the strike was no longer a local issue, but should be seen as a state or national one, and he and other Pennsylvania labor leaders debated leading a statewide strike that would take three hundred thousand employees out of their workplaces. Telegrams and money pouring in backed Pratt's claim that "the eyes of the civilized world [are] upon the City of Philadelphia." Working people in big cities like New York City, Chicago, and San Francisco supported the general strike, but so did those in smaller towns like Salt Lake City, Sacramento, Omaha, and Lansford, Pennsylvania. "Stand your ground against the man-eating sharks of Philadelphia," wrote the luminary of the left, Eugene Debs. "We are with you to the finish."[38]

Socialists particularly backed the general strike, seeing it as perhaps the opening salvo in a far larger clash. Philadelphia was witnessing "the first great general strike in the history of America," wrote the Socialist *Evening Call*, and its occurrence in Philadelphia, "the most representative American city," was causing a stir across the country. This strike, the paper continued, was "the greatest labor war in the history of the United States," a fight against the "uncrowned kings [who would] rule over us." The PRT, Socialists argued, had "robbed the citizens of Philadelphia for years, stolen its franchises and corrupted its government." Now was the time to fight back, not only by marching in the streets and clashing with the political establishment and their police—and thereby acting as "soldiers of the common good"—but also by demanding public ownership of the transportation system. "Private or corporate ownership of street car systems and other great means of transportation and production invariably leads to the oppression of the workers and to struggles between employers and employed," the *Evening Call* wrote. "It leads to the practice of extortion upon the public for the profit of the owners [and] political corruption and the domination of state and city by private property interests." The only solution, the Socialist Party concluded, was "the speedy establishment of public ownership and operation of the street railways and similar enterprises, which are necessary to the life of the public, but are now controlled for profit by a small portion of the people." Philadelphia's general strike would, Socialists hoped, go down as the battle "that capitalism will remember as long as the system remains that is responsible for such a struggle."[39]

Nationwide, the strike's opponents knew its significance as surely as organized laborers and Socialists. Because of a campaign apparently coordinated by NAM and related trade groups, hundreds of telegrams and letters poured into PRT offices from businessmen and white-collar professionals from Philadelphia and across the nation. They saw the general strike as a pivotal conflict between capital and labor, one that had national implications for labor relations, law and order, the structure of power, and—in their most overwrought moments—the fate of the nation. "The Union gaining victory in Philadelphia," wrote a Chicago brass manufacturing company, means "trouble all along the line." "Compromise," added Youngstown Sheet and Tube, "would have a wonderfully bad effect on labor throughout the entire country, for if a sympathetic strike won in Philadelphia, why not every where else." Employers particularly worried about the fate of the open shop, that working people would through their unions gain some control over the employment and production decisions of business. "The burden of maintaining the open shop has fallen upon your company," wrote the Citizens Alliance of Grand Rapids. "We trust that you will stand firmly for [this] American idea." "Upon your action this day depends the life or death of the open shop," concluded the Worcester Builders Exchange. "The entire country will pay the penalty if you recede."[40]

Although the future of labor relations mattered to many writers, for others the related issue of law and order and the structure of power hung in the balance. The PRT, wrote the National Sewing Machine Company, was defending "certain fundamental principles that are embodied in the Constitution of our country." Unions had a right to organize, but companies were under no obligation to recognize them, the writer continued. "The question must invariably be fought out sooner or later . . . and we trust that you will stand firmly for what is legally and morally fair." "In the interest of righteous industrial and commercial freedom for our country," added the president of NAM, "I beseech of you to stand firm by your patriotic conviction. . . . The principal [sic] at stake in your fight is one which lies at the very feet of our national life." "Law, liberty and the right of a man to work as and where he chooses is at stake," wrote another correspondent. The PRT, wrote a New York stove manufacturer, was led by "bold and courageous men" who were fittingly fighting for "American independence [in a city where] our forefathers signed the Declaration of Independence."[41]

For the most alarmist correspondents, nothing less than the fate of the nation was at stake, and they often described the strike in martial or religious terms. The PRT was "performing a patriotic duty," wrote a New Jersey engineer, waging "a gallant fight . . . against the dictation of the Union." Losing meant a "reign of terror and bloodshed." For the good of the country, wrote another man, the PRT had to "beat the thugs and outlaws that shed blood and destroy your property."

Only a half century removed from the Civil War, many writers tied this strike to that great conflict. "Remember Lincoln and don't give up the fundamentals of all liberty—Who in all time was denounced more than Abraham Lincoln? Who now more honored and revered?—He stood and strove for true *Liberty*," wrote one man. "I gave—man and boy—four years and three months of the best years of my life," wrote a Civil War veteran, "to uphold and maintain the unity of these states, but if the mob is to dominate in the land, my service and sacrifice were then naïve?" These strikers, wrote another man, had to be suppressed "just like the South." Luckily, in the eyes of a number of writers, Kruger and other PRT leaders were just the men for the job. "God is with you," wrote one Philadelphian. He would guide Kruger as he did Moses "with the same pillar of cloud by day and fire by night." "The salvation of our country," wrote another man, depends "upon such men of nerve and strength."[42]

The middle-class press agreed with the doomsayers and demanded that the authorities act. The *New York Times* called the strike "a declaration of war upon the community," and *Outlook* said the government must "suppress violence . . . boldly and firmly." If it did not, *Current Literature* warned, the strike would spread to "the shoe factories of Lynn, the carpet mills of Yonkers, the silk mills of Paterson, the steel mills of Pittsburg, the packing houses of Kansas City, the building operations of San Francisco and transportation in all our cities." Journals across the country echoed *Current Literature*, whose editors argued, not entirely accurately, that Philadelphia was witnessing "something entirely new to American history, the 'general strike.'" Revolution, observers believed, was in the air.[43]

For many reasons, then, Philadelphia's turbulent transit strike and the general strike it spawned touched a nerve for America's business class. Socialism, they feared, was on the rise, and events in Philadelphia showed what happened when "Anarchy" and "the mob spirit," in the words of one observer, were let loose. For the sake of business, law and order, and nation, the PRT had to win.[44]

With the pressure mounting, Republican politicians, unlike in 1909, refused to back the workers. Without a looming primary, political leaders denied the strikers an audience and let the PRT carry the day. Officials from the state senate to the president of the United States received letters urging them to stay out of the matter. Philadelphia businessmen told State Senator Clarence Wolf and George Earle Jr., who served on the PRT's board of directors, to "hold fast, be consistent . . . don't talk, and reap the fruits of your victory." "The company must have the privilege to hire and to discharge labor and the right to conduct its business without interference," another man wrote to Mayor Reyburn. "Your attitude and action during the trolley strike have received the hearty endorsement of businessmen. . . . This evil has reached such a stage that only surgery will cleanse the trolley system of the festering sore." Across the board, politicians heeded the advice they received. Earle said management had the right to fire whom it pleased. State Senator McNichol spent the strike in Florida, claiming he did not know what was happening. U.S. Senator Boies Penrose refused to get involved, as did President Taft, who said this was purely a local matter. The politicians' stance bitterly disappointed the workers, who had appealed to all of them for help. But with McNichol "walking out on his promise," as Pratt put it, there was little they could do. The lack of political support dealt a crushing blow to the general strike.[45]

With the politicians staying out, the AFL and its state branch decided not to expand the strike, which left working-class Philadelphians to fend for themselves. They continued the general strike for three weeks, but on March 25, at the urging of the Amalgamated, everyone but the transit workers went back to work. The general strike had proven the justice of their cause, Amalgamated leaders said, but remaining out represented too much of a sacrifice for their fellow citizens. As one observer put it, "The men certainly put up a fight, but [you have to realize] you can't buck your head against a stone wall." The transit strike nonetheless continued for nearly a month. On April 19 a committee representing the employees reached an agreement with the PRT that the company would take back all the strikers, and the 173 men who had been fired two months earlier would have their cases settled through arbitration. The strikers narrowly voted down the deal by a count of 1,265 to 1,258, but the Amalgamated's international leadership told the men they had to accept the agreement and go back to work. They did so, although many felt the union was led by "crooks" and "sellouts." Small groups still threw stones at cars and insulted drivers not wearing union buttons, but the upheaval was over. Within a few weeks the company reneged on its pledges, and many of the strikers found that they had lost their jobs.[46]

Most labor leaders' postmortems on the strike were overly optimistic at best. To Pratt, the general strike was a "decided success" that showed "the loyalty of the working class of people to the principle of our republican form of Government and proved conclusively that they can be depended upon to defend justice and right and the preservation of our constitutional liberty." The Amalgamated's international leadership argued that the strike represented "an advance agent" in the campaign to organize the working class, while the local union reported that "the future looks brighter than ever for Division 477 and the Philadelphia car men." Samuel Gompers offered a more balanced assessment, admitting that working Philadelphians lost but arguing that the PRT and its corporate and political allies had been exposed as caring about nothing but profits. The general strike, he said, "was the only way these wage-workers could protest against the disreputableness of Philadelphia's dominant elements." The IWW was openly critical of Gompers and his AFL, arguing that they had not stood tall in support of the general strike. Louis Duchez, an IWW leader, nonetheless hoped that the strike had taught "workers . . . and the whole State of Pennsylvania class consciousness and solidarity [better] than a whole trainload of literature." Most observers agreed with *Outlook* magazine, however, which argued that no one won. "Is this the best our civilization can do?" the editors asked. To them, the strike looked like a street duel as two "bad men" took pot shots at each other. Philadelphia, the nation, they argued, had to do better, had to act more civilized. The country could not tolerate open warfare on its streets.[47]

Although the Amalgamated surely suffered a terrible blow, any assessment that sees the PRT as a clear winner misses the mark. The strike cost the company $2.4 million and forced it to float a loan of $2.5 million more. In all, it took the PRT over a decade to pay off the costs of the strike. By that time, Kruger and other hard-line leaders of the PRT had been forced out in favor of Thomas Mitten, a man nationally known for rescuing transit industries from financial and labor-related stress. With Mitten in charge, the PRT continued some of its confrontational strategies in dealing with labor (a spy network to keep track of "troublesome" workers, for example), but overall the tone of labor relations in Philadelphia transit changed from one of raw violence to a more subtle domination of the workforce through a company union. Many of the issues that made work in the transit industry a miserable experience did not disappear, but the 1910 general strike marked the passing of the most violent time in Philadelphia's transportation history.[48]

THE AGE OF THOMAS MITTEN

The tumultuous general strike of 1910 had brought simmering class tensions so forcefully to the surface that many Philadelphians worried about another clash. Anarchy may have been, or at least appeared to be, on the march, but the frontal assault approach to labor relations employed by John Parsons and Charles Kruger had not helped matters. In fact, their actions had been found not only wanting but downright dangerous in an industry that knit together working-class neighborhoods across a metropolis. Never again, vowed Philadelphia's elite financial and political class, could PRT management put the city in such a position.

In October 1911 the PRT Board of Directors, including its city representatives, and the company's shareholders beseeched the eminent Philadelphian Edward Stotesbury of the banking house J. P. Morgan & Company to save their transportation system and by extension their city. "There is no enterprise which touches as closely the general welfare of a city and the convenience and comfort of its citizens as the local transportation problem. This is especially true in Philadelphia by reason of the close relations between the company and the city under the contract of 1907," their letter read. "We believe that nothing will do as much to rehabilitate the company in public opinion, and so add to the prosperity of the city, as a management which will be recognized as having as its first consideration the city's interest. We turn to you as the citizen who, in our opinion, can accomplish most along these lines, and we appeal to your well known civic pride to give this public service corporation, and through it to the city, the benefit of your personal and business interest and association." Stotesbury took over two weeks to consider the matter, undoubtedly to let the city fathers stew further too, before accepting the offer. In his reply, Stotesbury said he was doing this only out of "an earnest desire to advance the welfare of the general public as well as the business interests of the city." He did not place primary importance on the financial welfare of the company or its stockholders, but instead portrayed himself as the man who would save the city from the PRT and the PRT from itself, from its bad old ways of doing business. Stotesbury's single greatest demand was the right to bring in the man widely viewed as the savior of transit systems across the United States, Thomas Mitten.[1]

This chapter examines the Mitten era at the PRT, a period that lasted nearly two decades, from 1911 to 1929. During that time, Mitten cemented himself as a national figure, the man whose plan for industrial relations had putatively finally solved the labor question. Mitten vowed that through his plan—a combination of company unionism, welfare capitalism, and an employee stock purchase program—he would "humanize the capitalistic system" as he made "every employe a stockholder." No single part of Mitten's plan was revolutionary; in fact, he fit well, and simultaneously helped lead, the shift to a less violent, more contractual, negotiated labor relations regime emerging in the 1910s and 1920s. But Mitten's charisma and the particular way he amalgamated all the strands of his labor relations thinking made him a national figure. The fact that the Mitten Plan's success came on the heels of Philadelphia's violent transit strikes and the general strike of 1910 made Philadelphia and its transportation company a centerpiece in a nationwide discussion about the evolution of labor relations and how the United States could avoid the dangers of bolshevism. To many people, in Philadelphia, around the country, and even overseas, Mitten appeared to be a savior. He was, as the former secretary of the National War Labor Board W. Jett Lauck put it, a "far-seeing" leader, a "torchbearer of industrial democracy." He would bring peace, tranquillity, in our time.[2]

Mitten was, from all appearances, a true believer in his cause; yet there was another side to him and his plan. To implement his vision and "serve" his workers, he launched a system of social control that drastically limited their agency. Mitten broke the Amalgamated and ran it out of the city. He employed spies on the trolleys and subways. He locked workers into a stock purchase plan that tied their current livelihoods to their future retirements and put both in jeopardy when the company hit hard financial times. And while he implemented a plan that used his workers' own acquiescence to cap their earnings, he pulled so much wealth out of the company that it was all but bankrupt when he died. Thomas Mitten, the protector and savior of the working class, as he fancied himself, could implement his plans only by stripping his workers of their right to organize and leaving them nearly penniless when their investments proved worthless. Mitten, by personality and approach, was markedly different from his predecessors at the PRT, but the end result for his workers—particularly their financial security and their right to organize—was much the same.

Although Mitten's reputation rested largely on the labor relations plan that he developed, he was also a national figure because of his vision for urban transportation. To him, the PRT had to be a full transportation system, not just a trolley company. Mitten developed ventures that ranged from taxi and bus service to funeral cars and even Philadelphia's first airline. He was a true entrepreneur who thought in grand ways about what a transportation company could do for its city. His ventures, however, required the expenditure of scarce funds, and kept the company from making its basic services—the subway and trolley lines—more accommodating to the public's needs. Similarly, Mitten's commitment to private

FIGURE 8. Thomas E. Mitten, "Chief" of the Philadelphia Rapid Transit Company and an iconic figure in transportation and labor circles in the 1910s and 1920s. Photo courtesy of Historical Society of Pennsylvania.

enterprise and thus maximum dividends led him to limit the growth of the transit system if expanding it threatened to eat into scarce capital, which he believed should go to the company's stockholders. By the late 1920s, his expansion of the PRT's ventures and simultaneous restriction of the system's growth in its basic services brought to the fore questions about the value, utility, and limits of private ownership of a public service. The career of Thomas Mitten, with his labor relations plan, his commitment to private ownership, and his entrepreneurial ventures, opens a window into how Americans in the 1910s and '20s were thinking about two vital issues of the day: the labor question and the provision of public services in an urbanizing nation.

Thomas Mitten, the Working Class, and the PRT

Thomas Mitten came from bona fide hardscrabble roots. He was born into an aristocratic family fallen on hard times and immigrated as a boy from Sussex, England, in 1875. His family passed through Philadelphia, their "gateway to a new life," heading west to Indiana, where they tried to make a living as farmers. Locals looked down on the Mitten family as lower-class laborers, a view that made Thomas Mitten recoil and determine that he had to escape the agricultural life. Mitten's best opportunity, such as it was, arose on the Chicago Railroad, where he worked at the line's rural Swanington Station and made so little money that he sometimes had to live in a boxcar because of his poverty. Seeing the Chicago Railroad as a dead end, Mitten headed to Denver, often riding freight cars, to participate in the gold rush and eventually run a rail yard in that city. Mitten lost what little he had in the Panic of 1893, an event that made him suspicious of bankers for the rest of his life.[3]

Mitten got his big break when he moved to Milwaukee in 1895 to work at the city's transit company. He started as a line supervisor and quickly learned that the workers knew the system best and distrusted any novice leader. Mitten sympathized with the workers in the contest between what he called "the lords of the mahogany table and [the] serfs." But he also believed capital had certain prerogatives, especially the right to a fair return on investment, and urged management to fight back when the Amalgamated tried to organize Milwaukee's transit workers. Those workers struck in 1896, and most of the city backed them in what grew into a violent conflict, but Mitten pushed upper management to stand its ground, arguing that they could not afford what he called "capitulation." He took control of the strikebreakers brought in to squelch the uprising, telling them that they had "come to fight against the tyranny of a labor organization that has misled the employes on this property." He argued that he did not oppose organized labor per se, but that the Amalgamated had to learn strikes would not work. Increased pay, he said, came from smarter, harder, more efficient work. If the men would provide that, then he personally would "go to the officials and demand better wages."

Despite the tense conflict in the city—Mitten was reportedly nearly shot, and the business community called it a "reign of terror"—the transit company prevailed. Afterward, Mitten was torn: he was sure he had given management the proper advice, but he also felt badly for what he termed the "misled men" who had fallen for the promises of the union. He vowed to "never again beat men into submission ... never again take part in a war so savage and relentless with his own kind." Mitten's efforts secured him the office of general superintendent for the company, and he became known as a driven, austere manager (to subordinates, even his son, he was always "Mr. Mitten" or "the Chief"), but one whom the workers respected because he was always out on the lines. When he left the system, the workers honored him with a testimonial album that marked their "abiding faith [in his] discriminating firmness and great ability [that] place him in the front rank of the electric railway service to which he has devoted himself."[4]

Mitten moved on to Buffalo, New York, where he took over J. P. Morgan's International Railway Corporation (on whose board of directors Edward Stotesbury held a seat, offering just one example of how the nation's largest banking houses and its transit companies were interlocked), which gave him the opportunity to apply lessons learned in Milwaukee and develop his ideas about how workers and management might cooperate. In Buffalo, Mitten gave workers a small raise that demonstrated his commitment to them and promised that if they worked harder and more efficiently they would get what he called "a fair share of what they produce." Bankers, Mitten argued, never understood what he had learned from personal experience: that you must pay more and get more labor out of workers for the economy to prosper. At the same time, Mitten also established an employees' association to keep them from joining the Amalgamated. Many of the seeds of Mitten's ideas were thus sown: treat the men better, give them somewhat better pay, promise them even more if they work hard, develop an employees' association, and fight tooth and nail to keep out a real union. Mitten made Buffalo's transit company so orderly that J. P. Morgan & Company enticed him to move to their Chicago system, where labor conflict had flared for decades. There, Mitten implemented what he had done in Buffalo, but added two components: wages pegged to the cost of living and an acceptance of "reasonable" unions that would not demand the closed shop. Mitten once again settled a troubled system and earned the moniker "Morgan's Napoleon" from the local press. By the age of forty, Mitten had put two million dollars in the bank and run some of the nation's largest transit systems. He was ready to retire to study labor relations full time, but then Stotesbury called and Philadelphia presented what he hoped would be the culmination of his dream. Over his years in Milwaukee, Buffalo, and Chicago, Mitten formulated his basic ideas for labor relations. They continued to develop, no doubt, but when he left for Philadelphia he knew in broad strokes what he wanted to accomplish.[5]

Mitten made two demands before taking the job: he wanted the financial resources necessary to rebuild the PRT and the managerial control required to implement his labor relations plan. He received both with little resistance. Commentators widely understood that the company had exhausted its financial resources in paying construction costs for the Market Street Subway and absorbing the strikes of 1909 and 1910. By the end of fiscal year 1909–1910, the PRT had accumulated a deficit of $1.3 million, which precluded it from doing routine maintenance, buying new equipment, or paying dividends to stockholders. Working together, Stotesbury-Mitten Management, as their team was known in the early years, demanded that the city back a $10 million bond issue and that the company set aside 15 percent of gross earnings for seven years to cover maintenance and renewal expenditures. Mitten immediately used this money to purchase five hundred new cars and perform maintenance that had been deferred for years. Mitten had full authority to do so because he was named Stotesbury's direct representative in charge of operating and further developing the PRT and given the office of chairman of the executive committee (Stotesbury took the title chairman of the board of directors). Anytime that the company or the city challenged the management team's plans, Stotesbury threatened to tender his resignation. That proved a sufficient threat to dampen any opposition.[6]

Mitten knew that given the PRT's history, especially the difficult management-employee relationship, he would not be able to walk in and implement his plan by fiat. So he developed a multilayered strategy to win over the press, the public, and the system's workers. Mitten vowed that he could turn around the PRT, but only with the support of the city's newspapers. So he held individual meetings with ten different outlets, including the *Bulletin*, the *Public Ledger*, and the *Inquirer*. The politics of the paper did not matter to Mitten, but their reporting and editorials did. Many of the editors, predisposed to backing middle-class business values, adopted the line Mitten wished and vouched for his labor relations philosophies, attacked union organizers on the system, and supported higher fares if they led to better service. Mitten cultivated the public with stories in the press about how better transit increased trade and thus the tax base, which benefited all Philadelphians. He emphasized how under his management the PRT would institute a safety campaign so trolleys would no longer endanger the city's youth. He reached a mass audience by condensing his mouthpiece magazine *Service Talks* into a pocket-size pamphlet with a distribution that reached 425,000 and subsequently instituted corporate radio broadcasts in the 1920s. And he periodically had conductors and ticket agents sell stock to thousands of ordinary Philadelphians, the "car-rider-owners," as he termed them. Demand for PRT stock grew so great that in one ten-day period the company raised $10 million just by selling to teachers and housewives, doctors and shopkeepers. Reporters widely noted the brilliance of Mitten's program of selling stock to ordinary Philadelphians: it freed him from having to work with the banks while also convincing the public that supporting the PRT was in their financial interest.[7]

The third part of Mitten's strategy, directed at his workforce, was the most sophisticated. Mitten began by reaching out to the PRT workers still scarred by the last two years of strikes. Repeatedly, he told them that the era of unfair treatment at the hands of the old regime—the "Dark Ages," as he sometimes called it—had made them "righteously antagonistic" and that they had justly rebelled. Mitten understood their anger, but now he offered a new day. He vowed to treat the workers fairly and in exchange merit their trust. "I am bound to get you to believe in me, as you cannot stand out against the truth," he said. "Eventually you have got to come to it if I play the game square. If I don't, I cannot succeed. I have succeeded in everything I have undertaken. I expect to succeed here because I expect to play the game square." To emphasize this message and others, Mitten developed a series of internal communication organs, such as *Service Talks* and the *Employees Bulletin*, that emphasized how workers and management were in business together, that success for one meant success for all, that they all had to be committed to efficiency, honesty, and customer service. Mitten also had employees join management in writing work rules for the PRT. These rules, which governed everything from drinking and smoking on the job to the treatment of passengers and equipment, were to be fair to both the company and its workers and enforced with punishments upon which both sides agreed. "Discipline," Mitten told his workers, had to be "just and firm, and at the same time tempered with reason and common sense. Discipline never should be invoked as a punishment, but only as a corrective of faults."[8]

Although Mitten's first efforts targeted the employees on the job, he also reached into their homes and communities. The PRT held annual picnics at its suburban Willow Grove Park, about fifteen miles north of Center City on Broad Street. There, employees, their families, and guests engaged in track and field activities, played baseball, ate lunch, and listened to speeches by Mitten and PRT management that exhorted them to see the company as one big family. The famous minister Billy Sunday often came in to give sermons to the gathering, and Mitten's personal friend John Philip Sousa even wrote a song, "March of the Mitten Men," that was performed every year. It was a festive time, a "real lovefeast," in the words of one reporter, that promoted an allegiance between the workers and their leader, and by the 1920s it attracted as many as twenty-five thousand people. Under Mitten, the PRT also developed an education division that provided classes for workers on operating electric trolleys, dealing with the public, and other ways to advance their careers at the PRT. The courses focused mostly on workplace duties—Mitten once said, "Education to be worth its cost must be harnessed to the industrial machine"—but instructors did help employees find classes at other institutions that focused on English, economics, and other academic subjects. Finally, Mitten regularly appealed to the "wives, mothers and sweethearts of P.R.T. men" to influence them to be "absolutely loyal to the P.R.T." They should always ask, "What are we doing for Mr. Mitten, the Philadelphia Rapid Transit Company and our husbands or fathers?" Only they, Mitten once said in his inimitable style, could keep their men on the straight and narrow, loyal, responsible, and employed, avoiding the "sleepless pillow of the family in debt."[9]

The workers listened to Mitten with an open mind and tentatively decided to try to work with him. He had the track record of having risen in the industry, he had the support of workers at other systems around the country, he at least seemed to understand their plight, and he promised that they would all prosper if they worked together to make the company stronger. The PRT's workers, understanding the company's desperate financial situation and the need to provide a vital public service, acknowledged that cooperation might work better than continued confrontation. C. O. Pratt told the press that Mitten seemed "fair," that the Amalgamated agreed with Mitten on the need to tamp down labor troubles, and that he had "every confidence that the new management, under [Mitten's] direction, will work harmoniously and satisfactorily." The car men who had been fired for presenting their demands to PRT management in the past prepared a statement listing the improvements they desired (chief among these was a pay raise), and asked Mitten for a meeting. Mitten arranged several meetings that included two hours with officers of the local chapter of the Amalgamated, including President Peter Driscoll and Vice President H. A. Flynn, and other conferences with Pratt and Mahon. By all accounts Mitten listened attentively to the workers' concerns but would not commit himself to any action. Nonetheless, the press reported that both parties left the meetings in good humor, "and it was evident that relations between Mr. Mitten and the committee [union] are amicable." Mitten vowed that further meetings would take place as the company sought to address wage and working-condition issues, and he promised to give the employees a half-cent raise (to twenty-three and a half cents per hour) in two weeks on July 1, 1911. Given the conflicts that had transpired between PRT management and its workforce over the past year—indeed the past two decades—this was as auspicious a beginning as Mitten, the workers, or any Philadelphian could have hoped for.[10]

The Mitten Plan, the Early Years

The earliest iteration of the Mitten Plan, begun in August 1911 and lasting until 1918, was relatively limited and straightforward. It provided the employees a wage pool of 22 percent of gross passenger revenues, offered sick, death, and pension benefits, and put in place a system for collective bargaining. The central problem Mitten most needed to overcome, he later told a congressional panel, was a "lack of cooperation" between workers and management. His plan would lead to labor and capital "pushing together for a common end in a way which is not approached anywhere to my knowledge." Mitten's insight was that by setting the wage pool at 22 percent, he could argue to the workers that the harder they worked, the more efficiently they ran the system, the more the PRT could secure "increased wages and comfort of the men as well as increased returns to the Company."[11]

Most workers initially welcomed the plan, with Hugh Barron, the treasurer of the local Amalgamated, overstating the case to call it "the most far-reaching proposition ever made by any employer to his employees the world over. It is a beautiful theory of cooperation between capital and labor. Whether it is practical can only be decided after the plan is put into operation. But the plan is a worthy one." Pratt remained more circumspect about Mitten's plan, wanting to see it in action before offering such praise. He found a substantial backing for his position among many of the three thousand men who attended one Amalgamated rally. They were the ones Ida Tarbell likely quoted in her otherwise rosy article on the Mittens: "All agreed," she wrote, "it was good to look at, but probably 'bull—— meant to fool them into more work.'" Mitten responded by opening the company's books, and offering to do so whenever the workers wanted. "That is what got me," one man told Tarbell, "when he really allowed us to do that—insisted that we do it. I knew then it was all right."[12]

Tarbell's informant may have bought the plan, but Pratt grew more critical and pressed Mitten to accept the Amalgamated as the workers' representative. Mitten preferred that they belong to the Cooperative Association, the PRT's company union that was designed to implement the Mitten Plan, but publicly claimed he would happily accept an "outside" union if it met two stipulations. First, in an open election the Amalgamated or any other representative organization besides the Cooperative Association had to receive 67 percent of the vote. The Cooperative was the default winner otherwise. Second, no union could be a "troublemaker" but instead had to be "organized for co-operation with management to increase efficiency." Any employee who was a "troublemaker" or supported such an organization was in line for dismissal—"a day of reckoning," as one commentator called it. The high hurdle of a two-thirds majority and the requirement that a union cooperate with management signaled that Mitten's acceptance of unions, while oft stated, ran shallow.[13]

To head off Pratt's criticisms and give the workers a chance to express their preference for representation, Mitten held two elections, the first in November

1911 and the second in July 1913. Coming on the heels of the general strike, the elections garnered citywide attention from the press, and even reform fusion mayor Rudolph Blankenburg appeared before the workers to encourage them to back the Cooperative Association and Mitten Plan. In the first election, the Amalgamated received 4,276 votes out of 6,642 cast. That tally fell only 174 votes shy of the number needed for the Amalgamated to carry the election, but Mitten's stipulation of two-thirds support for the outside union put the Cooperative Association in charge. In the 1913 election, support for the Amalgamated fell, as 2,028 workers voted for the union and 4,320 voted for the Cooperative.[14]

The losses, especially the first vote in 1911, rankled a number of workers who found the two-thirds rule and other aspects of the elections unfair. The PRT, Pratt argued, had given pittance raises to make the workers think the company was on their side. It had constructed the ballot so that a yes vote was for the Cooperative Association and a no vote was for an outside organization. The ballots were far from secret, requiring workers to sign them and supply their badge number. Finally, Mitten banned union buttons, to make it harder for employees to declare their allegiance. Surveying the elections Mitten had constructed, Pratt called them "shams," continued to organize the PRT's workforce, and challenged Mitten to a debate. Mitten in turn labeled Pratt a troublemaker and warned the employees that "no man may expect to remain in the service who does not obey orders and who refuses to work in the direction I am now outlining." In three days, Mitten fired twenty-one men who had declared their allegiance to Pratt. Pratt then started making overtures to the IWW, gave a speech sponsored by the Socialist Party, and in an appeal for justice sent to the mayor summoned everyone's worst fears. Pratt referenced the labor conflicts of the last two years and threatened another strike that would involve "the community in a strife that will mean sacrifice and hardships."[15]

Pratt's post-election actions touched off a conflict within the Amalgamated that divided its Philadelphia membership for the rest of Mitten's tenure and left the union so weakened that it could not stand up to Mitten in the future. Mahon found the election results disappointing, but knew the rules Mitten had put forward and accepted the result. He abhorred the idea of another violent clash in Philadelphia and told Pratt and his followers in no uncertain terms that they could not follow such a course. Pratt, Mahon argued, relished strikes and tried to set himself up as a "dictator" during the 1910 general strike. Now, he was trying to split the Amalgamated by forming an independent "Carmen's Union" that had about one thousand members. Pratt and his followers had two choices: they could either renounce this heresy and come back into the fold with Pratt and his top loyalists subject to discipline, or they would all be removed from the Amalgamated, and the local's charter would be revoked. When International officers went to Philadelphia to investigate Pratt's group, they met a raucous crowd who "hurled at them . . . insulting epithets of every description." Feeling it had no choice, the Amalgamated kept Local 477 alive but removed Pratt and twenty of his top supporters. This internecine fight gutted the union (the "Amalgamated Association actually destroyed itself on the P.R.T. property," in the words of one student of the transit company's workforce) to the extent that it would never fully recover. Mitten never again felt the need to treat with the Amalgamated or even run regular representation elections. No outside union, he argued, "enjoyed a majority confidence of the employees"; only the Cooperative Association could represent their interests. This union factionalism not only undercut worker solidarity in the early 1910s; it set a pattern of intra-class conflict that while stemming from various issues through the years marred PRT workers' lives for the next half century.[16]

As Mitten gained control over the PRT workforce and cemented his plan, he liked to claim that it was all his own idea and unique in the United States, even the world. "I worked it out of my past experience, which showed the need of cooperation," he once told the federal Commission on Industrial Relations. "It occurred to me, let us establish that portion of the gross earnings which the company would pay, and then let both work together to get from it the very greatest wage possible for the men. It is the principle, not the percent paid—but the principle." To the Amalgamated, the plan seemed revolutionary enough that Mahon printed it in its entirety in the union's journal, with the suggestion that it might offer the next step in industrial relations.[17]

In reality, Mitten's plan both advanced and captured the broader spirit of labor relations of the time, emphasizing company unions and welfare capitalism that ameliorated the worst excesses of a capitalist economy by giving workers a voice, however moderate, in their work lives and some fringe benefits that went beyond wage compensation. Company unions, as they came to be known, began as employee representation plans (ERPs) in Europe, mostly Great Britain and Germany, in the last quarter of the nineteenth century. Company unions spread to the United States, beginning, Stuart Brandes found in his study of American welfare capitalism, in Boston at Filene's department store, where employees joined management in 1898 to form a committee to administer a medical clinic and insurance plan and then established the Filene Cooperative Association Council in 1905. Company unions spread slowly at first, then surged during World War I when the Great War threw American labor relations into turmoil. The United States had 225 company unions in 1919, 725 in 1922, and 814 in 1924. Economists estimated in the early 1920s that some one million American workers had representation from a company union. "It became an accepted axiom of progressive management thought in the 1920s," wrote Bruce Kaufman in his study of industrial relations, "that an employee representation plan was an essential ingredient of good industrial relations."[18]

Perhaps, but behind the plans companies had ulterior motives about controlling their workers. Management often implemented them to head off "worse" alternatives, especially strong unions in the AFL or IWW. They also regularly established them after bitter, violent strikes to assuage their workers. The Colorado Fuel and Iron Company in 1915, in the wake of the Ludlow massacre, offers a prime example, but so too does the PRT four years earlier. Almost all managers acknowledged that company unions advanced a cause dear to their heart: the open shop. As D. R. Kennedy, an adviser to several large companies that instituted these plans, once said, "After all what difference does it make whether one plant has a 'shop committee,' a 'works council,' a 'Leitch plan,' a 'company union,' or whatever else it may be called? . . . They can all be called 'company unions,' and they all mean the one big fundamental point—*the open shop* [emphasis in original]." Samuel Gompers understood this and blasted company unions as a "semblance of democracy," "a pretense admirably calculated to deceive." Whatever flaws company unions had more broadly, the Cooperative Association at the PRT was widely regarded as a shining example of the possibility of industrial democracy. It and a few other companies stood, wrote Lauck, "above all others as indicating a sincerity of purpose and as offering a basis for future constructive action. . . . The discovery of the theoretical worthiness of even one plan in itself gives assurance of the possibility of the general realization of industrial democracy in the coming years." The Mitten example notwithstanding, as the threat of union organizing faded in the 1920s, so too did company unions. By 1928 the United States had about four hundred companies that still had the plans. The PRT was one of those.[19]

In addition to developing the PRT's company union, Mitten also tapped into, and advanced, a national trend toward welfare capitalism, which fulfilled a similar objective of ameliorating the greatest excesses of capitalism and thereby helping dampen sharper manifestations of worker militancy. Welfare capitalism predated Mitten's ventures, starting in the second half of the nineteenth century in such Massachusetts firms as the Waltham Watch Company and Ludlow Manufacturing, and later at larger companies, including the Pullman Palace Car Company and Procter & Gamble. In the Philadelphia area, John B. Stetson, Henry Disston and Sons, John Wanamaker, and others all instituted welfare programs. President Theodore Roosevelt became a staunch backer, as did the National Civic Federation and all but the hoariest employers—all who wanted to put the days of raw industrial conflict behind them. To these industrialists, as Brandes put it, welfare capitalism "emphasized the development of the humane, rationalized corporation as the chief instrument of social progress." It was in some ways an "almost utopian plan." Welfare capitalism grew steadily through the first decade of the twentieth century, surging in World War I through the mid-1920s. A 1926 study showed that 80 percent of the nation's fifteen hundred largest companies had adopted welfare capitalism by that point.[20]

Versions of welfare capitalism varied from company to company, but all had basic components that involved employers spending a limited amount of their revenues to improve workers' lives and in the process gain some control over them. Early programs focused on housing and over the years expanded to provide entire towns (Pullman stands out in this case), education for workers, school systems for their children, medical care and health insurance, and death benefits. Larger businesses also ran recreation leagues and held annual picnics, all in the name of creating a "family" for their workers and building loyalty to the company. Mitten provided all these perks at the PRT, but he particularly supported employee stock ownership, which many companies grouped under the larger heading of "profit sharing." Few U.S. companies used this form of welfare capitalism as late as the 1880s, but by the 1910s some 250 companies participated (60 used a stock ownership plan), including U.S. Steel and International Harvester. The PRT was thus one of the first, but by no means the only to do so. By 1927, four hundred companies used stock ownership, including Bethlehem Steel and Firestone, where every one of its fourteen thousand workers owned stock.[21]

Whether welfare capitalism was good for workers has drawn mixed reviews. Selig Perlman and Philip Taft, in their *History of Labor in the United States*, argued that welfare capitalism "attempted to treat the worker as a human being and to approach his problems from his own point of view." It provided education and housing, a fair wage, and some security from unemployment, injury, and death. Brandes also saw welfare capitalism as improving workers' lives, although his assessment was more guarded. To him, "Welfare capitalism met the human problems of industrialization in a way which at best can be termed only minimally acceptable." Government investigators went further than Brandes. Company towns, one wrote, were little more than feudalism. "A man who in the struggle of life has no home to retire to; whose rug lies upon another's floor; whose fire is in another's grate," he continued, "can neither become great himself nor inspire greatness in his own family." Welfare capitalism, in this view, amounted to little more than another method of controlling workers' lives.[22]

Like company unions, welfare capitalism, for a variety of reasons, essentially played itself out by the late 1920s. Companies grew larger and management more distant in that decade. Many employers thus dropped educational, recreational, and religious strains of welfare capitalism, holding on to the most practical ones, such as medical coverage and pension plans. In addition, more workers owned cars and could live farther from their workplace. The most important blow, however, came from the Depression. In hard economic times, employers jettisoned these programs, which they saw as a fringe benefit rather than a key component of the social compact. As the historian Lizabeth Cohen put it, workers became "surer than ever that employers only valued welfare capitalism when it was convenient and cheap." The Mitten Plan may have been one of the most comprehensive, and most lauded, examples of welfare capitalism, but it followed a similar pattern and disappeared when hard times set in.[23]

But that was a future that PRT workers in the 1910s could not divine. In that decade and the early 1920s, they saw new developments in the Mitten Plan and rising wages that put them in the vanguard of U.S. industrial relations. PRT workers' wages stood at tweny-three cents per hour in 1910, climbed to twenty-five cents in 1912, thirty cents in 1915, and reached fifty-one cents by the end of the decade (these numbers outpaced inflation in the first half of the decade and matched it when the price increases of World War I set in). The PRT's total expenditure on wages grew by 151 percent over that period. These "voluntary increases," Thomas Mitten's son and corporate lieutenant A.A. Mitten trumpeted, fostered "a remarkable change in that intangible and invaluable asset to employer and employee alike—improved morale in the organization." Raises no doubt improve morale, but A. A. Mitten conveniently forgot that they were not always voluntary. This is not surprising, given that labor conflict at their transit company would have upset the worldview of both Mittens, but the younger Mitten's boast was historical revisionism at its worst.[24]

In reality, for several years after the Amalgamated splintered in 1912, Philadelphia's transit employees used their company union to discuss workplace issues with Thomas Mitten and accepted the PRT's corporate welfare largesse, but they again grew restive during World War I. In part, the city's transit workers were reacting to war-induced inflation that roiled the entire country and led to food riots in Philadelphia, New York City, and Boston. But they also caught the spirit of "industrial democracy," a term popularized by the Fabian socialists Sidney and Beatrice Webb of Great Britain in their book *Industrial Democracy*. As the historian Nell Irvin Painter has pointed out, "industrial democracy" meant many things to many people, but for workers it in essence meant having greater control over their workplaces, bringing the ideals of political democracy to the shop floor. The workplace could not be a dictatorship, in the view of advocates such as Walter Lippmann, because "without democracy in industry, there is no such thing as democracy in America." Industrial democracy of course fit the spirit of the time, coming in the midst of a war that President Woodrow Wilson announced was being waged so the world could "be made safe for democracy."[25]

The combination of economic pressures and working-class aspirations created a tumultuous, if brief, moment. By the end of World War I nearly three million American workers had joined unions, and the turnover rate jumped by 33 percent as people looked for better jobs. Companies across the nation reported that workers routinely engaged in low-level struggles for control in the workplace, and the number of official strikes averaged 3,746 from 1916 to 1918. A number of these strikes—in Kansas City and Waco, in Springfield, Illinois, and in Wilkes-Barre, Pennsylvania—took place on overburdened transit systems where management asked their employees to work ever harder, haul more people, even as the value of their wages declined. The greatest labor unrest in Pennsylvania, outside of the PRT, started during the war and came to a head just afterward with the steel strikes of 1919 and a strike at Cramp's Shipyard in 1921. The nationwide upheaval at times became so turbulent that employers worried a Bolshevik revolution was afoot. Workers were agitated enough that the federal government stepped in to calm labor relations. Federal authorities chiefly worried about maintaining production during the war, so in 1918 President Wilson established the National War Labor Board (NWLB) with former president William Howard Taft as its chair to arbitrate industrial disputes.[26]

The PRT's workers, then, were by no means alone, but their dispute does highlight these key trends in American labor relations. Pratt, who still held the loyalty of hundreds of Philadelphia's transit workers, contacted Mitten in May 1918 to ask that the PRT raise wages by five cents an hour to offset inflation, a proposal that would have cost the company an additional $4 million a year. Mitten responded as Pratt must have known he would: the PRT could not possibly afford this raise without a fare increase, and since the company had a union in place he had no intention of discussing this with Pratt or any representative of the Amalgamated. To rally support, Pratt held an all-night meeting of the car men where they affirmed their loyalty to the United States by singing "The Star-Spangled Banner" and appealed to the NWLB to mediate their case. Pratt argued, as he had in the past, that the workers' claims were not just about wages, but about their right to a greater say in their work lives. They had to be "recognized as human beings and given some voice in the matter, instead of being considered only as a mechanical appliance attached to the car." One thousand PRT workers then voted to strike if the company would not negotiate with them. The next day, thirty-four hundred men walked off the job, bringing the PRT nearly to a halt.[27]

The exigencies of war helped the workers achieve at least some of their demands and at the same time kept violence to a minimum. A few men did stone a car in Kensington, and police broke up a group of rowdy workers at Cramp's Shipyard. In case the city had a repeat of 1910, Superintendent of Police William Mills swore in an extra eighteen hundred policemen and armed them with riot sticks. But there was no way the federal government would let the situation escalate this time, not with Philadelphia being vital to the war effort. On day three of the strike, the NWLB announced that it would hold hearings to settle the strike and did so behind closed doors, with Taft mediating. Mitten protested that this was an internal matter that he could handle, but the NWLB did not care to hear it—one board member labeled Mitten "obstinate"—and in fact had made it clear in other cases that the prerogatives of management would take a back seat to both the war effort and the needs of a worker to have "a wage sufficient to sustain himself and his family in health and reasonable comfort."[28]

Mitten knew that the NWLB's involvement meant he had to tread lightly. He was never inclined to use violence to crush workers as the old guard had been, but now he acquiesced to a two-cent pay raise and vowed to give three cents more if he could increase the fare. He also welcomed back all the strikers and promised they could keep their seniority rights. Mitten's semi-capitulation ended the strike by the end of May, and commentators widely noted that he had done so to keep the NWLB out of the PRT's affairs, to "save the war labor board the bother of passing on the carmen's wage question," as the *Public Ledger* put it. In his letter to the NWLB, Mitten promised the board that the company's workers would be happy with their raises and that the PRT would supply "adequate service to the shipyards, navy yards, arsenals, and other plants engaged in the manufacture of munitions." Everyone at the company, he assured federal officials, fully understood that he had to "do his bit in helping the Government win the war." A few workers complained that Mitten still refused to recognize the Amalgamated or let them wear buttons to show their support, but the NWLB was finished with the case. War production was safe, and the board ignored the remaining grievances. Although federal support by no means led to the PRT's workers getting everything they wanted, it did compel Mitten to raise their wages not only in the moment but again in August so that the average increase was between five and nine cents per hour, depending on the employee's length of service.[29]

Higher wartime ridership meant the PRT could afford to pay the additional wages out of gross receipts, but the 1918 strike did reveal the first glimmer of how the company could pit its workers against the public on the matter of wage negotiations. This ploy became commonplace in the decades after World War II. Mitten was farsighted enough to know that good financial times in the transit industry were war-induced and would not last beyond the armistice. Ridership would drop, but inflation would continue to rise. Ellis Ballard, the PRT's chief counsel, told reporters that Philadelphians had to understand that the United States was "no longer a 'nickel country.'" He was right for many reasons: streetcar workers' wages across the United States doubled in the 1910s; materials rose at about half that rate; the average ride grew longer as cities expanded and residents moved to the suburbs; fares remained locked in by contract or local regulation at five cents; the rise of the automobile siphoned off many riders; and jitneys (unlicensed cabs) cut into the transit business. Mitten saw fare increases as the only way out, but the 1907 contract mandated that he had to get government agreement through the state Public Service Commission. The PRT's workers, he told the press, were right to agitate for a higher wage, but the company could not afford to make it permanent, especially if it also had to purchase new equipment to replace the cars used hard during the war. Mitten then asked his workforce to vote in favor of a fare increase to fund their raises. Pratt counseled against doing so, saying that wages, not fares, were the workers' concern. But by a count of 94 percent in favor they voted their support of the fare increase. In essence, Mitten got the PRT's employees to demand a wealth transfer from the riders to the workers without tampering with dividends or payments to the underliers.[30]

For the first time in any substantial way, city newspapers challenged Mitten. The press agreed that workers deserved a raise, but it should come out of the company's $6 million surplus. "Is another $4,000,000 necessary?" the editors of the *Public Ledger* asked in reference to the proposed fare increase. The city, they continued, was not responsible for ensuring dividends for stockholders, but should instead safeguard the interests of the riding public. If Philadelphia agreed to a fare increase, then the company should "surrender to the city the right and the power to control the sort of service the company shall give." On occasion Mitten showed hesitancy over any fare increase, fearing that it would deter short-haul riders and cut into profits. But he certainly did not reject the Public Service Commission's 1920 ruling that the PRT could charge seven-cent fares. The press grumbled about the increase, but the hike was final. This was the first of many arguments to come over the difficult mix of fares, corporate income, and employee wages.[31]

The Mitten Plan, the 1920s

In the decade after the First World War, Mitten increasingly staked out his own ground in the fields of urban transportation and labor relations and gained national acclaim for himself and the PRT. This was in part because of the further development of the Mitten Plan and Mitten's entrepreneurialism. But the difficult state of labor relations across the United States also played a role as they made the PRT look notably placid. Perhaps, many commentators argued, Mitten really had solved the nation's labor question. Others, especially in the company's workforce and the city government, grew increasingly skeptical.

The labor upheavals of the immediate postwar years troubled the United States. Overall, some four million workers—20 percent of the total workforce—went on strike in 1919 alone, leading historians to term the year the "Great

Unrest" or the "Second Great Upheaval" (the first being the upsurge of labor and political activism in the mid-1880s). The most famous strikes included a general strike in Seattle, the police strike in Boston, and a wave of strikes at steel mills across the United States, including plants in western Pennsylvania. The violence, whether directly associated with a strike or not, could be raw and frightening. Bombs shook Philadelphia, Pittsburgh, Washington, D.C., and other cities; open warfare broke out in Logan County, West Virginia.[32]

Although workers exhibited a strong activist impulse, they faced obstacles that they could not surmount. With the war over, the NWLB lost its backing in the federal government and folded in May 1919. President Wilson declared that "the question which stands at the front of all others in every country amidst the present great awakening is the question of labor." But consumed by foreign affairs and advised that he should distance himself from labor, Wilson also backed away from his support of workers. Many in the middle class looked on the violence in horror, worried that class war, communism, anarchy, had arrived. The *Los Angeles Times* in its comments on the Seattle strike captured this sentiment well: bolshevism, the editors declared, was a "right-here now American menace." The government's abdication of its role as arbiter in labor relations, coupled with middle-class sentiment, created a conducive atmosphere for a Red Scare that made labor organizing and activism not only difficult, but to many people downright "unpatriotic." Business both helped to create and took advantage of the antilabor times and developed what state manufacturers' associations in 1921 labeled the "American Plan." Advocates of the American Plan, such as the Philadelphia Chamber of Commerce, claimed that they wanted to return to a time when all Americans had freedom of contract, the right to choose their employer and negotiate terms as they saw fit. In reality, the American Plan was an unabashed effort to destroy unionism and return the nation to the open shop.[33]

With employers mostly united behind the American Plan, the middle class largely backing the antilabor offensive, and the federal government absent, unions lost members, lost strikes, lost ground, tried to weather the storm. The AFL lost one-fourth of its members in the early 1920s. A major steel strike fell apart. So did strikes in coal mines, meatpacking plants, and garment factories. As employers broke their union opposition and strikes commonly ended in routs, workers gave up their wartime-inspired higher ideals. Industrial democracy, in particular, was a casualty, replaced by an industrial relations system that sought to smooth over workplace tensions without surrendering any real control to the workforce. With an economic depression setting in in the early 1920s and inflation continuing apace—prices of basic commodities, food, and housing all shot up during the war and afterward—it is small wonder that workers felt battered just a few years after the war. Many workers in Pennsylvania, often employed in older industries, faced especially difficult times. Wages in the hosiery industry fell 15 percent in 1920, coal mines lost market share and shed jobs, Cramp's Shipyard closed in 1927, and Baldwin Locomotive began a downhill slide from which it never fully recovered.[34]

Given these tumultuous times, Mitten's second decade at the PRT looked then, and in some ways looks now, rather remarkable. He began the postwar period by reconfiguring the Mitten Plan in late 1918. With the NWLB still in place and monitoring the PRT, Mitten wanted to make sure the Cooperative Association looked like an independent union, and he also wanted to improve the benefits that the plan provided. Mitten vowed in the PRT company bulletin that workers would have "a free and independent vote for representatives for collective bargaining" and propounded a multilayered committee structure that provided for deliberation and appeals of all decisions. He also established a Cooperative Welfare Association that used company funds and worker dues to provide or enhance benefits such as life insurance, medical benefits, pensions, and a savings fund that paid 5 percent interest. The NWLB reviewed Mitten's new plan and pronounced that the "general intent and spirit of its provisions were entirely in accord with [the labor board's] own principles." Mitten was thus in the clear with the federal government. A simple review of the circumstances in which other workers found themselves was all it took for PRT employees to think the new Mitten Plan was a good one too: 99.55 percent joined the Cooperative Association by 1920. Everyone involved—in Washington and in Philadelphia—ignored, or at least chose not to care about, the fact that Mitten retained control of the association. According to the bylaws, the PRT's president and chairman of the board of directors served as chairman of the association and appointed the secretary-treasurer. Mitten set the agenda and controlled the purse strings.[35]

In addition to the benefits, Mitten also reconceived the compensation system. Claiming that "abnormal conditions due to war" had falsely elevated gross earnings, Mitten jettisoned the 22 percent formula and replaced it with a plan that based PRT workers' wages on the average of four comparator cities: Chicago, Detroit, Cleveland, and Buffalo. This scheme, in Mitten's view, proved too volatile over time, and in 1926 he put in place a "market basket" approach that paid the workers based on the average cost of common consumer goods in Philadelphia. By 1926, PRT employees who had made twenty-three cents an hour in 1910 earned seventy-seven cents an hour (inflation, of course, played a significant role in their rising wages). But the men did not actually receive the full seventy-seven cents. Instead, they were mandated as of 1922 to contribute 10 percent of their salary—what Mitten called their "wage dividend" or "added wage" generated by "super cooperation"—to a Wage Dividend Fund that then invested the proceeds in PRT stock. The employees' stake in the company soared: 100,000 shares (one-sixth of the company in March 1923), 152,000 shares (one-fourth in January 1925), 221,000 shares (one-third in December 1925). If the trend continued, by the end of the decade PRT workers would own a controlling interest in the company.[36]

For Mitten and his supporters this was an unmitigated good, or so they claimed. He would personally create "industrial democracy" for the PRT's workers, even if the term had faded from the American discourse. The financial benefits derived from stock ownership, Mitten asserted, finally gave workers the answer to their age-old question: "Why should we work more efficiently, when all of the benefit goes to capital?" To drive his point home, Mitten took to calling the company's workers "employe-owners" and routinely argued that the United States could achieve its true destiny—becoming the greatest country in the world because it supplied every citizen's needs—only when the working class had enough financial resources that it could hire the capitalists and choose its own management. Buying PRT stock, wrote A. A. Mitten, gave the workers "a more concrete lesson in the value of investing . . . and a steadier and happier employe is gained for the company and a better citizen for the community." "[By] using its head and its cash, as well as its skill and its brawn," the senior Mitten wrote, labor was writing "a new Declaration of Independence," forging a "movement for the betterment of mankind," "transform[ing] the whole industrial structure of the world." This "industrial utopia," he believed, would save the United States from the fate of a stagnant England or a Bolshevik Russia. Other companies may have had stock purchase plans, but none had the public relations service that Mitten provided for himself.[37]

Mitten saw his entire plan—the company union, the welfare capitalism, the stock purchase scheme—as his legacy, his gift to the world, but to him the employees' investment in and eventual ownership of their own company was his greatest achievement. That part of his plan was "grounded into your hearts, into your sensibilities," he told the assembled workers at the 1925 annual picnic. Many, at least in the early to mid-1920s, agreed. The Mitten Plan, motorman Ralph Nyman told the press, gave workers "peace and contentment to the fullest extent [because] co-operation always wins." Conductor Charles Cochran recalled the "chaos of strife" of the pre-Mitten era and likened the plan to a lily growing in the mud. The chance for employees to have a say in their workplace, to get ahead financially, he told the *Philadelphia Record*, was the industrial representation of "organized democracy," sung to "the music of laughter that means 'Liberty, Equality and Fraternity,' the trinity of justice." These employees were not alone: many PRT workers became so enamored of Mitten and his dealings with labor that they agreed to go to Buffalo to help put down a transit strike there in 1922. They referred to themselves as Vacationists (some called them "Storm Troops")

and vowed to defend the "Star Spangled Banner [against] the Red Flag of Anarchy and Lawlessness." With Philadelphia transit workers in town to operate the
vehicles, Buffalo's strike officially continued for several years but without a debili-
tating impact on the system. By the fall of 1922, "Mittenism was safely entrenched
and Bolshevism beaten."[38]

The Mitten Plan was so strong, had enough worker support, that the Amalgamated capitulated and struck a deal with Thomas Mitten. Brokered by Lauck,
the Mitten-Mahon Agreement of 1928 essentially stated that the Amalgamated
would stay out of Philadelphia in exchange for the PRT accepting arbitration,
the dues checkoff for its company union, and company funding of benefits programs. The Amalgamated never again had a significant presence in Philadelphia.
Many commentators saw the agreement as yet another example of Mitten offering the way forward. "It is good for industrial America to see men like Mitten
and Mahon working shoulder to shoulder," wrote the Philadelphia weekly labor
newspaper *Progressive Labor World*. "Co-operation always wins right results."
Critics, however, saw this as pure abdication of labor's independence, "the low
water mark of self-confident unionism," as Perlman and Taft put it. The editors of Philadelphia's *Trade Union News* castigated the Amalgamated for leading the way in labor moving "body and soul into the house of its enemy." And
A. J. Muste of Brookwood Labor College wondered if the United States was on
the road to becoming another fascist Italy, where "industrial and political dictatorship" reigned.[39]

Mitten may have trumpeted his plan, especially the stock purchase scheme
that made workers minor capitalists, to the world, secured the support of many
of his workers, and cowed the Amalgamated, but closer inspection reveals his
ulterior motives, or at least shows that the Mitten Plan helped Mitten as much
as the PRT's employees. By mixing up PRT workers' interests as employees and
investors, Mitten strengthened his control over them while bolstering the financial position of the company and, as it turned out, himself. Eliminating independent representation limited the likelihood of a strike; having workers buy stock
in the company essentially eliminated it. "Strikes and labor disturbances," Mitten admitted, "end when stopping work means stopping the worker's dividend
check as well as his pay envelope." Tying greater corporate profits to workers'
income also enabled Mitten to goad them into working harder. Continued opposition to one-man cars that eliminated jobs while making drivers also do much
of the work of conductors, Mitten argued, ran counter to workers' interests. So
too did anything less than full effort: "When we slack now, or shirk, or become
indifferent or careless, we work directly against our own interests." Failing to
give customers a good experience might lead them not to ride the PRT in the
future. "Walkers and autoists," as *Service Talks* cautioned, "butter no parsnips for

railroad men." On numerous occasions, the PRT's workforce voluntarily agreed to take less pay if doing so protected their investment: they delayed pay raises until a fare increase solidified the company's bottom line, and put more money into the stock purchase plan than was required. And even as they did so, they knew that they were not allowed to access their money until they left the PRT, and at that point the company, based on what was most advantageous for it, would decide whether they got the market value of the stock or their initial investment. Financial advisers, noted Henry Dennison in his 1925 article, "The Employee Investor," clearly understood the danger of employees putting "too many eggs in one basket without a real opportunity to watch the basket." Yet the PRT's workers, with Mitten's encouragement, did just that as they combined income and investments and thought of them as a significant part of their pension. Such thinking made some sense when economic times were good; it portended disaster when the company floundered.[40]

The employees' investments poured millions of dollars into the company, which made it appear more financially sound than it in fact was. Employees' money gave Mitten easy access to an interest-free line of credit, which allowed him to purchase equipment, extend trolley lines, and build a transportation empire without having to rely on investment banks, about which he maintained a healthy skepticism born of his early years in the railroad industry. The constant demand for PRT stock also kept its price, and the company's overall valuation, artificially high. Mitten used those valuations to argue how valuable his contribution was to the company and extract extraordinary fees from the PRT. A 1930 *Evening Bulletin* report found that between 1911 and 1929, the transit company paid Mitten and his son over $7 million in fees, with an average of $1 million per year coming in the last five years. Local newspapers noted that the president of the United States made $75,000 per year; PRT workers noted that their decent wages were not so decent as their Chief's. Mitten claimed he was worth every penny, represented a "fair return" on the city's investment, as he put it, because he had brought labor peace and "super cooperation" to the company.[41]

Many observers agreed with Mitten on the value, socially if not economically, of what he had created at the PRT. The *New York World* called the Mitten Plan one of the "Ten Decisive Ideas" in American history. Economists from Johns Hopkins, Yale, and the federal government lauded it, especially the stock purchase plan, as representing "the most enlightened thought in industrial relations." "Mr. Mitten," the Philadelphia press wrote, had emerged as "a national character." "[His plan] needs to be adopted by all the great industries—transportation, mining, textiles, iron and steel," the *Philadelphia Daily News* continued. "If these great employing industries adopt the plan, America is headed for a period of social prosperity and content without parallel in human history." Overzealous commentators groped for analogies: Mitten was the Babe Ruth of industry to some, the Abraham Lincoln of labor relations to others. He had, after all, led a "bloodless revolution" that would transition the United States "from the era of swollen fortunes and wage slaves to the age of ownership consistent with effort." Mitten gained such fame that John Wanamaker asked him to run his department store empire when he passed (Mitten declined); the Federal Electric Railways Commission had him testify about how his plan might help solve the nation's labor question; and a British delegation headed by Sir William McKenzie, president of that country's Industrial Court of Arbitration, interviewed Mitten to find out how he made his workers prosperous and content. Even famed Harvard University president Charles Eliot—no fan of the working class, labor unions, or immigrants—singled the Mitten Plan out for praise. It offered "the most hopeful sign of peace in industry." Eliot thought Mitten a bit naïve in his effort to turn workers into capitalists, but they found agreement in what Mitten termed their shared dedication to "devoting our lives to human betterment in industrial, educational or spiritual fields."[42]

The tributes may have represented the majority view of Mitten and his plan, but discontent lurked behind the scenes and mounted over time. Mitten could claim he was open to organized labor, but workers testified to the U.S. Commission on Industrial Relations that everyone knew there was no such thing as a secret ballot in his union elections and that Mitten's obvious end goal was to eliminate the Amalgamated and any other independent opposition. PRT managers went to employees' homes, pushing them to sign loyalty letters and making them vow to help the company purge any "undesirable employe." Mitten told the workforce that he was glad so many had signed to show their "cooperation [and] loyalty." He did not intend to "commence firing men wholesale," he added, but the threat was clear enough that Clarence Pratt testified that PRT workers felt "compelled to sign a statement." "This is hardly cooperation," Pratt concluded. The Cooperative Association's representation election victory itself was fraudulent, workers argued, a paper tiger that Mitten used as a "cemetery for grievances." To further keep the men in line, the company openly admitted to employing a spy ring that monitored employee behaviors ranging from theft to union organizing. The workers may have gotten better pay—and even that brought contention, as they lagged Chicago, Washington, D.C., and other cities—but by methods that often lowered the quality of their work life. Surveying the PRT, workers found that Mitten cut the workforce by 10 percent and forced through one-man car operation; the company increased its swing runs, which required employees to work as many as fourteen hours a day for approximately ten hours of pay; and station superintendents encouraged the "speedup" so trolleys covered their routes faster than advertised (and faster than some drivers thought safe).[43]

By the mid-1920s, the stock plan—Mitten's crown jewel—had begun to leave a growing number of workers and observers unimpressed, even skeptical. As Robert Feustel, an engineer who evaluated the PRT for the City of Philadelphia, wrote, "Super management and super co-ordination are the principal themes until it is little wonder that super assumptions were necessary in order to support the claims which were made." The Mitten Plan, he added, had taken on "the air of being some religious fetish rather than the presentation of facts." There was nothing particularly instructive here: Mitten worked the men hard, they were efficient but not paid all that well, and their "surplus" income was tied up in stock that Mitten controlled. Other observers were less kind: Mitten's stock plan had turned the workers into "rainbow chasers," one man told the Commission on Industrial Relations; the Mitten Plan was a "house of cards," a city official wrote. It all amounted not to the next step in the social progress of man, but to a subtler, less violent form of social control, of "dictation" by a "coercive plan" that suggested to some observers Russian Bolshevik practices. Mitten's methods of control may have been subtler than the violence of 1910, but over time many observers came to discern how his plan controlled the employees all the same.[44]

Mitten, the Transit System, and the City

Mitten's chief goal may have been to implement his labor relations plan, but he always thought big about what the PRT could accomplish. His plans brought him into conflict with Philadelphia officials who believed he built the transit system and his own reputation at the expense of his employees and the city's residents. From early in his tenure, Mitten argued that he had a civic duty to drag Philadelphia out of its "pot-bellied complacency" by transforming what he considered its "traditional" street railway company into a full "transportation company" that would build a larger, more prosperous city. But because the PRT had private ownership, Mitten always kept in mind that the provision of a public good required turning a profit. One of the conflicts that emerged early and often entailed city officials charging Mitten with diverting scarce capital to pay dividends rather than using it to upgrade service.[45]

This was a fair point to an extent, and it did raise again the bedeviling issue of a private company providing a public good, but it is important to note that Mitten did make a number of efforts to improve the PRT's service and even turn the company into a national exemplar. Mitten replaced the old streetcar fleet with lighter, more durable, and more modern Nearside cars designed by one of his lieutenants, Ralph Senter (who went on to become president of the Philadelphia Transportation Company). He rerouted the streetcars in the business district to rationalize the routes and timetables, making the system more efficient and user friendly. He greatly increased the PRT's electric power production and developed Willow Grove Park as a weekend destination so the company's cars did not sit idle two days a week. Mitten also used PRT streetcars to ship freight, newspapers, and the mail, and for a time even ran a funeral car complete with a flower compartment and benches for pallbearers. As automobiles grew in popularity, Mitten tried to adapt by building garages in Center City and park-and-ride lots at the end of rapid transit lines. In fact, the lot at the west end of the Market Street elevated in Upper Darby was reputed to be the first of its kind in the nation. To be sure, other urban transit companies ran similar side businesses, but none gained Mitten's national reputation for the sheer variety and volume of the services provided.[46]

Mitten's greatest innovations came in the form of the other types of transportation he provided to paying passengers. After the First World War, Mitten understood that to survive, the PRT had to compete with automobiles, had to accommodate the growing numbers who would only "ride on rubber." Mitten led the way in developing a gas-electric bus that provided greater comfort (his claim of "luxury" was undoubtedly overinflated) than the streetcars that to many Philadelphians seemed old, slow, antiquated. In 1923 Philadelphia began to transition a few routes at a time from streetcars to motor buses, which mirrored a number of cities—London (1904), New York City (1905), Cleveland (1912)—that got an even earlier start. Mitten expanded the bus service first into southern New Jersey and then developed interurban bus routes to New York City, Washington, D.C., and points in between that the PRT ran in conjunction with the Pennsylvania Railroad, until the latter company purchased the lines in 1936. The PRT, claiming that "every phase of City transportation should be brought under one direct management in order to conserve street space and make the largest possible proportion of transportation earnings available for expansion and improvement," also bought the city's four cab companies (Yellow, Quaker, Cunningham, and Diamond). By 1927, Mitten operated eleven hundred cabs, marking the first time, one commentator observed, that "the taxicab had been made an integral part of the transportation system of a large city."[47]

Mitten's foray into air service was his most adventurous undertaking of all. Traveling in Europe in 1926, Mitten encountered a government-subsidized, privately operated air passenger transport system that connected the continent's major cities. The network fired his imagination, and he determined to put together an airline that would connect Philadelphia to Washington, D.C., during his city's upcoming sesquicentennial. Mitten contracted with Anthony H. G. Fokker to supply three eight-passenger planes and persuaded the U.S. Navy to let his fledgling airline use its flying field. Inaugurated on July 16, 1926, the

PRT Air Service charged passengers twenty-five dollars for a round-trip ticket (including its most famous passenger, Will Rogers) and carried U.S. mail for ten cents a letter. The service lasted until November 30, 1926, ending along with the sesquicentennial. During that time the PRT carried thirty-seven hundred passengers, which amounted to nearly three-fourths of all commercial passengers in the United States that year. The airline lost money—$118,000 in less than five months—but demonstrated, Mitten claimed, that "air service in America has arrived. Its development under competent management will make for the most striking revolution in the history of modern transport." Developing an air transportation system was vital to the nation's future, he argued, but the PRT had lost thirty-two dollars for each passenger carried. The U.S. government would have to subsidize airlines, Mitten continued. Such a system was not "socialism," as his critics who clung to their belief in an outmoded, unadulterated laissez-faire economics claimed, but would instead be the modern manifestation of the model the United States employed to develop its world-class railroad system. At Secretary of Commerce Herbert Hoover's request, Mitten even loaned his best traffic engineer, R. Harland Horton, to the government to conduct an aerial survey that was to be used to build a two-thousand-mile network connecting Boston, New York City, Philadelphia, Washington, D.C., Atlanta, Birmingham, New Orleans, Nashville, Louisville, Indianapolis, and Chicago. Mitten had envisioned the country's need to develop an air transportation system, but the PRT simply did not have the resources to bring it to fruition. When the federal government, locked in a laissez-faire mind-set, refused to put resources into the system Horton surveyed, Mitten had no choice but to abandon it with regret.[48]

Ida Tarbell, writing in *American Magazine* in 1930, captured Mitten's grand vision: to make "something no city had as yet seen—a unified transportation system in which street cars, subways, elevateds, motorbuses, cabs, even flying machines cooperated in carrying people where they wanted to go, in the quickest, cheapest, and most comfortable fashion." Such a vision had its critics. To be sure, many Philadelphians took some civic pride in the national prominence of their transit company. But many grumbled that their transit fares subsidized Mitten's ancillary ventures. The cab companies may have furthered Mitten's transportation empire, but the press noted that the PRT mainly bought the companies because they cut into streetcar profits. Absorbing them allowed the PRT to charge a higher rate for a cab ride, which drove passengers to the subways, trolleys, and buses. The PRT's power, its near monopoly on all forms of public transportation, shut out smaller companies that tried to start lower-priced services that would have benefited Philadelphians. To city officials, a central question boiled down to whether a privately owned transportation monopoly, grand or not, benefited Philadelphia more than a number of companies each offering its own service.[49]

That monopoly gave Mitten great power in a growing debate with Philadelphia officials about whether a private company could supply a public good and do so in a way that best benefited the city and its people. Here, the strong commitment that Philadelphia's political and business class held to private enterprise often ran at cross-purposes with what was in the best interest of the city's people and its economic and geographic growth. It was during the Mitten era that valuation of the PRT and how its operations could best serve the metropolis began to assume a predominant place in the public and political discourse about transportation. That debate about the value of the company and how it deployed its resources, with permutations, lasted half a century, until the city bought the PRT and created SEPTA in the late 1960s.

City officials, with the support of the business community, led the charge to expand Philadelphia's transportation system. Mayor Blankenburg appointed A. Merritt Taylor as the city's transit commissioner in 1912, with the express task of developing improved transit facilities for the city and its suburbs by subway, streetcar, or other "modern methods." Taylor had served as president of the Philadelphia and West Chester Traction Company, the forerunner of the Philadelphia Suburban Transportation Company in Delaware County (commonly known as Red Arrow), and came in with a distinct and expansive vision of what the city's transportation system should become. Taylor built on a "Dream Metropolis" concept first circulated in 1911 that urged Philadelphians to make their city more like Paris or Washington, D.C., with monumental buildings and diagonal boulevards. To this Taylor added a delivery loop around Center City and high-speed lines on Broad Street, Woodland Avenue, and Kensington Avenue. These, to Taylor, took priority, while other lines that he proposed along Passayunk Avenue, Lancaster Avenue, and several other streets could wait. Taylor's concept—with an estimated cost of $57.6 million—was to build high-speed arteries across the city, on both radial and transversal lines, and feed them with streetcar capillaries. Not only would the system fuel the city's economy, grow its property values, and enlarge its tax base; it would also be good for its residents. "The people," Taylor argued, would "obtain wide and comfortable range of movement and enormous and valuable time saving." As the city grew, so would Philadelphians' opportunity to take new jobs, live away from industrial areas, and enjoy the region's recreational areas. Transit represented an economic as well as a social good, and it was the city's "direct moral obligation" to provide the best system it could.[50]

Taylor had the loudest voice and most expansive vision, but his was far from the only one. William S. Twining, who succeeded Taylor as transit commissioner in 1916, backed the construction of an expansive system. A hard-headed engineer, Twining understood, more than Taylor, the prohibitive expense of the full Taylor plan and backed away from parts of it. But in the end, he argued that the PRT did not do enough for the city, that service, not profits, had to "determine whether they [new lines] shall be furnished and operated." Business groups joined the chorus. The South Philadelphia Business Men's Association, for example, petitioned the city to build a Broad Street subway that connected South Philadelphia to Center City and then continued farther north. Other business groups, including the Philadelphia Real Estate Board, appealed for the construction of an elevated to Frankford that would "do justice to [their] business interests."[51]

Thomas Mitten, who had such tight control of the city's transportation system, listened to the arguments and mostly ignored them. The Market Street Subway, he knew, had nearly bankrupted the PRT before his arrival. Moreover, the company had invested heavily in surface lines and high-speed transit—subways or elevateds would only rob fare-paying riders from the streetcars. Center City suffered from congestion that seemed to grow by the day, and Mitten supported underground transportation there, but only if it received public funding and the city also provided some compensation for the dividends PRT stockholders were sure to lose when passengers headed underground instead of onto the trolleys. City officials were correct in their assessment of what rapid transit could do for property values: the valuation of land in West Philadelphia soared to $35,263 per acre after the Market Street line came through, while acreage in far North Philadelphia languished at $6,658 per acre without benefit of a subway line. No matter. With Mitten having such power, of all the proposed high-speed lines, only two were built in the era: the Frankford elevated line opened in 1922, and the Broad Street Subway in 1928. Twining, after feuding with Mitten for nearly a decade, stepped down in 1923. He was the last forceful transportation administrator the city had.[52]

Mitten's refusal to expand the system if doing so would harm his stockholders brought concerns about the public/private nature of the PRT forcefully to the fore. Public transportation was a public/private venture in Philadelphia, and the investment in high-speed lines, such as the Frankford elevated, emphasized that fact. As Mitten explained, the PRT acted as an operating company and leased Philadelphia's city-built lines. In the case of the Frankford el, the PRT paid $780,000 per year starting in 1927 for the right to operate the line. The company had to make that payment before it could offer any dividends to its stockholders, which led Mitten to pronounce that the PRT would enter such arrangements only if new lines supplemented the existing system. He could not justify supporting any expansion that harmed his stockholders' interests. Rankled by such thinking, Twining argued that the city-company relationship had become so "muddled . . . so entangled with government that growth has not kept pace with the public needs." Twining favored private operation of the transit system, but cautioned the PRT that it had to remember it provided a public service and was thus subject to government—the people's—control.[53]

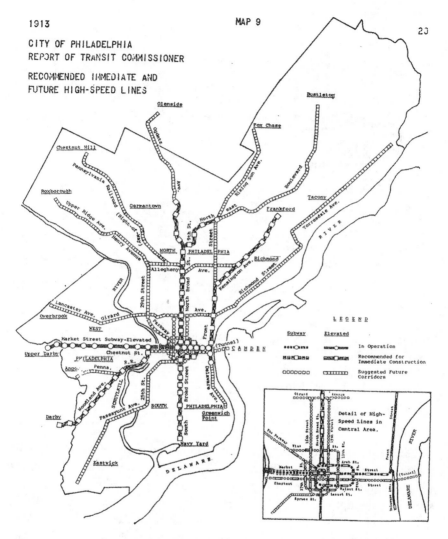

FIGURE 9. Proposed high-speed lines for Philadelphia, 1913. City planners proposed an expansive network of high-speed lines that would have made Philadelphia's transportation system the rival of any in the world. Such a system, which would have been expensive and potentially cut into PRT dividends, was never built. From Robert P. Sechler, *Speed Lines to City and Suburbs: A Summary of Rapid Transit Development in Metropolitan Philadelphia from 1879 to 1974.* Courtesy of Robert P. Sechler.

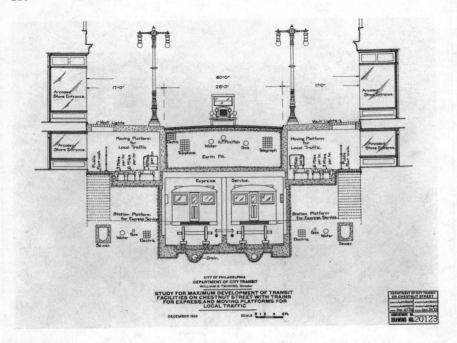

FIGURE 10. Chestnut Street proposed subway. Philadelphia city planners envisioned a cutting-edge system that if constructed would have provided express service, moving platforms, and easy access to shopping for riders. Photo courtesy of PhillyHistory.org, a project of the Philadelphia Department of Records.

Philadelphians generally agreed with Twining about the provision of public transit by private enterprise, but a strain of city politics, informed by the nationwide "gas and water" socialist impulse that sought public control of municipal services, did emphasize public ownership. As early as 1912, some city officials started discussing public ownership, although Mayor Blankenburg dismissed the idea because too much politics would be involved. Sure enough, cries of "socialism" rang out to beat back the idea, and Blankenburg instead brought in Taylor. The movement did not, however, die. In 1920, the city's Transit Commission argued that the charges levied by the underliers were so exorbitant that they prevented the PRT from building the best possible system for the public. The best solution, the commission asserted, was to exercise the power of eminent domain and take over the system. This proved controversial at first, but in 1928 Mayor Harry Mackey moved to begin the process of condemning the underliers, a move that concluded with the PRT's bankruptcy and reorganization in 1940. Even this process, however, did not result in public ownership.[54]

The key to the condemnation proceedings was the valuation of the company. The city and the PRT wrangled over that number throughout the 1920s and would in fact do so until Philadelphia took over the company in the 1960s. Valuation ultimately determined how much money stockholders would receive. Mitten claimed the PRT was worth $300 million; Mayor Moore said half that. Never one to shy away from a fight, or a moment of hyperbole, Mitten told his workers—a significant component of his stockholders—that the city was "trying to ravish us." PRT employees then agreed to buy more stock to keep the value of the company elevated. "Mitten is a smart man, a very able man," the mayor said. "Employees have been induced to buy stock [and I hope] they [do] not lose."[55]

Downfall

The mayor's fears for the PRT's workers proved prescient. In 1927, Mitten purchased from Albert Greenfield, a local banker and real estate developer, the assets of the Producers and Consumers Bank, a cooperative bank founded by Philadelphia's Central Labor Union in 1922 of which Greenfield had taken receivership. Mitten wanted the foundering bank's resources as a platform for his proposed Mitten Bank Securities Corporation (MBSC), which would combine the money the PRT's workers had invested in their company through the Cooperative Association with the accounts remaining in the bank. Mitten offered the account holders, all working-class people, a deal: they could leave their money in the bank and be made whole when it turned enough of a profit, or withdraw 60 percent of their funds. Most chose to stay. To complete the funding of the bank, Mitten exchanged the PRT workers' shares in their company for some $18 million in shares of MBSC. They thus went from nearly owning the PRT to owning stock in a financial institution emerging from bankruptcy.[56]

Now that he owned a bank (and named himself president and his son vice president), Mitten claimed he could complete his goal of "helping labor become capital." He exhorted PRT workers to put their life savings in MBSC and advertised the institution to the wider public as a bank for the people, the ultimate "democratic institution." Mitten vowed to revolutionize banking by putting "service first and profit afterward." MBSC opened divisions for workers and for women, established a small loans department, promoted banking by mail, and offered travelers' checks and other services usually reserved at the time for wealthier customers. All told, some twenty-six thousand people invested in MBSC securities, and many more relied on the bank's services. The Central Labor Union and AFL opposed Mitten's bank because of his refusal to allow organized labor at the PRT, but overall the bank received a warm reception in working-class Philadelphia, at least at first. Mitten, whose health began to decline in the mid-1920s, ran MBSC for nearly two years before turning the bank over to his son on January 1, 1929.[57]

Nine months after leaving MBSC Mitten died, drowning when his boat capsized at his estate in the Poconos. His death brought an immediate outpouring of sympathy from the working class and the wealthy. Some forty thousand people filed through the carbarn at Tenth and Luzerne to pay their respects, with some reportedly saying through their tears, "He was the best friend the working men ever had." Nearly seventeen thousand transit workers signed a bound volume to express their sympathy. J. P. Morgan sent flowers, and President Herbert Hoover extended a note of sympathy, as did the presidents of General Electric, the Pennsylvania Railroad, and many other firms. People seemed dumbstruck that their "Chief" had died. The press wrote articles about Mitten's Japanese valet standing silent guard over his body, about deer "crashing through the underbrush and shrubbery, to stare at the house where their friend's body lay." Rumors abounded that Mitten had committed suicide, been assassinated by his enemies, had a mask placed on a dummy corpse so he could sneak out of the country undetected. Those feeling less fanciful, or conspiratorial, discussed Mitten's $3.5 million estate and applauded how it would be used to support the continuation of his work with the Mitten Plan and MBSC.[58]

The good feelings soured four weeks later when the stock market crashed. After that, it became apparent what Mitten had done, and the PRT's workers paid the price for the next decade. To be sure, the Depression hurt working-class Americans across the board, but Philadelphia's transit workers had put all their eggs—income, investments, and retirement funds—in one basket, and the crash wiped them out. The value of their MBSC stock plummeted from $25 per share to $1.87, which meant their $18 million investment in PRT stock fell to approximately $800,000 by the time they got paid in 1941 (a loss of 96 percent). Several workers filed lawsuits, averring that Mitten had misled them, misappropriated their money, strong-armed them into transferring their investment in a business they knew to a bank that Mitten owned. They were the victim of "a fabric of trickery, constraint and compulsion." Their problems, the workers argued, were in part a consequence of the bad economy, but mostly of Mitten and his management team fearing that they, the workers, owned enough PRT stock to actually take control of the company. Mitten's slogan of "labor becoming capital" was a farce, they believed, a cruel one that robbed them of their life's savings.[59]

The city, increasingly at odds with Mitten, and now his son, sided with the PRT workers and began a two-year lawsuit against the company in 1929. The Mitten Plan, said the bombastic city controller and future mayor S. Davis Wilson, was "a farce and a sham." "The men have no voice in the affairs and management of P.R.T.," he continued, "and their earnings and labor have gone to aggrandizement of management at the expense of the men." City attorneys mocked the idea of "super cooperation," claiming that all Mitten provided was "super fees," and they demanded that his company, Mitten Management, which was now headed by his son, return the workers' PRT stock. The newspapers, finally out from under Mitten's spell, called on the city to end any contracts with Mitten Management, the PRT, or its underliers. Philadelphians had been treated like "suckers" long enough. The Common Pleas Court found it exceedingly difficult to sort out Mitten's financial relationship with the PRT and MBSC but ultimately removed Mitten Management from the PRT and ruled that his estate had to be turned over to the PRT to satisfy what debts it could. PRT workers' lawsuits continued through the 1930s, but there was little left to claim. [60]

In the end, Mitten appears to have been a true believer in his labor relations cause and in the value public transportation could add to human society. He had an expansive vision of the PRT and an entrepreneurial mind to match, and he made Philadelphia a national leader in multiple modes of transportation. He also thought he had found the solution to the labor question, a way to humanize capitalism at a time that bolshevism, at least in the minds of the business class, was on the march. Across the United States, commentators pointed to the PRT as the way forward in handling labor relations, and there was some truth to their hopeful views. Mitten and his plan did some good for some people: himself (he gained wealth and acclaim), the people of Philadelphia (they got twenty years of mostly uninterrupted transit service), and at times the transit workers (who thought they were taking control of the company through stock purchases and leading the way in revolutionizing labor relations). But in the end, the Mitten Plan could not hold as its flaws were exposed—the need to control employees' activities, the dependency on a growing market, the stock value crumbling in an economic downturn—and the workers it was supposed to benefit were actually grievously harmed by it. Mitten was certainly unwilling to use the raw violence that plagued the system in the years before his arrival, but it is debatable whether the PRT's workers were better off after his tenure than they were before. The hard times coming in the Great Depression would not make their lives any easier.

HARD TIMES AND A HATE STRIKE

The Philadelphia Rapid Transit Company's annual report in 1938 brought the stockholders a dire warning, one that they had heard all too often in the 1930s. "PRT must get out of the courtroom and back to the carbarn! 1938 passenger revenue was $31,738,599, lower even than in 1933, and $2,258,306 less than in 1937. . . . Unemployment and general business conditions have cut deeply into your company's revenue. But they are not the only factors operating against PRT. Inability to modernize the system, to buy new cars and buses, to offer quicker, quieter and more comfortable rides, has hurt us severely." To survive, management concluded, we must "end litigation, reorganize, modernize!"[1]

The Great Depression pushed the PRT into desperate circumstances, and management spent the 1930s trying to save itself and the company from bankruptcy. This meant repeated trips to court, penny-pinching on service, and continuing to use a company union under different guise to circumvent New Deal labor law. By the end of the decade, though, none of it much mattered. As was the case at transit systems in Chicago, New York City, and elsewhere, the Depression devastated the finances of the PRT and threatened to leave the company in ruins. Philadelphia's transit workers thus faced the hard choice of taking pay cuts and layoffs or organizing in a strong industrial union to fight for their livelihoods. It took them most of the Depression and World War II period to make their decision, but by the early 1940s they began shifting their position to abandon their company union for the first time in thirty years and support strong unionization, which opened the door for the Transport Workers Union (TWU, an affiliate of the Congress of Industrial Organizations—CIO). The TWU, which won a representation election in 1944, brought the first independent voice to advocate for the PRT workers' cause since 1910, and the union's negotiators immediately demanded a contract that would greatly improve employee pay and benefits.

It quickly became clear to PRT management that the TWU was solidifying its hold over the workers and was a formidable foe at the bargaining table. It was also clear that World War II had reshaped the job market in the city, forcing the company to hire women as drivers. African Americans also sought these jobs, but management and many members of the company's white workforce rejected them for most of the war. Recognizing the way racism divided the working class at the PRT, management employed a culturally pervasive language of racial discrimination in an attempt to undermine the TWU. Many of the PRT's white workers agreed with management about the need to keep the company segregated and in some ways even led the fight. This dangerous ploy by management in conjunction with like-minded workers resulted in the worst wartime "hate strike" in the United States, a strike that shut down America's third-largest war production center just eight weeks after D-day. It ultimately took President Franklin Roosevelt's deployment of five thousand heavily armed troops to the city to put down the strike.

In examining the Depression and World War II together, this chapter highlights the links among the PRT's difficult financial position, the "threat" posed by the TWU (workers would have called it "vigorous representation"), and the racial tensions of the war years. For a time, PRT management relied on lessons learned from Thomas Mitten. In particular, they used a company union to try to inspire worker loyalty and dampen their activism. But when the National Labor Relations Act (better known as the Wagner Act) forbid the use of a company union, the PRT had to open its doors to independent representation for its workers. When the TWU won its bargaining election, it became clear that the union would extract major wage concessions that management felt it could not afford. So PRT leadership shifted from using a company union to deploying the language of race. Many white PRT workers bought the arguments, advanced them, and thus inflicted great damage on the transit workforce generally as well as on black employees specifically. This use of racial division proved to be a short-lived but explosive foray that demonstrated both the supple nature of PRT management's continuing quest to control its workforce and the way the workers at times could be their own worst enemies.

The Great Depression Comes to Philadelphia and the PRT

"The Great Depression," wrote the historian Roger Simon, "fell upon Philadelphia as a vast tornado, flattening almost everything in its way." It changed Philadelphia's economy, the lived experience of the city's working class, the nature of local and national politics. The PRT and its workers, of course, could not escape this shattering event, and in the 1930s they suffered (although not as badly as workers in many other industries) as the economic downturn meant less business activity, fewer riders going to work or shopping, and less money thus coming into company coffers. The city had to have public transportation, but it was an open question for many years whether the PRT as a corporate entity would survive and how many transit workers would keep their livelihoods.[2]

At first, in the late 1920s, Philadelphians hoped the Great Depression might touch their city only lightly. In 1927 Philadelphia ranked third in the United

States in both wealth and the value of products produced, with the textile industry employing some ninety thousand people, and other sectors such as apparel, metals, chemicals, railroad cars, and electronic equipment providing work for tens of thousands more. Rather than being dominated by a single industry, then, the city had a diverse economic base that cushioned many residents from the worst of the Depression in its first few years. Yet business leaders worried that the city's industries had peaked. The proportion of the population working in manufacturing had dropped from about half in 1890 to under a third by 1930. Some of those workers found clerical positions in the city's burgeoning service economy, but in Kensington, South Philadelphia, and other dense working-class communities, unemployment could periodically reach 30 percent before the Depression even began.[3]

When the full weight of the Depression hit Philadelphia in the early 1930s, it devastated the manufacturing and financial sectors of the city's economy. Between 1929 and 1933, business activity fell by a third and manufacturing output by half. Major employers such as Baldwin Locomotive, Brill, Budd, Disston, and Philco either closed their doors or laid off thousands. Smaller concerns, especially in textiles and garments, had an even more difficult time. Some fifty banks, including Albert Greenfield's Bankers Trust Company, with one hundred thousand depositors and $35 million in deposits, and nearly two thousand savings and loan societies collapsed. Philadelphians, many of whom had lost their jobs, found their life savings swept away.[4]

Unemployment presented the greatest problem to most Philadelphians. A survey before the stock market crash in October 1929 found that 10 percent of the city's workers were unemployed, with the problem concentrated in industrial areas. That number grew to 15 percent in 1930 and hit its peak when 40 percent were unemployed and another 20 percent were underemployed in 1932–1933. Per capita income fell to half what it had been before the Depression began. Black Philadelphians were hit hardest, with some communities experiencing 50 percent unemployment. Women tended to have an easier time than men finding employment, chiefly because they often worked in service jobs or in the low-paying garment industry. Overall, unemployment was a persistent problem for a decade: As late as 1940, despite federal programs, 20 percent of Philadelphians were looking for work. Across the city, Philadelphians grew accustomed to soup kitchen lines and Hoovervilles springing up on vacant land.[5]

Mirroring transportation companies in New York City, Chicago, San Antonio, and most other cities across the United States, the economic downturn had a direct impact on the PRT's ridership, which reverberated through the company down to the workers. The Great Depression, to give two prominent examples, forced Chicago Rapid Transit into receivership and led to public control of a consolidated system in New York City. In Philadelphia, PRT numbers fell from a peak of 966 million riders in 1926 to 793 million in 1931, bottomed out at 627 million in 1933, and rebounded only to 637 million in 1938. Rides per capita in those years stood at 509 in 1926, 404 in 1931, 315 in 1933, and 310 in 1938. Revenues expectedly followed suit: $57 million in 1926, $45 million in 1931, $33 million in 1933, and $32 million in 1938. The taxicab division showed an even more precipitous drop in revenue, falling 72 percent between 1929 and 1933 (from $6 million to $1.7 million).[6]

Such numbers made finances and the company's future the central issues for the PRT in the 1930s. For a decade, company management, stockholders, employees, and city officials wrangled in Philadelphia's press and courtrooms over how much the company was worth, who should own it, who should run it, whether investors, including the company's workers, could get any restitution for their losses. Newspaper editors and city officials understood the unavoidable impact of the national economic downturn and fixated on the underliers as one of the most significant problems facing the PRT. The underlying companies were "thieves," according to one report; "hogs" in the words of the *Philadelphia Record*; vultures looking for the "last pound of flesh" in the *Public Ledger*'s estimation. The U.S. District Court cautioned the underliers that the PRT had enough money to meet its payroll and provide transportation to the city of Philadelphia but could no longer make payments on its standing debt. The underlying companies had to readjust their charges or the company would go bankrupt and they would get nothing. This was no idle threat: By 1933, the PRT had fallen behind by $645,000 in its payments to the underliers, and it did not look like the situation would improve. In response, the underliers offered to cut their payments by $1.8 million per year (a 20 percent reduction) until the PRT was on more stable financial footing, but to most observers that was simply not enough.[7]

The PRT's precarious financial position, exacerbated by the underliers, threatened the survival and arrested much of the expansion of a public service vital to the city's functioning. A 1930 municipal study reported that the system provided adequate service but had no reserves to build out its surface lines. The PRT had to halt its track renewal program and used money set aside for equipment maintenance and replacement to pay its funded debt. This, the company noted, was an untenable situation. The PRT did manage to replace many of its buses by the end of the 1930s (the *Inquirer* remarked that "the antiques" had certainly "earned a rest in whatever haven it is to which broken down old buses go in their old age") and in the process continued the transition from streetcars to rubber-wheeled vehicles. To try to offset the cost, management pushed one-man cars on the remaining trolleys and reinstituted "triple header" runs on buses, which paid eight hours but required drivers to report to work three times in one twenty-four-hour period to cover shift changes in the city's factories. Federal funds did help extend parts of the rapid transit system, with construction of the Delaware River Bridge high-speed line starting in 1934 (operation began in 1936) and the South Broad Street Subway to Snyder Avenue in 1937 (operation began in 1938). Other lines, however, were little more than federally sponsored "make work," because the PRT could not afford to outfit and operate the new tunnels. The Locust Street subway dig, for example, started in 1937, but the line did not begin operation until 1953, and a Market Street subway extension into West Philadelphia was sealed off until 1947, and service did not begin until 1955.[8]

In this context of financial emergency, the push for public ownership gained new legitimacy and momentum. Most observers no longer viewed proponents as Socialists lacking comprehension of the verities of the American economic system. Far from it: supporters of public ownership now numbered among the city's political leaders and opinion makers in the press. One study, in light of the fact that the underliers had received nearly $230 million since the PRT was organized in 1902, described city purchase as a matter of "social justice." The public's money in the midst of the Great Depression should no longer go to underlying companies that provided no service. A government report urged city purchase of the transit system because doing so would "greatly aid in carrying out the City's transit development program. It would facilitate the making of desired track changes and removals and thus contribute to the efficiency in use of the streets. The income and security of the City as landlord would be reasonably safe as a business risk. The proposed purchase would be a logical and important step in the unification and simplification of the City's transit system with ultimate relief from the heavy draft now made upon transit revenues by the Underlier rentals." More than anyone, Mayor S. Davis Wilson led the charge. Wilson knew the PRT had to be reorganized and believed the city had to take full control of the company. If the underliers were still to have any stake, their payments under his plan would be reduced to $2 million per year. The city, he said, was locked in a "rotten system . . . whereby the life's blood of the people is being drained from them." They were, Wilson added, "like those gunmen I'm chasing out of town." He then threatened to punch a PRT attorney in the nose.[9]

Threats aside, Wilson well represented the thinking of most Philadelphians who would no longer tolerate business as usual at the PRT, and the company thus faced great public as well as financial pressure to reorganize. In an attempt to stave off bankruptcy, the PRT finally agreed to a corporate reorganization in 1931 that resulted in the removal of Mitten Management (although A. A. Mitten stayed on as a corporate officer) and the creation of a new board of directors that included area business leaders such as John Gribbel of the American Meter Company, George MacKinnon of Stetson, and Herbert Tily from Strawbridge & Clothier.

The reorganization did not change the tide, however, and on October 1, 1934, the PRT filed for bankruptcy. Judge George Welsh of the U.S. District Court guided the company through the proceedings, and observers noted that he clearly did not want a public takeover. Over the course of some five years Welsh worked with city and PRT attorneys to set up a board of trustees to oversee the operation of the PRT, create the Philadelphia Transportation Company (PTC) as a replacement for the PRT, install Albert Greenfield as chairman of the board of the new company (five of the twenty-one members represented the city), and arrange a stock plan that swapped PRT for PTC stock. Welsh's plan also replaced the perpetual debt owed to the underliers with bonds that paid $1 million to $2 million per year, depending on company performance. The PTC began operation on January 1, 1940, and at that point the relationship with the underliers, which had so burdened the city, came to an end.[10]

Although the court's plan derailed Wilson's effort to take the transit system public, overall it met with approval. Critics claimed that the former underliers still received more money than they were worth, but even so the deal freed $3.5 million per year that the new company could use to improve its service and replace old equipment. Finally, management claimed that, after years in court, the PRT had embarked on "the road to modern transit and a greater Philadelphia!" The *Philadelphia Record*, a persistent critic of the underliers, told its readers that the PRT's reorganization was one of the biggest events of 1940 and that they should expect better service. "Exploited for decades, given worse and worse service for more and more carfare, the Philadelphia public has long had a square deal coming," wrote the editors. "It is promised they will get it now. We expect to see that promise fulfilled." The PRT's workers also backed the plan, hoping that it might arrest the company's decline, save their jobs, turn around their fortunes.[11]

As the PRT made its way through the reorganization process, Philadelphia's transit workers throughout the 1930s periodically took the company to court, hoping to recover some of the money they lost in their investments in company stock. They usually filed as a class (as PRT employees and/or members of the Cooperative Association) and asked that the courts remember their investment, often amounting to their life savings, in the PRT as the company reorganized. Unfortunately for them, the money was all but gone, and Judge Welsh finally had to tell the plaintiffs that he understood their plight but they had little recourse. "You see, they came in with eggs," he said. "They are now scrambled eggs. You never can get the egg back again. . . . I can't unscramble this egg. We will see if we can't make it into an omelet, that everybody can get a piece out of it." A tortured analogy can still capture the truth, and Welsh certainly did. By the time Philadelphia's transit workers finally got something back from the company, it was 1945 and they received ten cents on the dollar. The press painted the 1,250 men who showed up to get their payment at the Drexel Bank at Fifteenth and Walnut Streets as untutored, unwashed, bordering on pathetic. Their unjust and rather painful experience, the workers' attorneys argued and Mayor Wilson agreed, meant they had to have a greater say in the affairs of the company or they would be left vulnerable again. If the reorganized company tried to dominate them, to run its business without their input, Wilson warned that the city would have "a repetition of the terrible catastrophe of 1911." He meant the strike of 1910, but the mix-up of dates did not obscure his meaning.[12]

The quixotic fifteen-year effort to get something back from the company spoke to the workers' desperation. Not only did the men lose their life savings and retirement investments; in the 1930s they saw the PRT repeatedly respond to hard times with wage cuts and layoffs. Workers took a 16 percent pay reduction in 1932, another 7.7 percent in 1933, and the drumbeat continued through the 1930s. Layoffs did not cut as deeply as those in many industries, but they troubled PRT workers nonetheless. The company let 1,119 workers go in 1932, and Philadelphia's unemployment rate for street railway workers hit 10.6 percent that year. When transit employees complained, Ralph Senter, who had taken over as president of the company, cautioned them that their plight could be much worse: steel, steam railroads, and other industries had laid off one-quarter of their men or more. Plus, he reminded workers assembled at one meeting, they still belonged to the Cooperative Association and had agreed to peg their incomes to the price of a market basket of goods. A decline in prices during the Depression triggered a decline in wages. More, it was up to the men who had been elected as officials of the Cooperative to explain the situation to the rank and file, in effect to serve as the first line of defense as the company retrenched. In the early 1930s, the PRT's workers were willing to listen to Senter, accept the pay cuts and layoffs, and continue with the Cooperative Association that seemed to make them at least in part culpable for their own plight. But their acquiescence did not last.[13]

The New Deal and PRT Workers' Unrest

The political and labor relations context for workers in Philadelphia and at the PRT in the 1930s offered them a rare moment in the nation's history to organize along class lines. Mirroring working people across the United States, working-class Philadelphians forged a coalition that helped elect Franklin Roosevelt as president four times. Their politics was motivated by their disillusionment with local Republicans' meager response to the Depression and by their own support of federal New Deal programs. The Works Progress Administration and Civilian Conservation Corps, for example, employed tens of thousands of people to build the city's airport and refurbish its schools; the National Youth Administration provided health screenings for children; and the Home Owners' Loan Corporation helped thousands of Philadelphians keep their homes. The city's municipal offices, an anomaly in comparison to much of the rest of urban America, remained under the control of the Republican Party for another twenty years, until the liberal Democrat Joseph Clark won the mayor's office in 1952, but the 1930s nonetheless witnessed the beginnings of a significant shift in the city's politics.[14]

Federal programs that supported poorer Americans, as Lizabeth Cohen has demonstrated, were not just handed down from the White House, but were instead a result of working-class activism across the United States. In Philadelphia, that activism took many forms. Marches on City Hall, some led by the Communist Party, roiled the city, especially after conservative Mayor J. Hampton Moore used the police to repress them violently. Strikes, including sit-down strikes where workers took over the plant and refused to leave, occurred throughout the 1930s at Philadelphia employers such as Aberle Hosiery Mill and Exide. And the new Congress of Industrial Organizations, established in 1935 by proponents of industrial unionism who believed workers should be organized across an industry rather than by skill or trade, pushed a militant organizing agenda that brought tens of thousands of Philadelphia's workers into the fold in 1936 and 1937. By 1939, the CIO had 100,000 members in the metropolitan area, and unions overall counted some 250,000 people in their ranks.[15]

Such activism in Philadelphia and around the United States shaped national-level politics and helped push through new labor law that, often weakly, supported working-class interests. In the first years of the New Deal, Section 7A of the National Industrial Recovery Act (NIRA) guaranteed workers the right to organize without interference from employers. As the historian Peter Gottlieb has pointed out, the law at first gave workers optimism that they could form an independent organization that would advance their interests. But many corporations, including the PRT, circumvented this section of the NIRA by establishing or maintaining company unions. By 1935, when the Supreme Court ruled the NIRA unconstitutional, many workers had already recognized that the law did not do enough to help them form unions. It was the National Labor Relations Act of 1935 that finally put the federal government behind workers' right to organize their own unions and outlawed company unions. All of this activism and the laws that grew from it helped give working-class Americans the friendliest political atmosphere and presidential administration—although neither was flawless—that they had ever had at the national level. Transit workers and labor organizers at the PRT became emboldened by the political context and employed the state's machinery throughout the Depression and World War II eras to advance their cause at the transportation company.[16]

The first pronounced moment of worker restiveness at the PRT during the Depression came in a taxi strike in November 1933. After Mitten purchased the city's cab companies, he organized their workers in 1928 into a branch of the Cooperative, the Yellow Cab Cooperative Association. They lived with this organization for a few years, but pay cuts and layoffs led nine hundred of the system's cab drivers to join the Teamsters in October 1933. The men then contacted management to demand union recognition, a closed shop, the right to wear union buttons, the dues checkoff, higher wages, and free uniforms. The buttons especially mattered because they represented pride in and support of the union and had been banned by the PRT since 1912. The Cooperative, they argued, did not represent their interests, especially in lean times, and they expected with the help of the Teamsters to end their "virtual slavery." The company union's chief goal was "to keep them well in hand at all times," the drivers said. "It probably has the best methods of keeping its workers in hand—not the most brutal or the most vicious, but the most effective. It is against those conditions that [we] have revolted." President Senter told the men that they were free to join the Teamsters and then agreed to recognize the union and negotiate with the taxi drivers accordingly. He added, however, that under no circumstances would the PRT accept the closed shop or the workers' right to wear a union button. Senter's stance, particularly on the button issue, incensed the cab drivers, who said they would arbitrate all their other grievances but that the denial of the right to wear a button was a deal breaker. On November 26, the PRT refused to let any man wearing a button take out a cab. The drivers said they were locked out; the company said they had gone on strike. Either way, taxi service in the city shut down.[17]

Two decades after the last substantial conflict on Philadelphia's transit system, PRT management and workers once again clashed. The cab drivers, led by union president H. F. Galbraith, pledged that they would not instigate violence but would hold to their demands and not allow the company to intimidate them. "Stand pat on your demands," Galbraith told his union members, "and the public will be with you." Two months of newspaper reports made the case that the cab drivers did not necessarily instigate the violence, but they were certainly willing to participate in it. Almost all of the physical conflict took place between strikers and either men who crossed the picket line or against company property, chiefly taxis. On multiple occasions groups of a half dozen or more striking drivers forced cabs to the curb, assaulted the driver, and smashed the taxi. In one incident Joseph Donahue pulled over to pick up a woman walking with a baby in her arms. As he opened the door for her, five men jumped him and stole his cab. Donahue started walking toward the closest police station, at Twentieth and Buttonwood, when fire engines passed him. He turned to see his taxi ablaze. Elmer Romer, Frank Marone, William Hoskins, and dozens more drivers may not have fallen for the fake baby ruse, but they did receive beatings that put them in the hospital and served as warnings to others to leave the cabs in their garages or suffer the consequences. Some nights half a dozen cabs were burned, others pushed into the Delaware River, and police arrested 150 drivers in the first two weeks of the strike. The conflict, the staid *Evening Bulletin* told its readers, amounted to a "taxicab war" waged by union workers against the PRT and the good citizens of Philadelphia.[18]

That is, of course, one interpretation. But who was taking out these cabs if the men were on strike? asked the Teamsters. They were nothing more than "thugs and strike-breakers . . . imported by the company to operate the cabs" and were known for "carrying pistols and other weapons and were working under the leadership of a former pugilist." Senter repeatedly denied that the replacement workers were thugs, but the conflicts escalated nonetheless, and in testimony before the National Labor Board (NLB), representatives from the Teamsters pointed out that at least one of the new drivers was a notorious strikebreaker who had

FIGURE 11.   Low tide revealed PRT cabs sunk in the Delaware River as protest against the transit company during the 1933 taxicab strike. Photo courtesy of Historical Society of Pennsylvania.

been indicted for manslaughter. Senter also found a willing ally in Mayor Moore, who blamed the workplace conflict and any related violence entirely on the cab drivers and authorized the use of the city police to "suppress the first sign of disorder," which meant the liberal use of nightsticks as the wielders deemed necessary. Mayor Moore's "suppression of disorder" quickly became, in Galbraith's estimation, a "reign of terror" that unleashed "police provocation and brutality" on honest workers expressing their differences with management.[19]

As the conflict grew, the taxi strike became a matter of contentious public debate. Senter washed his hands of any responsibility for the violence and said the PRT would take back the strikers if it had room once the walkout was over, but that the company could not offer employment to anyone who had harmed vehicles or drivers. Some newspaper editors referred to the strikers as hoodlums who "terrorized" the riding public. Letters to the NLB called the workers "scum," "criminals," "communists," "the most vicious group of labor racketeers in the country." Observers appalled by the strike and supportive of the company argued that labor law and the Roosevelt administration promoted "this un-American, criminal-encouraging method of doing things." In the cab strike, a faint outline of the attacks that would be leveled against transit workers after World War II— Communist, criminal, un-American—began to emerge.[20]

Transit workers found supporters among Socialists, fellow workers, and politicians. The Socialist Party's Norman Thomas came to Philadelphia to rally support for the cab drivers by speaking at City Hall to one thousand strikers and their backers about the "shameful alliance of police and P.R.T. and . . . the recalcitrant attitude of P.R.T." Following Thomas's speech, Walter Shaw of the Bakery Drivers' Union pulled together twelve Philadelphia AFL unions engaged in transportation and encouraged them to all go out on strike in support of the cab drivers. Reports varied, but as many as twenty-five thousand drivers walked off the job for up to a week in a show of solidarity. Pennsylvania's maverick governor, the progressive Republican Gifford Pinchot, weighed in, telling the public that the police bore much of the responsibility for the violence because they refused to allow peaceful picketing. And S. Davis Wilson, the ever colorful city controller, made the most pointed comments from the political class. Senter was "incompetent," he said in a racist jab, "no more fit to run a taxicab company than a China man." Magistrates who jailed cab drivers, he continued, were clearly on the take for the PRT. Any close observer, he argued, had to see that the company had locked out its workers and then fomented most of the violence to make them look bad.[21]

With the turmoil brewing, cab drivers and Teamsters officials reached out to a federal government that they expected to offer support. Letters directly to President Roosevelt and National Labor Board chairman Senator Robert Wagner, several written in pencil and unsigned for fear of reprisals, asked that Wagner's board arbitrate the case. The NLB did so in late November and early December. After taking testimony from both sides, the board ruled mostly in favor of the cab drivers. Management and labor had to learn to negotiate, board members said, because "this is the day of cooperation. Those rugged individualists, you might say, are things of the past." "Labor," Senator Wagner told the PRT, "is getting some rights which I, for one, believe that labor should have had long ago. The laborer has been exploited . . . and we are lifting them out of that." The NLB then ruled that the PRT's Cooperative Association was "contrary to Section 7A of the National Recovery Act." The board, however, also ruled against the cab drivers' demand to be allowed to wear union buttons. All other issues were to be arbitrated, and the company was to put all men back on the job. The union accepted the NLB's decision (although it quibbled about the button issue), but Senter immediately rejected it.[22]

For many observers, Senter's defiance of the NLB revealed the politics behind the strike. S. Davis Wilson, as was often the case, captured the issue in vivid language. The PRT, Wilson wrote to Senator Wagner, was dominated by reactionaries who were determined "to evade their clear duty in this dispute and to defy the authority of the National Government." This was nothing more than "brazen defiance" of the president and his New Deal plans, and failure to enforce the decision would offer "clear encouragement to all those determined to hinder the restoration of industrial and economic stability throughout the country." "Nowhere in America," he argued, "has the Federal Government met with, nor is it likely to encounter, the character of bold and stubborn resistance to the President's recovery program which affects the relationship between capital and labor and all other emergency measures of the administration than in the City of Philadelphia and the State of Pennsylvania." The government had to take the PRT to court or risk rendering Section 7A a paper tiger. Several economists independently concurred with Wilson, writing to President Roosevelt that he had to act forcefully in the PRT situation, or the "National Labor Board [would be] seriously undermined." An open letter to the *Philadelphia Record* argued that whatever the company might say, "the real Philadelphians are with the cabmen who are striking for the principle of union recognition." "Mitten," the writer continued, "not only smashed the union but pulled off a fine lot of tricks on car riders, the city, and the stockholders of P.R.T. He was wise to make his sudden exit from earth before the rosy dream vanished. . . . I do not believe in violence—but a few broken windows and junked taxis aren't in it with the destroyed value in stock that P.R.T. saddled on its employes. A sunken taxi can be hauled out of the river but the destruction which men like Mitten accomplished can never be recovered." The cab strike, then, revealed a stark dividing line between PRT management and its supporters and the federal government and Philadelphians who wanted the New Deal machinery, and by extension unions, to succeed.[23]

Despite its intransigence, the PRT did succumb to federal as well as local pressure and accept arbitration before the NLB. The final agreement, reached on January 8, 1934, held that the PRT would accept either the Cooperative Association or the Teamsters as the cab drivers' bargaining agent, the company would put all strikers back on the job unless it could prove that they had engaged in violent actions, and the drivers' workweek would be reduced from sixty to fifty-four hours. The pact notably contained no mention of buttons. Senter agreed to the deal, and a day later the cab drivers voted in favor of it by a five-to-one margin. Galbraith and Wilson rejoiced at what the union leader called a "smashing victory for organized labor, not only in Philadelphia but in the entire United States."[24]

The "smashing victory," however, did not last long. A year later, the PRT decided to sell its cab companies rather than treat with the Teamsters. On the surface, the move made some economic sense, since the PRT had lost nearly $2 million on its taxi division between 1926 and 1935. More, the fleet of vehicles had been used so hard and maintenance so regularly deferred that PRT vice president Charles Ebert admitted the cars were little more than "wrecks." Wilson could not resist another shot at the company: Those cars were worth "about $8.50 each" he said; the PRT could "throw away the cabs and save money." In the end, the transit company sold off its taxi division for a paltry $298,000. The city, Wilson said, had to learn from this. No company should again be "permitted to establish such a monopoly," to have such control over a vital public service. Wilson may have been correct, but there was a second lesson as well, one identified by the U.S. District Court that was overseeing the PRT's reorganization. After the nasty cab strike of 1933, any sale, the court argued, had to ensure "that reasonable labor relations are assured."[25]

Although PRT management could sell off its taxi division, it could not for long escape larger changes in the nation's labor-relations regime or in the thinking of the company's workforce. The PRT, like many corporations in the mid-1930s that maintained company unions, kept its Cooperative Association in place for most of the decade despite the challenge levied by the taxi drivers. Ralph Senter and other members of the management team knew the New Deal state would subject them to greater regulation, and to meet the expected demands of the NIRA's industry codes, the PRT proactively reduced its trainmen's workweek from nearly seven days to five and a half, which curbed layoffs by spreading work hours to more people. It trumpeted at every opportunity how the Cooperative embodied the kind of labor-management cooperation—"cordial understanding," in management's terms—that the NIRA sought. And Boyd Garbutt, president of the PRT's Cooperative, assumed one of two labor seats on the seven-member executive council that administered the NIRA transit code.[26]

Although PRT management readily adapted to the NIRA, the law did not survive the 1935 *Schechter* decision in the Supreme Court, and Congress replaced its labor components with the National Labor Relations Act that same year. The Wagner Act remade labor relations broadly in the United States and specifically at the PRT in significant ways, particularly by guaranteeing workers the right to unionize, compelling employers to bargain with union representatives, and forbidding management sponsorship of company unions. PRT management at first questioned whether the Wagner Act applied to a local transit company, but passage of the Pennsylvania State Labor Relations Act, the "Little Wagner Act," on June 1, 1937, rendered the point moot. The PRT, by law, had to eliminate its company union. Within a week, the company posted flyers informing workers that the Cooperative had to be disbanded for purposes of contract negotiations and that it "stood ready to bargain collectively with whatever duly constituted union was designated by a majority of the employes for that purpose."[27]

The PRT may have publicly claimed it was ready to negotiate with a union of the workers' choosing, but it implemented a plan to make sure no outside labor organization would carry the company. Part of that plan involved continuing and extending the Cooperative's benefits so a company-friendly "union" would win a representation election. The PRT vowed to continue supporting its workers by offering life insurance policies, health insurance, and the pension plan, and publicly claimed it would pay another $400,000 toward those benefits. It also continued to sponsor company baseball and bowling leagues. At the same time, behind the scenes, A. A. Mitten met with Cooperative leaders about forming a new union, the Philadelphia Rapid Transit Employees Union (PRTEU). The PRT then paid PRTEU officers a salary of $6,000 and loaned the union $7,000 through the Mitten Bank Securities Corporation to finance its work. The PRT also promised to implement the dues checkoff (an automatic deduction of union dues from workers' paychecks) if the PRTEU won. Observers found the checkoff particularly suspicious, noting that "this is something which employers fight against tooth-and-nail. Yet without any struggle at all, it was granted to the company union."[28]

The July 1937 representation election was a smashing victory for the PRT and its new PRTEU over three other unions that campaigned for the workers' allegiance. Of 8,241 votes cast, the PRTEU won 6,551, or 80 percent. The Amalgamated, with Thomas Mitten gone, reneged on its 1928 promise to stay out of the PRT, but ran a lackluster campaign hobbled by many workers' memory of the 1910 general strike and received only 220 votes. The Brotherhood of Railroad Trainmen (BRT—an AFL affiliate), which had little experience with street railways compared to its work in steam railroads, made a better showing, securing 1,067 votes. And the fledgling Transport Workers Union (based in New York City and known for having a more radical politics) got 272 votes. Although this was certainly not an auspicious start for the TWU, it was the first real foray into Philadelphia for the union that became the PRTEU's chief competitor by the end of World War II. The results of the 1937 election disappointed organized labor, with the TWU calling the PRTEU an "unAmerican monstrosity" and the *Daily Worker* terming it "a notorious company union." But to most observers, such as Thomas Roberts, who interviewed a number of workers and analyzed the election in a 1959 dissertation, the PRTEU was the victor in a fairly contested election, albeit one on which the other unions spent little time or money. The workers had voted, Roberts argued, for an organization that was "little more than the Co-operative Plan under slightly different auspices." Until the PRT's employees cast off the PRTEU or any "union" related to it, he continued, "the Company [would] continue its domination of the carmen."[29]

The PRT's workers understandably thought at first that they were better off with the labor organization they knew. Many remembered, or had heard about, the violent strikes a quarter century earlier and knew the Amalgamated had abandoned the field in Philadelphia. The BRT had put little effort into the representation campaign. The TWU was still new—founded in 1934—and tinged with a radical reputation that made many wary. But it only took a few months, as the PRTEU's first contract negotiations with the company played out, for the transit workers to begin to doubt their choice. The contract signed on October 5, 1937, increased the PRT's wage bill by $1.1 million but overall proved to be a major disappointment to the company's workers. They received only a five-cent raise, which just matched the rare raise they had gotten in the worst years of the Depression. The company consolidated the workers' pension into the Social Security system, which meant the employees lost much of what they expected to receive from the PRT after retirement. And the PRT continued to refuse to pay overtime, provided no paid holidays, made drivers work fourteen-hour shifts when they were on swing runs, and implemented tight schedules that allowed little time for lunch or smoke breaks. Overall, PRT workers rightfully complained that their wages and working conditions lagged transit companies in comparable cities. Transportation workers in Chicago, Detroit, and New York City, for example, made 5 to 10 percent more than the employees at the PRT. By late 1938, PRTEU leaders and management heard rank-and-file rumblings that the new union, the extension of Mitten's Cooperative, had fallen "far short of fruition of its stated objective of a meaningful partnership of men and management." The TWU seized on this sentiment among the workers to charge "company domination and supervision of the incumbent union." The Pennsylvania State

Labor Relations Board investigated the claims and for the time being cleared the PRTEU of the charges of company unionism. But labor agitation at the transit system was just beginning to simmer, and this ruling provided only a momentary pause. World War II, for better and worse, brought labor relations at Philadelphia's transit company to a boil.[30]

World War II

Government spending for World War II revived the fortunes—at least for a time—of the company that was now the PTC. Unemployment, which had peaked at 40 percent in 1933, still stood at 20 percent in 1940. But over the course of 1941 the federal government invested $130 million in Philadelphia industries, reinvigorating the city's manufacturing base. By the end of the war Pennsylvania had received $11.7 billion in defense contracts (the sixth-largest total in the nation), with Philadelphia holding the lion's share of those dollars. Across Philadelphia, one out of every four workers found employment in the thirty-five hundred plants that produced war goods. Philadelphia's largest employers regained their footing: Baldwin Locomotive retooled to produce tanks, the Navy Yard and Cramp's shipyard made warships, Philco manufactured radar equipment, textile mills produced uniforms and blankets, and Frankford Arsenal rolled out countless weapons. Philadelphia, in the words of one commentator, produced everything "from battleships to braid."[31]

Defense work drew tens of thousands of people to Philadelphia. The city's total population grew by some 50,000 people during the war and jumped by another 100,000 between 1945 and 1950, topping out at 2,070,000, which was the high point in the city's history. The greatest growth came in the black population, which rose from some 250,000 to 376,000 people. African Americans represented 13 percent of the total population in 1940 and 18 percent in 1950, when they eclipsed the number of immigrants from foreign countries for the first time.[32]

Across the city, federal expenditures, the larger population, and the departure of some 180,000 Philadelphians for the armed services created great demand for workers, which overturned long-standing employment policies in area industries, including at the PTC. Philadelphia employers, who had accepted, implemented, and furthered gender- and race-based assumptions about who was "suitable" for different kinds of work, found themselves forced to turn to women and African Americans to staff open positions. Some three hundred thousand women were employed outside the home by war's end, with most beginning in service jobs. By 1944, however, women made up 40 percent of the manufacturing workforce. The great majority lost those higher-paying industrial jobs after the war, but the percentage of women in the workforce continued to rise in the late 1940s and 1950s.

African Americans similarly found employment in area industries. Midvale Steel, the Pennsylvania Railroad, and Sun Shipyard joined the PTC as key employers for the black community. Overall, African Americans working in manufacturing rose from fourteen thousand in 1940 to fifty-four thousand in 1944. As with women, war-era good times did not last. Many black Philadelphians doubled their pay rates during the war, but afterward they faced the typical problem of being "last hired, first fired," with Sun Shipyard, for example, laying off 80 percent of its black workforce in 1945–1946. Most black workers found no market for their skills after the war and returned to lower-paid manual and service work. A few found good jobs with the city government or in construction, but most were underemployed or unemployed.[33]

Those job-market problems lay in the future; during the war federal investment, coupled with population growth, revived Philadelphia's economy and by extension the fortunes of the PTC, which mirrored transit systems across the country in emerging from its doldrums at least for a time. The number of public transportation riders in the United States hit 23.4 billion people in 1946, more than twice the number in 1933. Ridership in Philadelphia surged from 672 million people in 1940 to over 1 billion in 1945. Gross revenues climbed commensurately in those years, from $34 million to $57.7 million. And PTC employment jumped from approximately ten thousand jobs to over twelve thousand, with those employees bringing back to their communities a total of some $18 million in wages in 1940 and $31.4 million in 1945. It should be noted, however, that the national number of riders in 1946 represented the apex for the industry, the war a respite from a longer secular decline: by 1960, only 9.4 billion people rode public transit as automobile ownership took off. As early as 1943, PTC officials feared this would be the case. In the company's annual report that year, they noted that "it appears that the high-water mark in wartime riding has been reached." Rush-hour demand was still high, but that would fade after the war when "normal conditions return [and] public transportation will again have to face the keen competition of the private automobile." Such pessimistic, and accurate, expectations set the tone for PTC management's approach to labor relations, especially with the TWU, during the war.[34]

With business booming, Philadelphia's transit company managers for the first time in any substantial way sought a source of labor beyond white males for driver and conductor positions. In 1943, the PTC carried 1.1 billion passengers, and its vehicles covered 109 million miles. At the same time, the company had twelve hundred men serving in the armed forces. To meet its unavoidable labor shortfall, the PTC first turned to women. The company ran a comprehensive advertising campaign that included newspapers and radio advertisements, posters, company publications, and door-to-door interviews that played on women's desire to work and on their patriotism. The PTC offered women preferential runs such as daytime hours only, limited their swing shifts, and promised equal pay for equal work. They received the same training as men and also had the opportunity to work in the shops, running drill presses and other machines. They would get training and good pay, the company promised, and at the same time carry out a "vital homefront task." "Carrying war workers to their jobs," one pamphlet told prospective drivers, "is essential to production and to Victory." Human-interest stories in the press followed the trailblazers in the industry, making the work sound as enticing as possible. "There's Glamor in the Cab," the *Evening Bulletin* told its readers in a feature on Regina Jewell, the first woman to operate a Market Street subway train. "Over 600,000 pounds of metal on wheels alive with 850 human beings," wrote the reporter, "that's a big load for a 23-year-old miss!" But she was up to the job. "My uncle first suggested I get a PTC job," Jewell said. "He's a motorman himself, but he was thinking of my landing some nice 'womanly' job, like cashier or office work. And you should have seen his mouth drop when he found I had done him one better. You see, he just runs a trolley car—and I'm handling high-speed trains!" Women responded to the call: By the end of 1943—the first year the company recruited female employees aside from its secretarial positions—over three hundred worked as drivers, and another eight hundred worked in the shop, maintenance, and clerical units.[35]

This change in the workforce did inspire some conflict. Many male drivers, once they adjusted to women operating the PTC's vehicles, found them to be competent, dedicated, and above all necessary. Some, however, worried that women posed a threat to their jobs, particularly voicing a fear that the PTC was using women to rob them of their seniority. Such concerns in the context of increasingly difficult labor relations with the company were understandable, but overblown. Most of the women said they had no desire to keep the job permanently. Jewell, for instance, said she had a fiancé in the army and planned "to quit her job, marry and raise a family after the war." PTC management also consistently assured its male workforce that women were on the line only "for the emergency period" and that no one would lose his position for serving overseas. The TWU, knowing how suspicious of their company many workers were becoming, played up the seniority issue to build support among the male workforce. "Both present employees and the new women employees, recently hired, are having their seniority rights disregarded," the union claimed. Setting aside certain runs for women violated the rights of more experienced men to select their routes. More, management had said these runs would not "be carried on the Depot seniority lists," which would allow the company to transfer women back and forth between carbarns as it saw fit. Experienced men and novice women, both would lose their rights to the company, and only the TWU, the union claimed,

FIGURE 12. The shortage of workers caused by World War II opened opportunities for women to work at the PTC. Photo courtesy of Historical Society of Pennsylvania.

could stand up for them. In the end, women's employment presented a challenge to the white male workforce, but it was a mostly limited one and in some ways overstated by the TWU.[36]

African American employment offered a much larger, more explosive issue that threatened to splinter the PTC's workforce. Transit company management and many white members of the workforce—opposed on so many other issues—both accepted and furthered a racist culture on the property. Philadelphia's transit company had run an openly racist shop since the industry developed in the late nineteenth century. African Americans could work as mechanics, messengers, and porters, but as the NAACP's *Crisis* put it, "regardless of training capacity, years of service, loyalty or desire a Negro could never hope to fill certain jobs, such as conductors, motormen, [or] bus drivers." "PTC," added the liberal newspaper *PM*, "has operated along Southern lines" for years, and white workers "simply accepted the fact that [driving] is a white man's job, and no Negro is going to get it." Company publications such as Thomas Mitten's *Service Talks* ran so-called "darky" jokes that made a mockery of black people, not to mention black workers. PTC management's use of employment policies that helped divide the workforce against itself along racial lines—what the historians David Roediger and Elizabeth Esch have termed "race management"—were common-place in the United States in that era, including in Philadelphia. In fact, the PTC was better than many local companies: as late as the end of the 1930s, according to the historian Walter Licht, Budd, Bendix, Cramps Shipyard, and Baldwin Locomotive employed thirty-five thousand workers but not a single African American. PRTEU leaders and many white employees cemented the PTC's institutionalized racism. African Americans could join the PRTEU, but the union backed the company's rule that blacks could not work as drivers or conductors. One union official bluntly summed up the PRTEU's stance, saying he would "see to it that no damned n——rs ever got to drive trolley cars in Philadelphia." Other unions with a presence at the PTC, aside from the left-leaning TWU, were as bad or worse. The BRT, for example, banned African American members. The Amalgamated had an official policy prohibiting racial discrimination, but federal officials argued the union did not live up to its promises. Racist practices, then, were culturally acceptable to many white managers and workers at the transit company, and management sought to use them to control the company's workforce, especially as the TWU gained support.[37]

The African American campaign for jobs at the PTC and the TWU drive to represent the company's workforce grew simultaneously. African Americans worked through the newly invigorated local chapter of the NAACP, the city's black press, federal agencies (especially the Fair Employment Practices Committee—FEPC), and the TWU to secure their rights. To black Philadelphians, work at the transit company represented both stable employment and the embodiment of America's wartime commitment to a democratic world. Their protests echoed the national Double V slogan that called for victory over fascism abroad and over Jim Crow at home. As the signs at one protest put it: "In Democracy, Freedom to Work Belongs to All," "PTC Sabotages the War Effort. Negroes Want to Work!" and "We Drive Tanks, Why Not Trolleys?" "We are trying," added the black rights leader Rev. E. Luther Cunningham, "to preserve democracy on the home front as our boys are doing on the firing line."[38]

The NAACP, led by Carolyn Moore and Theodore Spaulding, spearheaded the campaign. They both organized and drew on community activism that demanded that the PTC desegregate its workforce. This was an extension of a decades-old black political initiative that regarded public transit as a key site of citizenship. As the historian Robin Kelley has argued, streetcars and other public conveyances in the age of Jim Crow were "white-dominated . . . vigilantly undemocratic and potentially dangerous," a place where white transit workers acted as an arm of the segregated state, even as black citizens paid an equal fare and should have been entitled to equal treatment. Such politics found even greater purchase in Philadelphia, where as a quasi-public company the PTC received municipal support in exchange for operating the system. The NAACP worked with a committee of African American PTC workers (the Committee for Equal Job Opportunity of PTC Employees) headed by Roosevelt Neal and Raleigh Johnson that sought promotion to driver and conductor positions. Moore also organized two massive protest rallies—she called them "indignation meetings"—in 1943 that focused the black community on "the unfair labor policy of the Philadelphia Transit Company which undermines the morale of the Negro people, hampers the war effort and prolongs the war." "Jim Crow," they exclaimed in those meetings, "fights on Hitler's Side!"[39]

FIGURE 13.   Protest against Philadelphia Transportation Company hiring practices (November 8, 1943). With the ferment of World War II as backdrop, black Philadelphians carried out a series of demonstrations that eventually led to their gaining driving jobs at the PTC. John W. Mosley Photograph Collection, Charles L. Blockson Afro-American Collection, Temple University Libraries, Philadelphia.

The NAACP also pressured the federal government to act on behalf of black rights. The newly formed FEPC represented the government's first substantial commitment to securing equal rights, or at least more equal rights, for African Americans since Reconstruction. Black labor and civil rights activist A. Philip Randolph had used the March on Washington Movement to pressure President Roosevelt in the early 1940s to desegregate the military and end racial discrimination in industry. Roosevelt, concerned about African Americans marching on Washington to demand equal rights, issued Executive Order 8802 in 1941 to establish the FEPC. The committee had a limited budget and no law enforcement powers to back the investigations it launched into the war industries where workers submitted complaints of discrimination. Yet despite the committee's weaknesses, black Philadelphians found the FEPC, especially its local office, to be a valuable ally. Over two dozen African American job applicants turned away by the PTC filed protests with the FEPC, alleging that the company discriminated based on race, which violated Executive Order 9346 (the order that updated 8802). The FEPC held hearings in Philadelphia in the fall of 1943 and ordered company management and the PRTEU to cease their discrimination and promote African Americans to the position of driver. PTC managers and PRTEU leadership then took their case to Congress, where conservative, mostly southern, congressmen called into question not only the FEPC's decision at the PTC but the committee's very right to exist.[40]

PTC management could have ameliorated the problem by eliminating the company's discriminatory policies, but they chose not to do so, despite repeated requests. Moore wrote separately to Edward Hopkinson (chair of the PTC Board of Directors' executive committee) and A. A. Mitten (chair of the Committee on Industrial Relations) to ask that they intervene on behalf of black workers. She also wrote to Albert Greenfield, emphasizing that the PTC policy "serves as propaganda for the Axis." Moore included newspaper articles and photographs of the marches and demonstrations taking place across the city. Representatives from the Urban League met face to face with President Senter and Mitten, as did Johnson and Neal's Committee for Equal Job Opportunity. Each time they were rebuffed as Senter, Mitten, and others pointed to the so-called customs clause in the PTC-PRTEU contract, which forbade changing the agreement without the consent of both parties. The promotion of women, as Moore pointed out, obviously violated this clause. Mitten responded that he had worked with the union to get this change approved but that the company had no desire to extend the modification to black applicants, especially without the initial consent of the PRTEU. Mitten and Senter then directed Moore, Johnson, and the others to the PRTEU. As they expected, PRTEU president Frank Carney and secretary-treasurer Frank Cobourn offered unyielding support of segregationist policies.[41]

Management and its allies in the PRTEU, then, supported each other in preventing black promotions. Their actions stemmed from several reasons, many of which became clearer over time. The customs clause was the most obvious reason, but it only provided a pretext for the denial of opportunity for black workers. It followed a circular logic—each side said the other had to agree to a change—which would have led African American representatives to chase from management to union and back again indefinitely, if they had been so inclined. The racism of the era also provided a reason. The company had employed racist policies for generations and was reluctant to change, especially when doing so meant bowing to the pressure of black activists. White workers also enjoyed the restricted access to better PTC jobs that equated to steady pay and some job security just years after the Depression ended. More, these workers benefited from a racialized employment structure that provided what W. E. B. Du Bois termed a "psychological wage" that gave them a higher standing than the city's black population. This higher standing reverberated past the job market to the city's neighborhoods, which were equally segregated, including for PTC workers, who often lived near their carbarns. Many expressed grave concerns about African Americans not just driving "their" trolleys, but living in their neighborhoods, being around their wives and daughters. As one white driver said to reporters about protests against African Americans getting driving jobs: "Look at that row of houses over there. Suppose I and nine other whites lived in 'em and a nigger moved in right in the middle. What would I do? Get up a petition and have him tossed the hell out. That's all we're doing now, ain't it." The African American campaign for jobs, then, clearly provoked some of the ugliest racism that lurked all too near the surface of some members of the white working class, showing "the common man in his undershirt," as *PM*'s I. F. Stone put it.[42]

The racism of many of the PTC's white workers was undeniable, but it had such effectiveness because it was also a tool in management's hands as it sought to combat the rise of the TWU. Management's recognition of the power of white racism as an antilabor weapon, along with its desire to undermine the Roosevelt administration just months before the presidential election (which became apparent in the strike's aftermath), was the deeper reason behind the corporate racial intransigence. Led by Michael Quill, the Transport Workers Union began in New York City in 1934. It was known for its links to the Communist Party (although Quill was always coy about his own ties and grew increasingly anticommunist after World War II) and its radicalism that made it a talented and vexing opponent for management. Quill, who had been a member of the IRA while living in Ireland, joined John L. Lewis and other labor leaders when they took several industrial unions out of the more staid AFL to establish the CIO. The CIO more broadly, and the TWU specifically, led what the historian Joshua Freeman called "a virtual revolution" for the working class because of the demands they made and won on behalf of industrial workers. In the late 1930s, for example, TWU contracts in New York City raised workers' pay by 10 percent while cutting their hours to forty-eight per week. As members of the TWU, Freeman wrote, "New York transit workers began a long climb out of a state of special and intense degradation toward the full rights, privileges, and bounty of the modern working class." Quill, known as a great orator, not only knew how to organize workers and win better contracts at the bargaining table, but also how to work in the formal political realm, winning races for New York City Council, going toe to toe with the reactionary congressman Martin Dies's House Un-American Activities Committee, and supporting Franklin Roosevelt on timely occasions. Ralph Senter, who resisted a state minimum wage bill, opposed Democrats and other "radicals" in the government, and fought anything that would raise the cost of PTC's wages, knew the TWU was no PRTEU, and that Quill was no Carney.[43]

Despite the TWU's reputation, and in some ways because of its notoriety for radicalism, it took the union several years to gain traction at the PTC. The TWU lost representation elections in 1937 and 1942, the latter by a count of 2,500 to 500. Things looked bleak enough that the TWU contemplated withdrawing from Philadelphia, but Quill instead decided in early 1943 to send in one of his best organizers, the Irish leftist James Fitzsimon. Fitzsimon's politics helped him connect with Philadelphia's left-wing community, especially Saul Waldbaum, a fellow traveler with, if not a member of, the Communist Party, who had helped found the Philadelphia branch of the National Negro Congress (NNC) and served as attorney for the local TWU, the United Electrical Workers, and later Henry Wallace's Progressive Party. Fitzsimon found a similar ally in Arthur Huff Fauset, who led Philadelphia's NNC and supported the TWU in the editorial pages of the city's largest African American newspaper, the *Philadelphia Tribune*. And Fitzsimon also developed ties with the large Irish contingent in the company's workforce, cultivating a local leadership that included John McMonigle, Anthony Gallagher, and the Dougherty brothers. Several carbarns, mostly in Irish sections of West and North Philadelphia, developed reputations as "TWU depots," including Haverford, Cumberland, Woodland, and Jackson. PTC carbarn superintendents had always allowed the distribution of union material when the PRTEU was in charge, but as the TWU gained influence, the company banned all union activity. Fitzsimon correctly interpreted this as a good sign, that the PTC was getting nervous.[44]

Two events propelled Fitzsimon's campaign: the reopening of the button issue and the poor PRTEU-PTC contract of 1943. Once Fitzsimon made headway in some of the carbarns, he encouraged workers to wear TWU buttons, a direct challenge to the decades-old prohibition instituted by Thomas Mitten. The PTC immediately ordered all workers to remove the buttons and said that no driver would leave a carbarn while wearing one. Over one hundred drivers, including ten women, defied the company in November 1943 and tried to drive their routes while wearing TWU insignia. The PTC pulled them off the line, preventing riders from getting to their jobs at war industry plants such as Sun Shipyard, Baldwin Locomotive, and Westinghouse. Fitzsimon scored the PTC in the press: "This is not a strike; it's a lockout. If the company would rather insist on enforcing an antiquated regulation laid down in 1912 than take defense workers to their work, that's up to them." "This," he concluded, "is purely a matter of union discrimination." Cowed, the PTC agreed to submit the button issue to arbitration before the War Labor Board and let the workers return to their jobs the next day. The conflict over buttons showed the TWU had real teeth, Fitzsimon argued, and it did bring hundreds of workers into the fold.[45]

If the button issue showed the TWU's strength, the 1943 contract highlighted the PRTEU's weakness. Coming less than a year after the PRTEU's sweeping victory, the 1943 contract looked to PTC workers like nothing short of a betrayal. Management and the union negotiated the contract under the auspices of the War Labor Board, which granted the union the right to a higher wage than the Little Steel formula allowed (Little Steel limited workers in most industries to modest pay raises during World War II). The PRTEU asked for a ten-cents-an-hour increase; management offered four cents. After six months of negotiation, the PRTEU took what the company offered and failed to earn for its workers even a penny more. It also did nothing about swing runs or the forty-eight-hour workweek. For many workers, this was the last straw. Perhaps, they believed, the TWU was right: the PRTEU really was a "useless parasite." Hundreds of enraged PTC employees immediately told the company to halt its deduction of their union dues, and within a month several thousand resigned from the PRTEU.[46]

Fitzsimon's organizing, the defense of workers' right to wear a button, and the bad contract put the PTC's workforce in a state of flux and led the TWU to call for another representation election. Fitzsimon knew he had to capitalize on recent events and put together a campaign that focused on the TWU's commitment to bread-and-butter issues. The TWU, Fitzsimon repeatedly argued, "alone will be able to get [PTC workers] the things they want: higher wages, better working conditions, job security and a decent pension." "One Industry—One Union," was Fitzsimon's motto; "The Bigger Your TWU Majority—The Better Your Next Contract!" his mantra. TWU organizers reminded workers how the company had brought violence upon them, their families, their communities in 1910, how it had foisted "the curse of company unionism" on them for another thirty years after that. CIO president Philip Murray came to Philadelphia to urge PTC workers to vote for the TWU, to assert "the right to be free from the tyranny of a dominating boss—free from any discrimination against you by the company and a company union undermining you." The Steelworkers, Textile Workers, and Amalgamated Clothing Workers unions all lent their support and contributed money to the TWU's cause. Playful, yet trenchant, songs about the workers' plight and the hope inspired by the TWU emerged:

> The company chisels here and there, it's really not surprisin'
> The "Company Union's" sound asleep, and grievances are risin'.
> Who wants to sit in on report and loaf away the day,
> When factories offer steady work and guarantee the pay.
> . . . .
> So, stroll around the property and listen to 'em hum,
> "If we had a decent Union," and "The delegate's a bum."
> Remember you're the voter and the whole thing's up to you,
> Stand up and fight for what is right, SUPPORT T.W.U.
> A platform man working for P.T.C.[47]

In the end, Fitzsimon's plan bore fruit by November 1943, when the TWU could claim a membership of some six thousand workers.[48]

Other unions, especially the PRTEU, the Amalgamated, and the Teamsters (the latter two in the AFL), fought back against the rise of the TWU. Their first two strategies—violence and red-baiting—brought only limited results. Race-baiting, however, proved more fruitful. AFL unions, one commentator noted, were ready to use any strategy to prevent "the C.I.O. [from gaining] a foothold on the P.T.C. property, and thereby extend[ing] its growing influence in Philadelphia." PTC management would have agreed with that sentiment wholeheartedly: significantly, it did nothing to arrest the violence that took place on its own property. The AFL brought in 150 "thugs" (in Fitzsimon's words) who circulated through the carbarns, destroying CIO equipment and literature and intimidating the employees. "I'm not a talker, I'm a fighter," Edward Crumbock of the Teamsters told AFL supporters. "I want to bust some f——ing skulls, and I will before this is over. . . . This is a fight between the AF of L and the CIO, but we'll win; I don't mind going all the way. There will be some bloodshed before this is over." The AFL men then proceeded to put one TWU member in the hospital, beat up several others, and throw paint and gasoline through the window of another man's home. In the face of these attacks, Quill argued that the AFL resorted to violence because it was "completely without program." He then asked for Republican mayor Bernard Samuel and the PTC to provide police protection. None came.[49]

The Amalgamated and the AFL more broadly also initiated a Red Scare campaign against Quill and the TWU that started slowly but set a tone for years of labor conflict after World War II. Tellingly, the anticommunist attacks came not from the company but from within labor's ranks. James McDevitt, president of the Pennsylvania Federation of Labor (the AFL's state affiliate), labeled Quill a "Communist tool." "Red Mike's" union was "Communist dominated," according to the AFL. The only way to protect Philadelphia was to vote for the Amalgamated and keep the "Communist councilman from N.Y." out of their city. Quill scoffed at the accusations. "They charge us with Communism," he said, "and we charge them with rheumatism of the mind." "The 'red' scare," the TWU's Philadelphia local said in one flyer, "just won't stick. . . . PTC workers are too smart for that stuff!"[50]

That may have been the case when it came to the red-baiting campaign, but race-baiting proved more powerful. Again, the PTC did nothing to quell the unrest in its carbarns. The PRTEU warned transit workers that a vote for the TWU was a vote for "Negro Supremacy on the Company." Only the company union would "keep the 'Niggers' off" the transit lines. AFL sound trucks at carbarns across the city announced that "a vote for the [CIO] is a vote for Niggers on the job." The TWU responded cautiously. The union, especially as a member of the CIO, was known for its egalitarian principles, but Fitzsimon and other TWU leaders understood the volatility of the situation. Coming out too strongly in favor of black rights could antagonize many members of the white workforce and jeopardize all the TWU's work. So the TWU invited an African American employee named Groves to serve as an organizer, and Fitzsimon met with the NAACP, but did so quietly. The TWU also issued statements offering a sharp class-based analysis of how the company was using race to its advantage. The PTC had dominated its workers for decades, the TWU argued, and workers had to ignore the race "smoke-screen." "The company," TWU officials continued, "wants the employees to fight among themselves over Negro employment, and to oppose the government." Doing this would "prevent the employees from winning the election and thereafter from negotiating a good contract with higher wages, good working conditions and a sound pension plan. THAT WOULD SAVE PTC MILLIONS OF DOLLARS." The PTC was no different from "other reactionary employers [who] practice racial and other forms of discrimination in order to have a supply of CHEAP LABOR." "Without unity of all workers," they concluded, "we cannot win a good contract this year and keep winning better contracts every year thereafter." The PRTEU was in bed with the PTC, doing its bidding, the TWU argued. Both were "cornered rats" that knew they were in trouble. Still, TWU supporters at the PTC worried that the racist attacks were taking their toll. Grant Dinsmore, TWU chairman at the Woodland carbarn, warned that "our position on the Negro issue . . . will ruin us unless altered." "We are being blamed," he continued, "for being friends of the Negro and are not getting any credit for being friends as much of the white workers and of the country. Our prestige has gone down and . . . where we would have won by a 90 percent vote a few months ago, we would just barely win now." Regardless of how strong the TWU's support was, black Philadelphians knew the union was the best ally they had at the PTC. As the *Philadelphia Tribune* told its readers: "No self-respecting PTC employee who has a single drop of colored blood in his veins will vote for [the PRTEU or the Amalgamated.]"[51]

In the March 1944 representation election, the TWU emerged victorious, tossing out thirty years of company unionism and putting in place the union that represents Philadelphia transit workers to this day. The voting was, to the surprise of the TWU, heavy but peaceful, representing the culmination of "one of the city's most bitter factional labor fights in recent years." Of eight thousand votes cast, the TWU won 4,410, the PRTEU 1,786, the Amalgamated 1,677, and "no union" 147. The TWU's total was only 55 percent of the vote, but the remainder was split between the PRTEU with 23 percent and the Amalgamated with 20 percent. Overall, the PRTEU and Amalgamated performed better in the transportation division than in the maintenance division, where the bulk of the PTC's African Americans labored. More, those carbarns with the weakest support for the TWU (Allegheny, Richmond, and Luzerne) were in areas of North Philadelphia undergoing racial change, and the TWU's support of black rights would have looked to many white workers there like a "threat" to their neighborhoods.[52]

The voting pattern of the election did not bode well for racial tranquillity at the PTC in the coming months, but for the time being congratulatory telegrams poured in to the TWU. Irving Potash, a Communist leader of the Furriers Union, applauded the TWU for striking "a mighty blow to the forces that seek to divide the workers and the American people." The United Office and Professional Workers of America called the TWU's victory "one of the most heartening things that has happened to the American labor movement [because the union] is going to be able to play an increasingly vast and important role not only in New York but throughout the country." That union's Chicago affiliate asked that the TWU organize its local bus drivers next. To the New York Council of the CIO, the TWU's Philadelphia campaign represented "a victory not only for those immediately involved, but for the entire labor movement, since it is such a conclusive refutation of all those who sought to defeat CIO by means of the well-known tricks of red-baiting and other shabby schemes." And Max Yergan of the Communist-dominated National Negro Congress especially applauded the TWU's defeat of "the company union and the AF of L union [that] had resorted to the outmoded practice of Negro-baiting and red-baiting to seek the support of their fellow-workers." "That TWU was able to clarify the issues and win the election on a platform of full equality for all workers," Yergan concluded, "is a signal victory for all the democratic forces in the nation." Such congratulations from the CIO and the Left highlighted their animus toward the AFL and company unions and the need to combat racism in the working class, but notably failed to implicate the PTC for its part in the ugly politics of the representation election.[53]

The reason management perceived a need to divide its workforce and undermine the TWU by any means available became more obvious once the union made public its heightened contract demands. All told, the TWU had forty-two demands, divided among wages, working conditions, and a pension plan, that it wanted management to meet. Among its many points, the union argued that PTC workers made a vital contribution to the war effort and should be compensated accordingly, with a raise of at least fifteen cents an hour (the PRTEU had won one-quarter that amount). They also deserved time and a half for any work over eight hours per day and that same rate for working on scheduled days off or doing snow and cinder work. All employees should receive two weeks of vacation time, plus six holidays per year. The union would have the right to arbitrate any decisions about schedules and assignments, and discipline was to be a joint procedure handled by management and the TWU. Finally, the company had to reinstate its pension plan, contribute $50 per month for all employees, and no TWU member could be forced into retirement. This wide-ranging set of demands increased the PTC's wage bill and limited or eliminated long-standing managerial prerogatives. Management knew that, unless it could remove the TWU, these demands could cost the company some $3 million per year and force it to surrender much power to its workforce.[54]

In the months following the TWU's victory, racial animus grew to alarming levels in the PTC's carbarns, with the company doing nothing to combat it. Supporters of the PRTEU, in an attempt to pull the TWU apart, employed racist rhetoric that connected black advancement with the federal government and the CIO. "Your Sons and Buddies that are away Fighting for the Country," one letter read, "are being Stabed [sic] in the Back on the Home Front, By The National Association For The Advancement of Negroes, And The F.E.P.C. which is a 100% Negro Lobby, that was Created by Executive Decree, and financed by The U.S. treasury." African Americans, the letter continued, were using the war to "gain Control of All The Jobs and Everything Else that Belongs to the White People." Other leaflets told workers they should support "a white supremacy movement for the protection of [their] jobs" and that they had to protect the jobs of their friends in the Army not only from blacks but from the CIO. "Your buddies are in the Army," read one notice, "fighting and dying to protect the life of you and your family, and you are too yellow to protect their jobs until they return. Call a strike and refuse to teach the Negroes; the public is with you, the CIO sold you out." These charges stoked many white workers' anger, particularly after it became clear the company would in fact promote African Americans. "They [the CIO] betrayed us," one driver said. "We won't stand for Negroes running our trolley cars."[55]

Management put this racism to use. It was obvious that internal divisions could weaken the TWU broadly; but on a more concrete level, PTC leaders knew that the federal government could invoke the War Labor Disputes Act (commonly known as the Smith-Connally Act) to abrogate any contract and toss out the Transport Workers Union if a strike took place and it looked like the union was leading it. So management used its workers' racism to try to undermine the CIO union. First, managers posted notices that promised to upgrade African Americans on August 1 and insisted on an escape period that gave disgruntled white employees three weeks to stoke their anger and leave the union by August 23. PTC leadership then let former PRTEU officials hold meetings on company property, where hundreds of, although by no means all, white workers swore they would strike before working with African Americans. They made good on their threat on August 1, walking off the job and shutting down the nation's third-largest war production center just weeks after D-day.[56]

The strike threw the city into turmoil and threatened to touch off another race riot like the one that rocked Detroit in 1943. Strike leaders Frank Carney from the PRTEU and James McMenamin from the BRT used racist attacks at every turn to inflame white workers. The strike, McMenamin claimed, was about "the white race keeping *its place*." African Americans had a "standard of living . . . very far below the standard of operators"; they would infect whites with "bedbugs." "We don't want Negroes and we won't work with Negroes," Carney told a cheering crowd. "This is a white man's job." Their attacks on black workers bled into recriminations about federal "meddling" in local affairs, especially the FEPC order to upgrade African Americans, and charges that the CIO was a foreign organization bent on destroying local rules and values. PTC workers, as they had in past strikes, found much support in the city. One letter to Governor Edward Martin made the point clearly, charging that federal agencies were forcing white Philadelphians to submit to African American demands. The black worker, the letter stated, "is demanding, and the government is right behind— . . . that he be given jobs—the best, including the operation of our local transit lines. We submit to your good judgment—is this fair! Is it right!" Other Philadelphians complained that "the will of a few politically minded officers of the . . . TWU . . . [would prevail over] the will of 6500." A number of white drivers claimed that the CIO had lied to them, that the TWU's talk of black equality was supposed to be a ploy to get more votes and was to be abandoned after the representation election. "I'm getting out of the CIO right now!" one man told the press. "What we ought to do," said another, "is start our own independent union and put in a clause to keep the Negroes out." The last statement was music, no doubt, to management's ears.[57]

FIGURE 14.   Many white transit workers walked off the job rather than accept African Americans as drivers. The 1944 hate strike was one of the lowest points in Philadelphia labor history. Photo courtesy of Historical Society of Pennsylvania.

TWU leaders and many members, along with the union's black and liberal allies, opposed the strike, showing white PTC workers' racial views were far from monolithic, but their efforts were to little avail. TWU officers immediately saw several dangers. "The race issue and the strike [could] split the CIO union wide open," as *PM* put it, "with the leadership going 100 per cent against the strike, and the rank-and-file moving with the ex-company union leaders." Knowing this, TWU leadership sent some 250 stewards to the PTC's carbarns to encourage drivers to be "loyal patriotic Americans" and return to their jobs. A few sporadic back-to-work efforts grew from the stewards' work, but nothing sustained. TWU officers also feared the federal government would intervene to put down the strike and perhaps throw out the union for fomenting it. To head off any blame, they sent telegrams to federal officials assuring them that they were "unqualifiedly against strikes or other interruption of work in wartime." Liberal observers focused less on the danger to the TWU, more on the danger to American society and the nation's black population. The National Negro Congress and Urban League both wrote to applaud the TWU's "democratic and uncompromising stand [for] real democracy in action." Unions from cities as far flung as Buffalo and Oklahoma City wrote to support the TWU and the FEPC's directive, to decry working-class racism. So too did the leftist Thomas Jefferson Club of Oklahoma City and the National Lawyers Guild, the latter of which argued that the strike was nothing less than "the denial of the American principles," "a stab in the back which gives aid and comfort to the enemies of our country." "The issue," the guild tellingly noted, "appears to have been raised by persons no longer recognized as representatives of the employees and condoned, if not encouraged, by representatives of the management."[58]

Federal officials immediately understood the danger the strike presented. Industrial plants from Frankford Arsenal to Midvale Steel reported high absentee rates that slowed the production of all manner of matériel, from flamethrowers to fighter planes. Rear Admiral Milo Draemel warned that the strike might "delay the day of victory." More, in the midst of a putative war for democracy, the strike raised searching questions about the kind of society America stood for. As the NAACP's Walter White put it, stories like Philadelphia's strike or Detroit's race riot allowed the Japanese to say, "See what the United States does to its own colored people; this is the way you colored people of the world will be treated if the Allied nations win the war!" On August 3, two days after the walkout commenced, President Roosevelt ordered the secretary of war to assume control of the PTC and authorized Major General Philip Hayes to deploy five thousand armed troops to the city to break the strike. On August 7, under threat of being drafted and shipped to the front, the PTC's strikers returned to their jobs. Federal power broke the strike and assured that African Americans would get their promotions.[59]

To the strike's observers, the racism of working-class Philadelphians was the most obvious issue—"Their Lot Isn't Any Better in North," read one headline in the *New York Daily News* lamenting African Americans' plight—but for many analysts during the strike and afterward, the manipulations perpetrated by PTC officials and their Republican allies in Philadelphia's government merited equally close scrutiny. The two groups combined anti-CIO sentiment with reactionary politics to do all they could, through acts of commission and omission, to exacerbate the strike and thus damage the TWU and the Roosevelt administration. The harm to war production, and the distinct chance of a shattering race riot, did not seem to unduly bother them. PTC management knew on the night of July 31 that a strike would be called the next morning but did nothing to try to stop it. In fact, it brought in Louis Campbell—a man with ties to strikebreaker Pearl Bergoff and J. Howard Pew, the right-wing owner of Sun Shipyard—to stir up the workforce and reportedly buy off reluctant employees with company money. On the morning of August 1, dispatchers told workers reporting for duty to go home, that trouble was brewing and it would be "healthy" to stay away. The company then stopped selling tickets and turned the power off to the subway and elevated lines, ostensibly to protect its equipment, although it had not done so in previous strikes. Management also canceled a meeting of the executive committee of its board of directors and refused to confer with the TWU. At the same time, it left the carbarns open and allowed McMenamin and the other strike leaders to use them for headquarters. Corporate officials even refused to ask the federal government for help and rejected the opportunity to join the TWU, FEPC, and Red Cross in a radio program that urged workers to return to their jobs. Management's stance was, in the understated words of one confidential report, "curiously detached." Perhaps. A PTC worker provided a more accurate assessment: "Listen mister," he told a reporter, "that was no strike; it was practically a lock-out."[60]

The response of elected officials, including Mayor Bernard Samuel, who sat on the board of the PTC, had been equally obvious, equally egregious. Republican governor Edward Martin never uttered a public statement about the strike or provided any assistance. In Philadelphia, where local authorities had used state power to break strikes many times, Mayor Samuel refused police protection for workers who wanted to return to their jobs. In fact, policemen attended strike meetings but stayed in the background and were observed drinking and playing cards with the strikers afterward. The difference in police behavior was striking. "Those who have had the privilege," wrote the *Pennsylvania Labor Record*, "of watching Philadelphia's police force, the flower of half a century of Republican discipline, go to work on pickets, can do naught but marvel at how the virus of liberalism and tolerance of the new G.O.P. infected the city fathers." Samuel also refused to meet with TWU leadership or give the War Production Board airtime to urge drivers to return to their jobs. When the NAACP asked for permission to send two sound trucks into black neighborhoods to call for calm, Samuel refused the request without comment. He even failed to convene the City-Wide Inter-racial Committee because it would "add to the confusion."[61]

The underlying anti-Roosevelt, anti-liberal politics of the strike were all too clear. A. A. Mitten produced a poster that would have rescinded the order to promote African Americans because doing so would stop the strike that had "crippled every war industry in the Philadelphia area." Frank McNamee of the War Manpower Commission (WMC) rejected Mitten's overture, leading him to tell the federal official, "It's your baby and it's on your lap." Orville Bullitt, a local investment banker who worked for the War Labor Board, met with Carney and McMenamin to discuss overturning the FEPC order, a move that commentators thought was motivated by his desire to help PTC's management and its investors. The WMC rejected this proposal, too, and soon after Bullitt resigned from government service to return to his banking practice. Following these efforts, the PTC told the public "[We are] in a helpless position in that, through compliance with a government order, [we] brought about the stoppage of work by [our own] operators." Were it not for the federal government, the spokesman implied, the system would be running as usual. The strike was the fault of Roosevelt and his intrusive liberal administration.[62]

In the weeks after the strike, race relations organizations, newspaper reporters, and to a lesser extent a grand jury hammered PTC management and their political allies for the way they had used racial animosity to further their labor and political agendas. Working-class racism undoubtedly played a role, but "the real roots of the strike are to be found in the long history of company-dominated employees' associations," wrote the American Council on Race Relations. The company was desperate to break "the only union that has ever gone among P.T.C. employees with its hands free of the shackles of company domination." "The company, its stooge employes' outfit and subversive groups playing Hitler's game," wrote the left-wing news service Federated Press, had hatched the "plot to break the union and damage the Roosevelt administration." The country, wrote one magazine editor, had narrowly avoided a "Philadelphia Putsch" that threatened to spread racial clashes to other cities and further the cause of political reactionaries. Only a federal investigation could get to the bottom of the strike, the TWU's Fitzsimon argued, and he and the CIO called on U.S. Attorney General Francis Biddle to investigate the transit strike and hold accountable those "plotters and perpetrators" who caused it. A grand jury did convene on August 9 and took testimony for almost two months. Its final report found fault with the TWU and federal directives but also charged the company with a lackadaisical attitude, one that at best offered "the weak excuse—'What's the use?'" In the end, twenty-seven PTC employees received fines of $100 for violating the Smith-Connally Act, but no one from the company's management or the city's political class was indicted. The grand jury "fizzled," in the words of the Federated Press, leaving "the question of who was behind the strike . . . unanswered." Everyone actually knew the answer to that question, the news service continued, but "the Republican-dominated grand jury that indicted the workers last October white-washed management."[63]

The grand jury may have downplayed management's role, but PTC leadership did not escape the strike unscathed. The work stoppage cost the company $800,000, and by the end of August Ralph Senter stepped down as president, after having served in that capacity for seventeen years. The PTC offered Senter the briefest of valedictories, writing in its annual report that he had "resigned because of ill health." The Federated Press called his presidency "another casualty of the transit strike," and Senter died in Eustis, Florida, four years later after battling an "extended illness." Charles Ebert, another Mitten protégé who had worked for him in Buffalo and Philadelphia since 1900, assumed the PTC's presidency. A. A. Mitten, whose standing at the company had been sharply damaged when the Mitten Plan came asunder, stayed on a little longer, resigning because of his health in 1948. After nearly forty years, the Mittens were out of Philadelphia's transportation system.[64]

In the weeks after the strike's end, the PTC's worst fears about the TWU came true. On August 9—a mere two days after the strike ended and the same day as the grand jury began its investigation—management signed the proposed contract that increased its wage bill by some $3 million per year. All told, the TWU drove a hard enough bargain that the percentage of corporate income paid to PTC workers went from 45 percent in 1943 to 55 percent in 1945 to 58 percent in 1946. The company had to backdate the wage increase to February 1944 and agree to let its workers wear TWU buttons. And it had to agree to a contract that had no explicit statement of management's right to discipline or discharge its employees. Because the company signed a one-year contract, it found itself in the position of having to negotiate every year, with the threat of a strike looming in the background. As the TWU gained its footing, it pressed even more issues in the first year: first came a battle to stop the expansion of one-man car operation, then for the dues checkoff, then increased pension benefits. Of course wages and working conditions mattered, the TWU argued, but it had a broader agenda. Working with the CIO to improve the overall lot of working-class people meant a "living wage" for all, a stronger economy that would "turn the wheels of this country [and provide] a good job for our veterans." "America must go forward!" the TWU told its constituents. "No more breadlines—no more apple stands. Good wages mean a peaceful, prosperous America. This is your fight against depression." The months after the 1944 hate strike set the context for years of negotiations to come.[65]

Despite the TWU's strong advocacy for the PTC workforce, the internecine divisions continued. Andrew Kelly and other former leaders of the PRTEU and supporters of the Amalgamated started a withdrawal movement against the TWU. The PTC happily supported the effort, supplying typewriters and note cards for workers to produce anti-TWU literature. "The Wreckers," as the TWU called them, found notable support. A petition drive accumulated some two thousand signatures from employees who wanted to leave the TWU, with their support of the PRTEU and racial animosity being the leading reasons for wanting to withdraw. The TWU faced real danger here: "We beat the company in an election," one union official said, then "on the colored issue they took the union right from under our feet." That may have been true in the short term, but in the long term the TWU, chiefly by securing favorable contracts, cemented its position as the representative of the PTC workforce and a worthy foe of management. In part, the TWU also solidified its place by keeping its greatest antagonists from the strike—James McMenamin, Frank Carney, and others—barred from transit jobs. The management of PTC—a company that had blacklisted strike leaders from previous stoppages—was willing, however, to take them back.[66]

Race, of course, did not disappear as a divisive issue for PTC workers, but it did fade remarkably quickly as black and white employees both realized the strength of the TWU, and many white workers recognized how they had let their base views allow them to be manipulated by management. African Americans obtained immediate returns from their support of the TWU. The 1944 contract did not have explicit nondiscrimination provisions, but it did do away with the customs clause that had prevented African American employment in driving jobs. The contract also secured additional overtime pay, a $400,000 fund to make up for wage "inequities and inequalities," and a night-shift premium, all for the maintenance division where most African Americans worked. In addition to the contract specifics, the TWU maintained a forthright stand in favor of African American rights. Any contract the TWU signs, Quill vowed, would provide "full opportunity for all our people irrespective of their race, color or political beliefs. . . . We stand for complete unity and equal rights for all." The TWU's *Transport Workers Bulletin* likely overstated the case when it said PTC workers "cheered the announcement that the company will post notices ending all discrimination against minorities," but in October 1944 the membership did elect five African Americans to office—four to the executive board and one as vice president. Maxwell Windham, the vice president, claimed he was not surprised by the results. Certainly a number of white workers harbored racist sentiments, he said, but "only if a fellow figures the strike last August was just on account of this race business will he be surprised. . . . I know the race business was just a phony, a trick to bust up our union." A white union member said much the same thing: "I voted for Windham because he was on the administrative slate. I know Windham was active in organizin' the union, and I don't see why I shouldn't vote for him on account he's a Negro. . . . Look, don't get me wrong. I ain't no nigger-lover. But I'm a union man, and we can't have no strong union if we get started fightin' among ourselves about things that just don't make no difference."[67]

Soon after the hate strike's end, then, the strength of the TWU in advocating for all the company's workers, its stand especially on behalf of black employees, and the growing number of African American workers at the PTC helped make the issue of race lose its potency to divide the workforce. Over the years, to be sure, black workers occasionally levied charges of discrimination against the PTC, but seldom against the TWU. No postwar strikes featured racial discrimination as a significant issue, and photos from rallies showed mixed-race crowds as early as 1946. The *Philadelphia Tribune* in 1955 evaluated how far African Americans at the company had come in just over a decade, and the results were heartening: the PTC employed four hundred black drivers; twenty black women worked as porters ("porterettes" they were called) in the depots; African Americans were too numerous to count in the maintenance division and on the repair crews, where many experienced black workers had been named foremen; and Arthur Thomas served as the only African American public relations consultant working for a transit company in the United States. PTC vice president David Phillips as well as the union deserved credit, the paper noted, for coming together to make sure the hate strike in 1944 was an aberration and a quickly forgotten one. Herbert Northrup, in his 1971 study *Negro Employment in Land and Air Transport*, found that the PTC employed 941 African Americans in 1950 (6.3 percent of the workforce), with that number growing to 1,019 (9.7 percent) in 1960. Only the transit systems serving New York City, Chicago, and Detroit employed more African Americans.[68]

The way race faded as an issue among the PTC's workers highlights the role that management and its allies played in fomenting the hate strike of 1944. This is not to absolve white workers who participated in the strike of their racist impulses or to say that they were manipulated into expressing sentiments that they did not hold. But it is to say that the hard times of the Depression, the reorganization of the company, and the rise of the TWU led management to believe that it was pressed to the wall financially. Its backing of other unions and employing red-baiting techniques did little to slow the Transport Workers Union. Race, however, offered a powerful, if dangerous, tool that could divide the working class against itself. If the Roosevelt administration had failed to back black rights and had invoked the Smith-Connally Act to abrogate the TWU contract proposal and toss out the union, then management's gambit would have worked. But the extraordinary use of military power on behalf of African Americans cemented their access to jobs at the PTC, averted catastrophic racial violence, and helped give the TWU firmer footing in Philadelphia. On one level the 1944 hate strike highlighted Philadelphia's war-era racial tensions, but on another it had longer effects, opening employment for African Americans at the PTC and ensconcing the TWU as a viable adversary of the company's management for decades to come. Luckily for all Philadelphians, race faded as a divisive issue among the company's workers, and management shifted to less explosive, and often effective, tactics that depended on winning public opinion to management's side.

LABOR RELATIONS AND PUBLIC RELATIONS

"The primary purpose of an urban transportation system," Mayor Joseph Clark told his Philadelphia radio audience in May 1953, "is to carry as many people as possible as cheaply and as comfortably as possible, wherever they want to go within the City in the ordinary conduct of their daily lives. If a city transportation system does not fulfill these requirements it isn't doing its job. You are just as much entitled to cheap, efficient, safe, comfortable, rapid transportation as you are to police protection and the collection of your trash." Clark went on to say that he did not really want the city to own or operate the PTC, but that the company had to overcome a number of thorny financial problems for Philadelphia to get the service it deserved; otherwise the city would have no choice. The PTC's chief problem, Clark argued, was overcapitalization, which meant there were too many bonds in the market and thus too much money being paid to security holders in comparison to the value of the company. That outflow of capital resulted in few cash reserves on hand and limited access to credit. The PTC thus could not purchase much-needed new equipment or maintain its existing stock. The company's only substantial source of liquidity came from fares, and here the PTC was locked in a vicious circle of raising fares, which led to declining ridership, which exacerbated the cash-flow problems, which again prompted the PTC to increase fares. It was a terrible financial situation for a company so vital to the welfare of the city of Philadelphia, and one that increasingly occupied the attention of the city's political leaders, its newspapers, and its populace.[1]

Increasingly tense labor relations between the PTC and the TWU were symptomatic of and contributed to the company's difficult financial position. In the decade after World War II, the company and its workers engaged in a running conflict over wages and working conditions, but also over employee discipline, fares, and service. In a period when organized labor at many companies, most notably General Motors, challenged basic managerial prerogatives, PTC officials confronted a difficult opponent that, in their view, threatened to undermine not just their bottom line but their right to manage. And for the PTC the challenge was particularly daunting, because unlike many industries in this era across the Northeast and Midwest that chose to relocate rather than negotiate with stronger unions, the transit company could not move. Other Philadelphia-area concerns such as SKF, GE, and B.F. Goodrich moved to the suburbs, while textile and apparel producers left the state and eventually the country. All told, Philadelphia suffered as much as any metropolitan area from deindustrialization, with fifty thousand jobs leaving the city in the 1950s and another two hundred thousand disappearing by 1985. But without this option, PTC managers had to be inventive in their thinking, and in the decade after World War II they developed an appeal to the public that emphasized the company's broad commitment to providing for the city's needs, as opposed to what they portrayed as the TWU's narrow interest in looking out for itself. The PTC, company officials argued, was virtuous; labor leaders (and they almost always stressed TWU officers rather than rank-and-file workers) were "unreasonable." Demands by labor leaders for higher wages and reduced hours, management argued, meant that the company, which always sought to "prevent a strike, with all its hardship on the people of Philadelphia—workers and shoppers, merchants and industries," was persistently left with "no alternative except to seek an immediate increase in fare." Any hardship the public faced, be it higher fares or a strike, came about, the PTC asserted, because of the unreasonable union.[2]

The TWU, of course, had a different view and likewise made its appeal to the public. In the eyes of PTC workers and union officials, the company made plenty of money but chose to spend it on dividends and managerial salaries rather than improved service and workers' wages. PTC workers sought a raise, a "modest demand, in view of the high cost of living," as the TWU put it in a 1945 open letter, because they worked hard to provide a vital service to the people of Philadelphia. "We are working to the utmost of our ability to give you the best kind of transit service," the TWU said. "If you're not getting the service you deserve today, we think you are entitled to know where the fault lies. We say it rests squarely on the shoulders of the PTC. More cars could be taken out of the barns, improved schedules could be devised, a more efficient utilization of manpower and equipment could be brought about. [But] PTC in effect says to its workers and its public—they can both fly a kite. It's an attitude of the workers be damned. The public be damned." Transit workers, in this view, were ordinary Philadelphians trying to do a good job and receive decent pay for their labor. They sought only what was fair.[3]

As both sides competed for the public's allegiance, the contest at the PTC—often as bitter, certainly as bombastic, but never as violent as those in the past—played out in strikes, politicians' speeches, newspaper editorials, and public discourse. Surprisingly, transit company management never employed anticommunism, a common cudgel of the period used to smear organized labor and other progressive causes, in an overt way. PTC officials instead persistently spoke of the demands of an "unfair" union, the "danger" posed by the "outsider" Michael Quill, the "threat" to their rights as managers. Philadelphians, they intimated, were not safe with the TWU representing the city's transit workers. This chapter, by focusing on the discourse around labor relations at the PTC, highlights how management employed an array of tools and terms beyond anticommunism to challenge the TWU, yet in the late 1940s and early 1950s such broader efforts still at times fueled charges of communism among the union's critics, some of whom were union members. In a period when Philadelphia's political class for the first time regularly sided with the transit workers, PTC management's rhetoric, and the way some Philadelphians adapted it within the context of the Red Scare, turned much of the population against the workforce. This realignment in some ways not only reshaped how Philadelphians perceived labor relations between the transit company and its workforce then, but also set a tone, provided a vocabulary, for how many people think about workers and their employers in the transit industry, and in public service work more generally, to this day.[4]

Postwar Challenges

A, if not the, central problem confronting the Philadelphia Transportation Company in the decade after World War II was developing and operating a transit system for a growing but suburbanizing metropolitan area while coping with declining ridership and financial constraints that threatened to cripple the company and erode its service. Philadelphia, of course, mirrored the nation in its newly suburbanized postwar population pattern. The city's population, which stood at 1,931,000 in 1940, hit its peak of 2,072,000 in 1950, then declined to 2,003,000 in 1960 and 1,949,000 in 1970. The metropolitan area, however, climbed throughout the period, from 3,199,000 people in 1940 to 4,824,000 in 1970. The percentage of those people living in the city proper fell by a third over that time, from 60 percent to 40 percent. From 1940 to 1950 alone, the populations of the largest suburban counties exploded: Bucks went from 108,000 to 145,000, Delaware from 311,000 to 414,000, and Montgomery from 289,000 to 353,000.[5]

The decline in population density across the United States, as one federal study pointed out, consistently resulted in higher costs per passenger and thus led to transit companies reducing their levels of service. There were exceptions to this rule: Cleveland and Toronto built new rail transit systems in this period, and New York City, Boston, and Chicago expanded their rail lines in restricted ways. In Philadelphia, consulting engineers advised the city to enhance its subway system or face "stagnation, closely followed by decline and decay." The city acted in a limited fashion, opening the Locust Street Subway in 1953 (it had been started with WPA funds in the 1930s and then abandoned), extending the Broad Street

line north to Fern Rock in 1956, and adding express service to the Broad Street
Subway in 1959. But overall, the industry's decline in service was steep. Retrench-
ing where possible, companies across the United States discontinued less profitable lines, abandoned tracks, and converted from streetcars to buses. The nation
hit its high point in miles of rail for transit at 44,835 in 1917. That number fell
gradually to 37,699 in 1929 when the Depression hit. Bad economic times cut
the miles of track that companies operated nearly in half, to 19,671 in 1940. That
number held fairly steady through the war, settling in at 18,151 in 1945, but then
the drop was precipitous: the entire nation only had 6,251 miles of track in service by 1955. So obvious was the decline that no company manufactured or purchased streetcars in the United States between 1952 and 1970. This was the case
despite the fact that the PCC car was developed in 1936 and known for its size,
comfort, and efficiency. The PCC cars were modern and rider-friendly enough
that Philadelphia, as well as Boston and San Francisco, used them into the 2000s.[6]

Some cities shrank the route miles of their systems, and the PTC did that on
occasion, abandoning such lines such as Olney to Sixth and Oregon, and Twenty-
Second and Snyder to Front and Market. But for the most part the reduction in
streetcar lines in Philadelphia was the result of conversion to buses and some
trackless trolleys. The PTC still operated 1,898 streetcars on fifty-eight different
lines in 1947, and its fleet of 1,612 trolleys in 1953 made it the largest remaining streetcar system in the United States. Nonetheless, the pattern was clear:
a company known for its streetcar system under Thomas Mitten just two decades
ago had 980 buses by 1949, a number that equaled more than half of the city's
trolleys. The conversion project, as was often the case, gained momentum in 1955
when National City Lines took over, but the PTC was making this shift before
NCL ever bought a single share of stock in the company.[7]

To boosters of the transition to buses, this was the next step in providing
a modern, cost-effective transportation system. "Street railways," read one report
on several transit systems, including the PTC, "should be replaced as rapidly
as practicable with modern buses." This was "the chief job of modernization,
which is essential both as financial salvation for the company and as service provision for the public." Buses were portrayed as modern conveyances, akin to cars
with rubber wheels on asphalt streets. They had greater maneuverability to steer
around tie-ups in the road and could thus help solve the problem of congestion (the average commute time in 1951 was 28 percent longer than in 1945),
which was exacerbated and too often caused by lumbering trolleys, the "antiquated, unattractive, slow-moving, crowded, and uneconomical transportation
facilities which have not kept up with technological progress and with changing
requirements of the traveling public," as one report put it. Moreover, routes could
be developed without worry about rails or overhead wires. As the population

dispersed, PTC management could more easily establish, alter, or abandon routes that used the public way and thus had little fixed technology and limited fixed costs. The PTC, under headings in its annual reports such as "Modernization and Service Improvements," heralded each purchase of more buses as a cost-saving effort that would allow the company to compete with the automobile and at the same time save millions of dollars in track and equipment maintenance costs.[8]

Critics of the transition—people who rode the streetcars and worked on the lines—understood the PTC's economic circumstances but did not want service to plunge just for the sake of the bottom line. Although the automobile was becoming a mainstay of American life, many Philadelphians did not own a car. Studies in the 1950s and 1960s found that people who continued to live in cities, as opposed to moving to the suburbs, were far less likely to own an automobile. Seventy-five percent of American households in 1960 owned at least one car, and that number moved up 4 percentage points over the next decade. At the same time, only 42 percent of people who made under $3,000 per year and 55 percent of those age sixty-five and over owned an automobile. Although New York City had by far the largest percentage of population that did not own a car, at 41 percent, Philadelphia, Chicago, Boston, and Pittsburgh formed a group of cities with more than one-quarter that did not. Given the way race shaped American suburbanization, and the related income disparity between suburbs and city, the provision of transportation in the urban center increasingly became a matter of meeting the needs of poorer people, a point that the PTC at times recognized, if obliquely, as when it told the public that its main job was to "furnish an essential service and fulfill a basic human need."[9]

Despite the PTC's occasional recognition of its social obligation, many Philadelphians questioned whether the company's new equipment furthered that mission. No form of transportation is perfect, but riders consistently thought the PCC streetcars were modern, comfortable, and performed adequately. They found buses, on the other hand, small (depending on configuration, about 20 percent fewer seats than a PCC car), underpowered, and malodorous. Mervin Borgnis, a PTC bus driver in the postwar era, recalled the difference between a trolley and a bus on the same route: "When the trolleys ran the line, the tie-ups were more frequent. However, once they were in motion they could pick-up the throngs that awaited their tardy arrival. With these buses with 35 seats and a narrow aisle, even when they ran on time they could not pick up all the people in rush hour. If a coal truck held up the bus, it quickly filled to capacity. After such a tie-up they left as many people standing on the corners as they picked up." A rider described the experience: "They crawled along at 15 miles an hour. A bus would pull up to our school and the kids would line up. Going to school we were jammed in like sardines. We kids called them Toonerville trolleys, but the smell of gasoline would have choked you." Adding to these experiences the fact that many of the new buses did not even have air conditioning, it is not too difficult to see why many of the riders—poorer, older, with fewer resources and often minorities—did not see the substitution of buses for streetcars as an unabashed example of the march to modernity.[10]

The PTC's workforce also looked askance at the change in technology, often seeing it as a means to lay off workers and assert greater control over those who remained. The most obvious manifestation of this was the operation of buses, which required one worker to drive and take fares, whereas trolleys used a motor-man and a conductor. The company could keep paying the higher wages it nego-tiated with the TWU, but to a significantly smaller workforce. In fact, the number of employees at the PTC declined throughout this period, from just over 12,000 in 1945 to 11,200 in 1952, to 10,300 in 1954, and 8,100 in 1956. The issue of con-verting from two-man cars to buses and what that did to employment prospects at the PTC became a matter of persistent labor tension for years at the com-pany. Maintenance of the vehicles also had implications for the workforce. The PTC correctly argued that keeping hard-used streetcars repaired and in service was expensive. But an internal report suggested that doing away with streetcars would also help bring maintenance workers to heel. Many of these workers were older (76 percent of the Kensington shop was over age fifty-seven), resistant to change, and recoiled at disciplinary action. The men had a reputation for loafing and winning grievances against their superiors. The company, the report argued, had to get better control of how maintenance employees performed their work and how supervisors evaluated them. Doing so would save the PTC some $500,000 per year. None of this could happen, the report concluded, without a change in labor relations led by management. Shifting to buses would allow PTC managers, or so they believed, to challenge maintenance workers' seniority and workplace rights, since many job classifications would disappear, to be replaced by positions specific to buses. The choice of technology, then, was not simply management's prerogative, but had significant implications for the PTC's workforce. It was, to the workers, a matter of contention, debate, negotiation.[11]

Suburbanization, the growth of automobile culture, and what most Phila-delphians regarded as a decline in service combined to have a devastating effect on PTC ridership levels. This was a common occurrence across the country. Annual public transit ridership nationally peaked at over 23 billion rides during World War II and immediately after, fell to 17 billion by 1950, and was down to 9.4 billion by 1960. Likewise, annual trips per capita in the United States peaked at 178 in 1946, dropped to 115 in 1950, and stood at 52 in 1960. Although inter-city passenger rail service went through a similarly steep decline, which led to the creation of Amtrak in 1971, some sectors of the transportation industry, such as taxis, did remain healthy. But overall, in Philadelphia the trend was obvious and seemingly inexorable: from the over 1 billion passengers carried during World War II, the number fell to 790 million in 1951, 705 million in 1953, 596 million in 1956.[12]

Transit company management, employees, and union leaders scrambled to respond to the desperate situation. Around the country some transit labor leaders, including Michael Quill in New York City and the Amalgamated Transit Union (successor to Mahon's Amalgamated), periodically responded to the decline in ridership by pushing for no-fare transit supported by public funds. The movement never gained much support in Philadelphia, largely because the company remained in private hands and the overarching view was that riders should pay their own way. Instead, the PTC embarked on advertising campaigns urging people to take public transportation to avoid the hassles of traffic jams and parking. The PTC segmented its campaign, telling the public generally that "Transit is a Bargain *Everywhere*!"; businessmen that the PTC offered "the easiest way to get to town!"; and women, through the perspective of one smiling but fearful female driver, "I like to drive . . . but NOT downtown! It used to be fun when traffic was lighter and parking easier—but no more!"[13]

Center City businesses, fearful of what suburbanization and traffic congestion were doing to their trade, joined with PTC management to support public transit. Working in tandem, they developed "shoppers' special" routes—as many as six thousand trips per week—that brought women residing in outlying areas to downtown stores during off-peak hours. They experimented with "ride-home-free-on-PTC" campaigns that gave shoppers discounts on transit fares. PTC management formed two-man teams to meet with local industries, real estate agencies, and community organizations to try to build support for a greater mass transit system. And they urged the city to prohibit street parking in Center City. The ultimate goal, the PTC and downtown corporate leaders agreed, was to make the price of driving so high, the cost of public transportation so reasonable, that suburbanites would either help pay for public transit facilities with taxes and parking fees or be "persuade[d] to consider public transit as an alternative." The TWU, despite its contentious relationship with management, signed on to these efforts by helping to promote "friendly service" campaigns and lobbying the city to extend its Broad Street Subway line.[14]

To Philadelphia's political leaders, these efforts were reasonable but far from sufficient. Public transportation, argued one study in 1950, was a "sick industry." It had a multitude of problems related to suburbanization, finances, and public perception, all of which resulted in a traffic problem that choked the city. "Our streets and highways are clogged by today's traffic," read a report by the Urban Traffic and Transportation Board (UTTB), a city advisory board created in 1954 to examine the region's transportation needs. "Our public transportation system has not been able to keep its riders; and unless heroic measures are taken, the future will bring greatly intensified congestion." Traffic volume, the board continued, had doubled from the late 1940s to the mid-1950s, and automobile registrations had risen by 70 percent. Demographic estimates showed the region's population growing by 29 percent by 1980, which would put traffic in the Central Business District beyond the "saturation point." Already, the report continued, "millions of dollars worth of time and gasoline are wasted yearly in the traffic tangle," and matters would only deteriorate until people chose to live and businesses to locate in other regions because they "are providing the best modern facilities for transportation." Philadelphia, argued the UTTB, was "becoming mired in a traffic problem so intense that the region's livability, its efficient functioning and its competitive power to attract population and industry [were] all seriously impaired." The city's future hinged on solving its transportation problems.[15]

Urban planners in Philadelphia, like their counterparts in many American cities in the age of the automobile, saw a solution, although not *the* solution, to Philadelphia's transportation difficulties in the construction of more expressways and parking lots. Led by Edmund Bacon, often known as the "Father of Modern Philadelphia," city planners and their political patrons rebuilt much of Center City, including Penn Center, Market East, Penn's Landing, Society Hill, and Independence Mall. At the same time, the city witnessed the construction of the Schuylkill Expressway (I 76), the Vine Street Expressway (I 676), and the Delaware Expressway (I 95). All told, the UTTB drew up plans to build three hundred miles of expressways (seventy-five miles within Philadelphia's city limits), another four hundred miles of arterial highways, and parking structures across Center City. Not all these were built—construction of the Crosstown Expressway was famously halted by protests in the 1960s—but they did signal the city's commitment to the automobile.[16]

Although certainly evident, the zeal for highway building in Philadelphia was tempered by the recognition that expressways alone could not solve the city's transportation problems. "It cannot too strongly be emphasized," read one report to Mayor Samuel, "that if Philadelphia is to develop its outlying sections, increase its population, add to its realty values, and continue as a thriving, progressive metropolis, it must take prompt action to expand its high speed transit system." Building new expressways was an expensive and necessary enterprise—a 1955 study estimated that the metropolitan area needed to spend $1.1 billion on expressways and highways alone—but, the report to Samuel continued, "We must not lose sight of the fact that improved highways breed more traffic; that as more and more motor traffic converges on the city, there will follow more and more interference with the movement of many hundreds of thousands of our people who travel to and from their places of employment and the shopping centers in street cars and buses and trackless trolleys. It is this inevitable increase in interference with the mass movement of people in public vehicles on our streets that will drive population out of Philadelphia at a time when we need to hold what we have and attract new citizens." The UTTB called for multiple improvements to make the public transit system modern and attractive: better cars and faster service on the Market–Frankford elevated line; extension of the Broad Street Subway into northeast Philadelphia; greater coordination with the suburban commuter railroads; funds from parking and automobile registrations earmarked to help pay for public transportation. These plans meant an investment of some $350 million in public transit—a sizable sum, but only one-third of the amount planned for highways. Failure to spend this amount would be even more costly, the UTTB cautioned. "The high costs of congestion now being incurred because of poor transport facilities [make it a] false economy for the community, the transport user, and the industry to attempt to provide needed mobility with antiquated mass transportation equipment and obsolete highways." The time to act, wrote the city's Bureau of Municipal Research, was now. Delay was tantamount to "letting the house burn down while you calculated how to put the fire out."[17]

Action may have been necessary, but developing a system this large, this expensive, this essential to the city's future required greater resources than anything a private company, even the largest privately owned transit system in the world in 1953, could marshal. For the first time in Philadelphia's history, a substantial and sustained push for municipal ownership emerged in the city's political class. Joseph Clark Jr. and Richardson Dilworth, liberal war veterans from prominent Pennsylvania families, led that class in the 1950s. Both of them running as Democrats, Dilworth won races for district attorney and then mayor, while Clark served as mayor and then U.S. senator. Together they formed a team that swept out the corrupt Republican machine in 1952 and brought a liberal politics, a sense of noblesse oblige, and a faith that government could do great things for the people of the city of Philadelphia. The Philadelphia Renaissance that they led no doubt created problems by destabilizing working-class communities, but it nonetheless helped to modernize the city; and public transportation, along with the urban renewal efforts of Edmund Bacon, was a key component of their efforts.[18]

Clark and Dilworth made their arguments about the necessity of developing public transportation in a language that focused on economic interest rather than social concerns. One mayoral report argued that the city had invested tens of millions of dollars in public and private money to revitalize Center City in the decade after World War II, but that money would not "pay the dividends it should . . . unless the physical redevelopment of our center-city area is accompanied by the modernization of our public transportation system." Modernization of the subways and elevateds alone would cost $100 million. The PTC could not afford that, Dilworth argued, in part because a private company did not have access to the kind of credit that a city did, but also because the PTC kept "draining all of the cash out of the Company by the payment of dividends [which] shows that it would not spend the money for these capital improvements even if it had it." The PTC did in fact pay out stock dividends totaling some $250,000 per year for much of the 1950s. Here, Clark made a softer argument than Dilworth: dividends to him were not "draining cash" but were instead part of the price of capitalism, but they could not come at the expense of the city's interests. Regardless, the city, Clark and Dilworth believed, had the money and the vital interest at stake to act. The lease agreement begun in 1907 ended in 1957, which gave Philadelphia the right to purchase the PTC in just a few years. Neither Clark nor Dilworth was fully ready for city ownership and operation of the transit system in the early to mid-1950s, but they instead shared a plan with the UTTB that they hoped would split the difference between, or perhaps blend, private and public interests. The city would buy the PTC and its equipment, and then enter into a leasing agreement with a private company to operate the system. The municipal government would therefore "retain financial control and thus the initiative and the right to determine management policies respecting services and fees." In the postwar years, the chief question became how much the company was worth—the city initially claimed a valuation of $50 million, the PTC claimed twice that—and it took more than a decade of wrangling to finally determine the value. Whatever the final cost, in Dilworth's view it was a "cheap price for saving our entire center-city area."[19]

Not everyone agreed with city ownership, of course, and certainly not at first. The *Evening Bulletin*, for example, argued that city ownership would not relieve the need for higher fares and would only help the company's bondholders when they sold their property. Local engineer and business owner Kern Dodge wrote that the very idea of public ownership of the company reminded him of the "wild, unbridled waste and inefficiency of the New Deal." To PTC president Charles Ebert, any discussion by the city of public ownership was "hostile and destructive," "unjust and wrong," a "serious harm to the people of Philadelphia." Such sentiments may have been strongly felt but were certainly out of step with the city's political leadership.[20]

To Clark, Dilworth, and city planners, an improved PTC, likely under city ownership, was the first step in what emerged as a larger plan for enhanced transportation in the metropolitan area. "The heart of the city," Dilworth argued, "must be preserved as the principal place of business, shopping, entertainment, hotels, restaurants, etc., for the entire urban area served by the city. . . . In order to accomplish this, there must be an over-all approach." Highways, he said, had to be built, extensions of the subway and elevated lines constructed, and commuter railroads integrated into the system. To accomplish this goal, the city had to take control of the local transit system, but also develop an authority that would expedite bringing "people by the thousands rapidly and comfortably into our center-city area so that area can once again grow and expand. In short, today it is just as essential for a big city to furnish rapid mass transportation as it is to furnish water and sewage services." The UTTB made Dilworth's proposal more concrete, recommending in 1955 that "a regional transportation organization be created as soon as possible to develop a comprehensive transportation system." The board pushed to have the agency coordinate five counties in Pennsylvania (Chester, Bucks, Montgomery, Delaware, and Philadelphia), plus, if an agreement could be worked out, the three New Jersey counties of Burlington, Gloucester, and Camden. Thinking even more expansively, the board proposed bringing in Mercer and Salem Counties in New Jersey and New Castle County in Delaware if possible. At the least, Philadelphia and its Pennsylvania suburbs had to have a coordinated transportation system.[21]

Such a system, a regional solution to a metropolitan problem, required a broad commitment to expanded political authority and a redistribution of public resources. Philadelphia had taken the first step in that process by creating the Urban Traffic and Transportation Board in 1954 (in some ways a correction to the city's abolition of its Department of City Transit under the new Home Rule Charter of 1951). Next, the city had to overcome the political divisions that gave counties autonomy over their policies with little concern for their neighbors. Everyone in the region had to recognize "the continued interdependence of city and suburbs," as the UTTB put it. That meant coming together to make policy, but also rehabilitating public transit with "a sharing of revenues from other parts of the total, integrated transportation system." "The public," the UTTB continued, "must assume financial responsibility for the cost of improvements and extensions of railroad commuter service that cannot be covered by revenues, but which reduce the need for other public expenditures." Some of the costs could be offset by a more equitable distribution of state funds and an infusion of federal aid to the Philadelphia region, but in the end it would take a substantial local public commitment and a redistribution of resources to fix the Philadelphia area's transportation problems. Given the way suburban counties jealously guarded their prerogatives and the racial politics of suburbanization, the board strongly recommended that "any plan for regional transportation organization should infringe as little as possible upon the existing political structure." As people became more accustomed to regional authority and it proved its ability, such a political structure would become stronger and hopefully lead to similar approaches to water supply, sewage disposal, control of air pollution, and recreation. In the end, the UTTB saw the possibility of using transportation to recast metropolitan politics. "It would be less than adequate," the board asserted, "if action taken in establishing a regional transportation agency should be taken without a view to the ultimate potential requirements for regional government." That dream of a new metropolitan framework foundered in the 1960s mostly because of the way race often shaped the politics of suburbanization. Nonetheless, in the decade after World War II, the region's new suburbanized geography, the rise of the automobile, the ascendancy of a liberal political worldview committed to government activism in city hall, and the need to revitalize Philadelphia combined to make a regional authority, at least for transportation, possible. SEPTA (the Southeastern Pennsylvania Transportation Authority) was not born until 1963, but in the 1950s it was at least a glimmer in the eyes of Clark and Dilworth.[22]

Labor Relations at the PTC—a Larger Context

Finances, always a central issue in labor relations at the PTC, continued to play a key role at the company, especially in the context of the strain put on the industry by suburbanization and the growing push for public ownership. For the first time in its corporate history, management at Philadelphia's transportation company felt buffeted by strong labor on one side and political leaders on the other. The TWU, and organized labor more generally, was at high tide in the postwar period. Unions hit their peak membership rate nationwide of approximately 35 percent in the mid-1950s, and the TWU's Local 234 was the largest CIO union in Philadelphia, with 9,422 members on its tenth anniversary at the PTC in 1954. The local even had enough members and resources to open its own school that year, offering courses in labor history, public speaking, parliamentary procedure, collective bargaining, and arbitration. Michael Quill was an indefatigable and savvy adversary throughout the period who knew how to win at the negotiating table and attract the attention, if not the support, of the public to the workers' cause. In prior decades, almost without exception, PTC management could have counted on Philadelphia's political class—machine politicians, allies of the business elite, devotees of laissez-faire platitudes—to align themselves with the company. No longer. Now, Clark and Dilworth pushed the company to be more responsive to city needs, more willing to expend its resources for the public interest. Public ownership, a regional system, a public authority to coordinate transportation were not just on the table; they were all a distinct possibility.[23]

PTC managers saw threats all around and developed and deployed a new method for trying to control their workforce and stave off liberal designs. Labor relations at the city's transit company in the decade after World War II became a matter of public relations, an effort to influence the opinion and win the allegiance of ordinary Philadelphians. This change, where the terms of who would work for whom and under what conditions—in other words, the labor question—was a matter of debate rather than an arena for violent conflict as in earlier periods, signified both a less brutal period in American labor relations and a moment when many people outside the labor movement saw class relations as an issue of public concern to be debated and determined. Politicians such as Clark and Dilworth, PTC officials including Charles Ebert, and TWU members from Michael Quill to rank-and-file transit workers all engaged in a vigorous debate about the value of public services, the rights of workers, and the needs of the public. Ironically, suburbanization, which had so hurt the PTC's finances, helped the company in this arena, as the region's new geographical organization pulled apart working-class communities and increasingly led Philadelphians to see TWU leaders and members as unreasonable and perhaps un-American. Both management and the TWU sought to bring the public to their side of the argument; overall management did so more successfully, although their efforts were spotty, uneven, incomplete. They nonetheless helped set the contours of the debate on the transportation company, its workforce, and by extension public service employees, for decades to come.[24]

In confronting a stronger union and more organized workers, PTC management experienced what businesses across the United States faced in the postwar decade. In those years, millions of coal miners and meatpackers, teachers and steelworkers, transportation employees and autoworkers walked out. In 1946 alone 4.6 million people went on strike in what the U.S. Bureau of Labor Statistics called "the most concentrated period of labor-management strife in the country's history." Unions presented a far-reaching set of demands that went beyond pay and hour issues to include organization of foremen, control of the shop floor, regulation of production, pricing of products, and deployment of technology. In Pennsylvania, organized labor was at its greatest strength in history, with more than 1.5 million workers belonging to unions in the 1950s, and they participated in strikes at Westinghouse, GE, GM, and steel mills across the state. Philip Murray, born in Scotland and raised outside Pittsburgh, rose through the United Mine Workers of America to become president of the CIO. Even public service workers finally found representation in the American Federation of State, County, and Municipal Employees, which especially grew in size and power in Philadelphia and proved to be a boon particularly for many African Americans who made a better life for themselves through these union-protected jobs. Labor's power manifested in formal politics, where union votes helped elect Clark and Dilworth to municipal offices in Philadelphia, put George Leader in the governor's mansion and Clark in the U.S. Senate, and pushed through the state legislature a higher minimum wage and a more capacious version of the federal Fair Labor Standards Act. Such political activism mirrored CIO efforts through its Political Action Committee (PAC) and other initiatives to influence American politics and was actually more successful in Pennsylvania than across much of the United States.[25]

The TWU in Philadelphia benefited from and helped advance organized labor's gathering strength both by the causes it espoused and the people it endeavored to bring into the union. Wages and hours of course mattered, but TWU leaders understood that the labor movement was strongest if it fought for something larger, and that meant supporting the CIO's social agenda. At mass meetings, public rallies, and in informational materials, Local 234 backed campaigns for affordable housing, decent medical care, and better schools. It called for an increased minimum wage, stronger labor legislation, and full employment. And it pushed for a United Nations to promote world peace and a "world federation of trade unions" (an apparent reference to the International Labour Organization, a UN agency that dealt with labor issues). "You have secured through the Transport Workers Union higher wages, many improved working conditions and a pension plan that many of us thought was impossible to secure," the TWU wrote. But the CIO offered the working class a broader vision. "Your dues pay for much more for you and your family and your neighbors all over the country. . . . The CIO has fought and will continue to fight for a program which benefits all Americans directly. These are vital fights conducted by TWU as part of the great CIO. Our fight on PTC is not a narrow fight. It is connected with all labor. We could not have organized on PTC if CIO hadn't paved the way, thru the sacrifice and united action of all CIO members coast to coast."[26]

TWU leaders knew they needed a strong organization to win the changes they sought, and they thus developed a multifaceted effort to solidify and expand their union. The local chapter routinely held mass meetings to discuss contract negotiations, organizing campaigns, developments in labor law, and other issues. Local officers made a special effort to reach out to veterans, including establishing a "Local Veterans Committee" and participating in CIO-sponsored veterans' rallies that demanded low-cost housing and better hospitals and schools. The union also sought to increase women's involvement, both in membership in the union—especially among the cashiers, where most women workers were still concentrated—and in an auxiliary for spouses of male employees. The TWU vowed to do away with gender-based pay differentials that compensated cashiers with wages as much as 30 percent below male conductors, obtain access to phones and water coolers for all cashiers, and increase vacation time. "Well girls," the union asked at the end of one of its pamphlets that urged women to join the union, "what are you waiting for? TWU stands on its record!" And the union gave its support to other Philadelphia-area workers engaged in conflicts with management, most notably members of the United Electrical, Radio and Machine Workers of America on strike at GE in 1946. "The entire labor movement of Philadelphia must be united in determination against the Fascist-like police brutality stemming from the arrogance of the General Electric management," wrote the TWU. All unions had to unite to fight for "decent wages, fair treatment, and all the other things that a true Democracy owes to its people."[27]

To the TWU, securing what a democracy owes its people required political activism. That first and foremost meant supporting the CIO PAC financially and through information campaigns. "Political action costs CIO lots of money," the TWU wrote. "We ask that every member of Transport Workers Union, Local 234—just as all CIO members all over this country are doing—contribute $1.00 to help elect the kind of men and women to public office who will pass laws to protect our jobs and our standard of living." Money, however, was only the starting point: the TWU encouraged its members to canvass their neighborhoods, urging people to register and vote. In its flyers, the TWU argued that the PAC fought for the rights of the little guy, "the veteran, the farmer, the small businessman [and] the working man," anyone who understood the danger "reactionaries" posed to the country. The Transport Workers at times claimed the PAC was nonpartisan, but overall the union made it clear that the great objective was to defeat the GOP, the party that enjoyed "sticking it to the little suckers who have been taking it for years." A strong PAC, in the TWU's view, would mean an end to local and state Republicans ignoring the troubled housing situation, disregarding labor law, torpedoing fair employment legislation, and using the police to break unions. Nationally, only the CIO could put together the resources necessary to elect "a good Congress who will vote for your best interests." The resources that Local 234 could marshal were obviously too small to significantly impact the U.S. Congress, but the union's activism did help elect Clark and Dilworth and developed allies on the city council as well.[28]

Businessmen, largely politically conservative, understood the tenor of the times, especially the strong organization of the working class and the New Deal state apparatus that continued to protect labor's rights, and reached a certain accommodation, a level of understanding, with and about unions. This accommodation was certainly not the supposed "labor-management accord" built on a fanciful social compact between workers and management that provided decent wages and benefits because they were good for business and created a stronger society. Historians Nelson Lichtenstein and Jefferson Cowie have cast grave doubt on that historical interpretation, if not entirely laid it to rest. No, this was businessmen best represented by NAM—led in part by Pennsylvanians J. Howard Pew of Sun Oil and the Mellon family of Gulf Oil—tolerating unions because they could not destroy them; viewing them as adversaries who stirred up working-class grievances to foster class solidarity and oppose the interests of business that employers believed offered the country the true path to a brighter future.[29]

At best, these businessmen believed, unions could be penned in with all but wages and hours taken off the table and labor leaders used to control rank-and-file restiveness. At GM, for one famous example, management rejected Walter Reuther's demand for a 30 percent increase in wages without a commensurate rise in the price of automobiles. Such demands, GM argued, were un-American and Socialist, a gross infringement on corporate prerogatives. Perhaps, but if that approach had been implemented at the PTC, Philadelphia would have avoided the inflationary wage-price spiral that plagued the city's transit system. GM did ultimately give some ground on pay, but only as a quid pro quo for control of prices being taken off the table. That settlement sketched the boundaries for the "legitimate" focus of labor negotiations across American industry. Also at GM, as well as Ford and other companies, management drove a hard bargain, signed detailed legalistic contracts, and then demanded that the union enforce them. Labor officials, hemmed in by their increasingly bureaucratic roles, could fall into the trap of disciplining their members, serving as the right hand of management rather than as the leaders of the rank and file. The TWU, under the leadership of the firebrand Michael Quill, avoided that fate more so than many unions, although restiveness and dissension certainly tore at class solidarity at the PTC. Business interests also used their political power to control labor, most significantly through the federal Taft-Hartley Act of 1947 that banned direct union contributions to candidates in federal elections, eliminated the closed shop, ended secondary boycotts, allowed states to pass right-to-work laws, and required loyalty oaths that purged Communists from organized labor. Pennsylvania's legislature followed the national lead, enacting a "Little Taft-Hartley" law that banned strikes by state and local government employees and eliminated secondary boycotts. PTC management sought to varying degrees to adopt all these national trends in business efforts to shape labor relations: restricting bargaining to wages and hours, pushing the union to assume the role of policing the contract, relying on political and legal allies to support the corporate position. The TWU in turn contested the PTC all along the line.[30]

Vying for Public Support at the PTC

In the decade after World War II, the TWU staged three strikes, in 1946, 1949, and 1953, that grew out of a running contest over wages, hours, working conditions, and one-man car operation, the last of which represented a larger conflict over managerial prerogatives to control the use of technology, distribute resources, and determine the size and deployment of the workforce. These strikes were annoyances to Philadelphians but not particularly long, large, or explosive like the conflicts in 1910 or 1944, so rather than tracing each job action, this section analyzes in composite the underlying issues that occasionally resulted in these formal strikes. The debates over the years of course had permutations, but over-all PTC workers argued that they worked long hours, made lower wages than the national average, had a hard job, and should not be subject to technological change by fiat that would make them operate buses and trolleys individually. Management pointed to the industry's financial condition, argued that transit workers had stable jobs that paid decent wages, and maintained that the decision about one-man operation was a company prerogative. The drumbeat of labor strife left Philadelphians from every walk of life increasingly exasperated with the shutdowns, fare hikes, threats of walkouts, and perpetual war of words in the press. Both sides had their work cut out for them in winning the public to their side in a setting so fractious that one wag in the *Inquirer* dubbed PTC the "Perpetual Turmoil Company."[31]

The first issue on the board after the war was, expectedly, wages. PTC, municipal, and federal reports all showed the same thing: that Philadelphia transit employees, who earned $1.43 an hour in 1950, made substantially less than similar workers in other cities. Chicago and Pittsburgh paid $1.65; Boston and Detroit $1.60; New York City (depending on the company), Washington, and Baltimore $1.55; San Francisco and Cleveland $1.50; Los Angeles $1.47. Only Saint Louis, of the twelve largest cities in the United States, paid as little as the PTC. The average yearly take-home pay for all PTC employees came to $3,300 (almost $3,600 for operators; just less than $3,000 for maintenance workers, many of whom were African American). Such wages, a city fact-finding board observed, were lower than those in comparable service industries in the Philadelphia area, such as trucking and railroad employment on the suburban Pennsylvania and Reading Railroads. PTC employees, the board concluded, deserved raises that demonstrated their similar worth to workers in other large cities as well as equivalent area industries and also helped them keep up with postwar inflation.[32]

Michael Quill and the TWU seized on this data and local political officials' support to demand higher wages throughout the late 1940s and early '50s. The union, after regularly opening with a demand to as much as double workers' pay, usually negotiated down to a number that amounted to approximately a 15 percent annual increase. To make clear to Philadelphians what was at stake, the TWU developed a public relations campaign that used television and radio spots, pamphlets, flyers, and speeches. Unlike company officials, Quill asserted in one such speech, "PTC employes are honest workers and don't steal money. They have to live on their wages and their wages are not sufficient to take care of anybody." Workers across Philadelphia, the TWU argued, could understand PTC employees' plight. "All organized labor," their flyers read, "is fighting for a decent wage in keeping with the high cost of living," and the company symbolized management throughout the United States with its "attitude that the employes be damned."[33]

The company might charge PTC workers with being "Reds," Quill argued (management hinted at that but in fact did not), but there was nothing more American than the flag and fighting for what you were owed. "We are saying to the management of this company [that] they have ridden on our backs long enough. Let them get off and walk for a little while and let them come across with the things we are entitled to." This notion of fighting for better pay as true "American" behavior became a recurring theme in rank-and-file discourse and also on the picket line where strikers, often in their World War II uniforms, carried the American flag and placards that read "Veterans Deserve Decent Wages," "PTC Veterans Right in Line," and "Great Wartime Profits for PTC: What about Us?" These wage campaigns were ultimately about getting what was fair, a "decent living wage" that would support working-class families, and Philadelphia's city

FIGURE 15. Women on the picket line. Although most female drivers left their jobs at the PTC after World War II, women took to the streets to support their husbands' contract demands in 1946. Photo courtesy of Historical Society of Pennsylvania.

council, increasingly in Democratic hands, agreed that the local government should be on the side of workers who simply sought a "living family wage." At times, the TWU argued, unions in the pursuit of justice had no alternative but to strike, and the public understood this. "PTC has flatly refused to bargain with us," wrote Local 234 president J. B. Dougherty in an open letter to Philadelphians. "PTC has notoriously over the years not treated its employes well. The company's attitude has been a viciously anti-union one." The company's stance, he continued, was irresponsible, and he had confidence that "you will support us in our modest demands, in view of the high cost of living [and] that Philadelphia's citizens prefer a spirit of labor-management cooperation between PTC and its workers." He concluded by asking the public to write to President Ebert, "urging upon the company an attitude more in keeping with the interest of the community and nation." Quill cautioned the public not to fall for repeated PTC propaganda about its poverty; if it could afford dividends, it could afford to pay its workers what they deserved. With this last line of argument, Quill tapped into a common belief, one dating from the heyday of the transit industry when entrepreneurs made millions, that the company sat on "hidden reserves" generated by "improper bookkeeping." PTC officials made it a point to immediately beat back such "false and absurd" charges when they heard them because, true or not (and the union had little evidence of actual malfeasance), management understood the "insidious . . . effect [such allegations had] on the minds of riders and PTC employes."[34]

Hours and working conditions also figured prominently in PTC workers' grievances. As late as the early 1950s, operators still worked an average of forty-four hours in a six-day workweek. Quill made the forty-hour five-day week a signature part of his efforts, and Mayor Clark supported the cause, arguing that the company should agree to what it must know is "the proper work week in American industry." Extra men (operators who had to report for work in case a regular driver was absent) and those who worked swing runs had it even worse. Extra men had to show up for work at 3 a.m. on weekdays as well as one day each weekend, not knowing if they would get runs or not, although the TWU did make sure by contract that they got the five-day week in 1949 and were paid for forty hours regardless of how many trips they made. Drivers on swing runs continued to work the two rush hours with a hole in the middle of the day, essentially being on duty for fourteen hours and receiving pay for eight. These schedules, combined with overloaded vehicles and lax maintenance standards, the TWU argued, put operators and the public at risk. "Safe and Sane Schedules!" became the union's rallying cry, and rank-and-file workers said, "Those who make these schedules should be made to work them." The increasing number of rules about fares, operation, equipment checks, and more, all of which were taught in the strikingly named "Indoctrination Room," made the job even more onerous. Management logically argued that it had an obligation to make sure equipment was safe for workers and the public—although drivers often complained about inadequate maintenance—and fares were collected honestly, but many workers believed the rules were really in place so the company could "prove" employees were wrong whenever an incident occurred. In protest, workers engaged in "rule book operation" (essentially a slowdown), whereby they refused to take out a vehicle until the schedule was precisely accurate and every vehicle check had been performed. Once on the road they stopped at every intersection, which purposely caused them to run late.[35]

As with the campaign for higher wages, the TWU pursued public backing. The union put together teams, often made up of veterans in uniform, to go into neighborhoods, telling residents, church groups, and small business owners their side of the issues. The TWU knew it had to obtain "their support, financial and moral." Flyers told Philadelphians that the "Safe and Sane Schedule Campaign" meant "better service for the people [and] better working conditions for the men and women." PTC workers, the union informed the public, had to keep to impossible schedules while working as many as seven hours straight without a break. "We ask you," one flyer read, "to imagine the condition of a PTC employes' nerves at the end of the day after operating on a split second timing for nine or ten hours at a stretch." "We ask you to give us your support," the flyer continued, because our campaign "will benefit you too." Now was the public's chance, the TWU asserted in a final effort to win the public's allegiance, to join with PTC employees in pushing the company to improve service and provide a safer ride.[36]

The city's last major development in transportation technology provided the third significant point of contention between labor and management. The shift to one-man operation, which meant rebuilding trolleys so they did not need a conductor, buying PCC cars that were designed to be run by a single operator, or substituting buses for streetcars, had significant implications for the PTC's workforce. The conversion to one-man operation took place across the United States in this period, with Chicago, Cleveland, Detroit, and other cities more than doubling the percentage of track using single-man cars between 1948 and 1951. Without fail, one-man operation eliminated transit workers' jobs, as the lines generally required half as many employees and the switch to buses additionally reduced the need for workers to provide car maintenance and track and electrical equipment repair. The switch often pushed out more senior employees because they lacked requisite preparation to operate one-man equipment, and most transit companies balked at retraining older workers nearing retirement. Companies, including the PTC, could thus reduce labor, equipment, and maintenance costs all at once, and use the end of trolley service as an excuse to abandon

FIGURE 16.   Men on the picket line. Philadelphia's transit workers emphasized the veterans in their ranks when they campaigned against the PTC in the years after World War II. Photo courtesy of Historical Society of Pennsylvania.

less profitable lines. As one PTC driver put it, when older workers left, "they took their runs with them!" Jobs fell hundreds at a time as the PTC performed the conversion, and the company saved as much as $2 million for each 10 percent cut in the workforce.[37]

The TWU responded to the conversion efforts with protests and appeals to friendly politicians and the public. On several occasions the union threatened to go on strike over breach of contract, as they termed the layoffs. Outside of the formal negotiation process, rank-and-file workers threatened to stage wildcat strikes if the company put employees out of work. Flyers, to cite one example, urged operators to walk the picket line at the Luzerne carbarn to "show the board of directors of P.T.C. that we are not going to be pushed around any longer." The union also staged demonstrations at City Hall to support passage of an ordinance banning one-man cars in Philadelphia. Workers backed these demonstrations with radio appeals and with petitions circulated among the public outside City Hall and all members of the city's CIO unions. In their petitions, radio spots, and public statements, TWU leaders argued that they were of course looking out for their members' interests, but those interests mirrored the public's. Too often, one-man conversion represented nothing more than corporate penny-pinching, and management readily turned unfamiliar equipment over to operators with insufficient training. The PTC saved a few dollars at the expense of TWU members' jobs, but also at the expense of public safety. One-man cars, the TWU's Andrew Kaelin told the city council, placed "an inhuman burden of work" on the operators, which meant "the service to which the public is entitled must suffer." "Further extension of one-man cars," he added, "threatens the safety of passengers, motorists and pedestrians." "The union," local TWU official P. J. O'Rourke told the press, had to speak out "against PTC and its officers whenever they cut the service to the riding public of the City of Philadelphia."[38]

The PTC delivered a multipronged response to the TWU's arguments that involved building pro-PTC sentiment among the workforce, the business community, and the larger public, using scare tactics on occasion, and publicly rebutting the TWU's arguments in contract negotiations. The company's efforts began, as the journal *Bus Transportation* put it, with "an aggressive advertising and public relations program designed to 'give 'em the facts,' and win back community goodwill." Their campaign involved running pro-company pieces in newspapers and on the radio, posting placards on buses and trolleys, producing boosterish movies, making presentations to schoolchildren, meeting with civic, business, and church groups, and reaching out to the black community. They also encouraged a "Glad-to-Have-You-Aboard Club" among the employees to emphasize "friendlier, more courteous service to the public." All told, the company spent over $100,000 per year on its public relations campaign, and observers widely noted how effectively it "contributed to improvement of the company's standing in the community," especially among businessmen, who constantly worried about the impact of transit strikes on Center City commerce.[39]

In addition to trying to build public support, PTC officials appealed to their workforce to recognize the benefits of the company in an effort to win the allegiance of at least some of the employees. Management produced a number of pamphlets with titles like "PTC Must Be a Good Company to Work For" that reminded workers of their steady job, pension, and fringe benefits that more than offset any shortfall in their take-home pay. Turnover at the PTC averaged less than 1 percent per year, management claimed, at a time that the national average in all industries was 4.5 percent, and more than six thousand employees had been with the company for ten years or more. PTC could claim this record, management continued, because it was a "stable company, considerate of our people." "There need be no fear of your future as PTC employees," management concluded with little accuracy but more than a hint of defensiveness. Regardless, 87 percent of returning veterans wanted their jobs back, and the transit company tried to smooth the reintegration process by hiring counselors, offering free medical exams, and promising that time in the service would count toward seniority and pensions for these "valued and vital members of PTC."[40]

At the same time that the PTC was selling to its overall workforce the advantage of employment at the company and promoting its welcoming atmosphere for veterans, it was also casting the TWU, and especially Michael Quill, as outsiders who cared nothing for Philadelphia, even endangered the city. In doing so, they played on some of the old sentiments from the war-era TWU organizing campaign about the CIO union as an interloper. PTC officials also touched on, but never explicitly used, newer red-baiting arguments driven by the postwar anticommunist scare that portrayed union leaders as alien, people with hidden agendas not to be trusted. In part, this approach demonstrated an effort to separate rank-and-file workers—that is, local Philadelphians—from TWU leadership. To spearhead this campaign, in 1945 the PTC brought in William MacReynolds as vice president for industrial relations. MacReynolds had worked in similar positions for Inland Steel and the Packard Motor Company and had a reputation as a "tough guy," someone who would stand up to big labor. As it played out over the years, the TWU came to see him as a strong but fair negotiator, a reputation that got MacReynolds fired in the mid-1950s when National City Lines took over the property. NCL even chose to do away with the Industrial Relations office.[41]

The PTC's campaign against Quill and "his" TWU was unrelenting. Time and again, Ebert, MacReynolds, and other corporate officials referred to Quill as a New Yorker, an outsider who did not care about Philadelphia. Following Quill's radical agenda, Ebert warned, would be "a breach of faith [by] the responsible operating employees of the Philadelphia Transportation Company" who he knew had a better sense of "decency, fair play and responsibility" than their labor boss. Quill's demands for better pay, fewer hours, and a halt to one-man conversion were invariably "unrealistic and unjustified," his contentiousness nothing less than a "war of nerves" that held Philadelphians hostage. Ebert claimed to find Quill's program shocking; to him, it meant "the officers of the union [were] bent on using tactics which may well destroy the company" and bring upon the city nothing but "suffering and loss." Comments occasionally wandered from this script, as when local transit workers were described as "husky union members" bent on intimidating others, but for the most part PTC officials such as Albert Greenfield voiced the belief that union leaders enflamed issues while "no deep-seated animosity exists between . . . management and employees."[42]

Rank-and-file workers heard management's hard charges against Quill and TWU leadership, and some accepted parts of it and even extended it as they leveled critiques of their own. Both Quill and local workers allowed their rhetoric to devolve at times into stereotypical types of red-baiting common to the postwar period. Ironically, although management set the context, the workers did this to themselves. Much like in the Mitten era, internal union discord was in some ways as damaging as any management offensive, and commentators including the *Evening Bulletin* noted how TWU divisions, sometimes called "factionalism," benefited the company in negotiations.[43]

To some Philadelphia transit workers, management was correct in its charge that Quill was an outsider who did not have the city's, or their, interest at heart. Quill lived in New York City and made no attempt to hide the fact that to him his city's transit industry was paramount. He developed a reputation for flying into cities for contract negotiations or union elections and then leaving just as quickly. Other cities, it seemed, mattered to him most because their wages and working conditions could set a pattern to which New York City transit officials would wish or have to conform. Philadelphia having low wages—the city often ranked in the twenties for transit workers' pay—was thus problematic ("a very embarrassing position," as Quill put it), more so because it served as an example for New York City than because PTC employees were being exploited. The executive board of Local 234 repeatedly complained that Quill ignored their advice, disregarded local workers' concerns, and used his Philadelphia-area ally, Andrew Kaelin, to control the local union. Quill, who could use his sharp tongue to the workers' advantage when sparring with management, turned it on the Philadelphia local, calling anyone who opposed him "drunks, jobseekers, or thieves." Such behavior understandably infuriated PTC workers, who on occasion defied their leadership and wrote in letters to company officials, politicians, and the newspapers that they did not want to quarrel constantly with management, that Quill was making them do it. The international TWU in Philadelphia, as one anonymous worker put it, had "no following [and thus] no leadership." Quill was just "making moves to make the Union membership feel that he is doing something for us," but if the rank and file could vote, the company would not have any strikes because "we are satisfied to work—and not 'pound the pavement' as Mike would have it." Quill should "get back to New York and let us work." "No following" was clearly an overstatement, but the rancor in the Philadelphia local nonetheless highlighted how in the eyes of at least some PTC workers the international union came across as an outsider using their concerns to advance ulterior motives.[44]

The divisions, the factionalism, played out most damagingly in charges and countercharges of Communist sympathies. Quill insisted that he had never belonged to the Communist Party, although he acknowledged that he attended Communist summer camps and met frequently with William Z. Foster and other party leaders. Most observers understood that in the 1930s and early 1940s Quill was firmly in the Communist orbit, but he avoided directly addressing the issue with the comment, "I'd rather be called a Red by the rats than a rat by the Reds." In 1948, as anticommunism grew in the United States and Taft-Hartley reshaped the political landscape, Quill, in the words of the *Evening Bulletin*, "disavowed all Communist sympathies and declared war on Red influences." Quill's "war" meant removing a number of his longtime allies from leadership positions in the International TWU, including Secretary-Treasurer Douglas MacMahon, Director of Organization John Santo, and Vice President Robert High, the last of whom was from Philadelphia's Local 234.[45]

Quill's treatment of his former allies, especially High, sparked much of the conflict in Philadelphia. Wages, hours, and one-man operation figured in elections for union leadership, but a significant part of the campaigning revolved around ferreting out who was a "Red," or perhaps more accurately, using such charges as weapons in an internecine political battle. At times Quill's opponents accused him of being a Communist, a betrayer of "honest, militant American Trade Unionism," and he felt obligated to have Philip Murray vouch for his loyalty to the CIO and his vigorous opposition to "those who seek to divide our organization from within." But most of the charges came from Quill and his loyalists, men who said they wanted "honest CIO trade unionism—not Communism." McMahon and his allies were "members of the Communist bloc" who sought to control the union and "oust Brother Quill from the Presidency of the TWU." Anyone who voted for Kaelin's opponent was a dupe who fell for the "RED package." Many rank-and-file workers, weary of the charges and countercharges, signed resolutions condemning the Communist Party but also demanding that Quill and McMahon halt their internecine "political fight [that] disrupted and weakened" their union.[46]

Such admonitions went unheeded throughout the late 1940s and early 1950s as internal union politics became a bruising affair. Kaelin, for just one example, was charged with being a champion of the Right. His defeat in 1951, in the words of Quill opponent Paul O'Rourke, finally signaled "a victory for the rank and file." A few years later, Quill engineered O'Rourke's ouster in what the press called Quill's elimination of the "last of his big TWU enemies." He also installed Kaelin as a vice president with the TWU International to make sure he had a loyalist on board to back his efforts. Each side charged the other with financial abuse, leading to expensive and embarrassing audits of the union's books that ultimately revealed no wrongdoing. At one point the executive board of Local 234 even considered censuring Quill for interference in Philadelphia's affairs, but ultimately decided not to. The TWU's conflict became so pronounced that the Amalgamated even tried to get back on the property in Philadelphia after a two-decade absence. The AFL union lambasted "Red Mike" in a number of flyers but found little traction with the workers. Quill, after referencing the 1910 strike, scoffed that clearly "the Amalgamated Association is still remembered in Philadelphia in the cries of the orphans and the tears of the widows it created." With the union already facing a difficult opponent in the PTC, these internal conflicts both weakened it in its negotiations with management and gave the public greater fodder for opposing the transit workers if not for siding with the company.[47]

Management, having done the groundwork to hopefully develop goodwill in the city and among some of the company's workers, while sowing dissension in the ranks or at least happily watching it grow on its own, achieved a mixed record of success in the debates and related contract negotiations about the specific issues of one-man conversion, wages, hours, and benefits. But in the larger battle for public opinion, as evidenced by the voluminous correspondence sent to the TWU, city politicians, and local newspapers, there was little doubt the PTC came out ahead, especially as management persuaded the public to believe pay raises were ineluctably tied to higher fares, and thus greater transportation costs were the fault of the union.[48]

Throughout the era, the PTC maintained that the TWU was overstating the problem of one-man conversion and that, regardless, any such decisions were management's prerogative. In opposing one-man operation, MacReynolds charged, Kaelin was disingenuous, Quill irresponsible, even "inhuman and unconscionable." The PTC, MacReynolds argued in the newspapers, had no intention of using one-man conversion to reduce its workforce. The company merely wanted to use one-man vehicles on Saturdays during the summer because ridership had fallen by one million passengers per day since World War II. The company, President Ebert vowed, would return to two-man operation in the fall. No official commented on PTC annual reports showing the company shedding thousands of jobs. TWU arguments about the safety risks presented by one-man cars were equally untrue in the PTC's estimation. The company's superintendent of surface transportation, W. J. Mack, wrote in an open letter that the PTC would never ask its workers "to sacrifice the public safety, or your safety, to meet any schedule," and MacReynolds told the press that PTC statistics showed one-man lines actually had better safety records than two-man lines. In using "deliberate distortions and exaggerations," MacReynolds claimed, Kaelin showed "a complete disregard for the interests of the public and the employes he represents." The company, MacReynolds concluded, was "reasonable and justified" in making these "proposed economies," and it, unlike the TWU, had the best interests of the public at heart. Overall, he said, it would be "a major Union boner" to continue this fight. The TWU did keep up its opposition to one-man conversion throughout the early 1950s, but without much conviction. One-man cars remained part of negotiations, but the union at various times considered and then decided against striking over one-man cars as a singular issue. Instead the union decided to negotiate higher pay of ten cents an hour for operators of one-man cars rather than strongly contest what management felt was its prerogative. The PTC essentially won this fight and generally had free rein to continue the conversion process that saw 70 percent of surface rail lines operated by one-man cars by 1952.[49]

In negotiations over hours and benefits, the PTC found a harder road. The Fair Labor Standards Act of 1938 had established an eight-hour day and forty-hour week as the standard work time for many types of employment in the United States. Workers, although not all covered by the FLSA, came to expect, or at least desire, that standard, and politicians who sought their votes, such as Mayor Clark, backed them. Pressured by the TWU and liberal politicians, the PTC grudgingly granted shorter hours, first promising, in 1949, the five-day week when possible, and then, in 1954, fully agreeing to provide it. Extra men secured the five-day week at the same time, but their work remained irregular. Management also agreed to the dues checkoff that it had previously given the PRTEU and to improve fringe benefits, including uniform allowances, three weeks paid vacation, six paid holidays, fifteen minutes of paid time for lunch, and more money in the pension system. In return, the TWU agreed to even more one-man conversions, the savings from which offset the company's increased expenditures.[50]

Of all the issues at stake, the greatest struggle took place over wage increases. In this case, the workers won the battle, but ultimately lost the larger war for public support. For wages, as opposed to one-man operation, the TWU made it clear that it was willing to go on strike. And in the postwar context of great inflation, working-class power, and liberal ascendancy, the PTC could not hold the line on pay rates, especially when even a short strike like the four-day walkout in 1953 cost the company $300,000. Repeatedly, the TWU either through strikes or negotiation won higher pay: twelve cents per hour in 1946, thirteen cents in 1947, fifteen cents in 1948. All told, the TWU secured a 42 percent pay raise between 1945 and 1948 (from 95 cents to $1.35 per hour) and more than doubled PTC workers' pay between 1940 and 1950. Although PTC employees were constantly in danger of slipping back, the raises finally brought them up to the average pay of the twenty largest transportation companies in the United States. With ridership dropping, each hike challenged the company's bottom line. One cent cost the PTC approximately $323,000 per year. TWU contracts added $5.5 million to company expenses in 1948, $2.3 million in 1949, $1.3 million in 1950, and $2.5 million in 1951. The raises meant the PTC had to spend 59.9 percent of gross revenue on labor costs in 1947, 62.6 percent in 1951, and 66.3 percent in 1953. Fearing that payroll expenses would spiral completely out of control, the company finally won a two-year contract in 1951 that locked in wages in exchange for a basic agreement on cost-of-living increases. This, Quill crowed, is "our greatest victory since the inception of the TWU with the PTC."[51]

To PTC managers, Quill's "victory" was only a stopgap. Such raises were unsustainable, they believed, and they tried different strategies to defeat the union and get the public on the company's side. Attacks on Quill and his out-of-town union were of course a standard approach. Union leaders, Ebert argued in the press, had "no sound argument in justification of their position [on pay raises] other than the flat statement that 'we want it.'" "Directed by the Union's New York leader, Michael Quill," he continued, the TWU had a "strangle-hold on Philadelphia." The PTC was always ready to negotiate in good faith, to protect "the people of Philadelphia—workers and shoppers, merchants and industries," from the "extreme hardship of a strike" that Quill and his "ruthless, selfish . . . Union leaders" would bring. Ebert also tried to repurpose the TWU's "living wage" argument in service of the PTC. The company acknowledged that decent wages contributed to the purchasing power of working-class communities and thus enhanced Philadelphia's economic vitality, but, he said, the workers had to understand that the PTC "must have the right to collect 'living fares' if we are to give adequate service and keep abreast of developing Philadelphia." The company wanted to "serve the people and business interests of Philadelphia," but the TWU's exorbitant demands left the PTC on "starvation rations . . . like an underfed hired man [who is] unable to adequately perform needed services."[52]

Those strategies were a start, but what worked best, what brought much of the public to the company's side, was the argument that transit workers' increased wages compelled the company to seek higher fares. Philadelphians and the PTC were trapped together in a wage-fare spiral that the PTC did not want, but that the TWU forced upon them. The company, not the union, was on the side of ordinary people, management argued. Every contract negotiation led to Ebert making some variation of the statement "We simply don't have the money." The TWU's wage demands, and to a lesser extent inflation-inspired higher material costs, combined with declining ridership meant the PTC, as it publicly argued, had no choice but to seek fare increases. There was no way, the transit company told the public in 1945, that a business could meet "the increasing pressure of higher wage costs against a fixed sales price for service rendered, not changed since 1924." "Rate relief" became the company's mantra, and officials constantly appealed to the public to understand its plight. "No one blames a worker for attempting to secure what his services are worth," wrote the PTC in one pamphlet. But the public had to understand the implications of the TWU's demands: "Another wage increase would mean another *fare* increase [emphasis in original]." "The real victim of [the TWU's] destructive tactics," the company continued in another flyer, is "the public. It is time that common sense prevailed and steps be taken to permit PTC, like any other business, to set its prices at a level that will meet its higher costs. . . . We want our riders and business men and civic leaders to understand all the factors in this critical situation. This is not just our problem as a transit company. If you use PTC, if you run a business that depends on PTC for customers and employes, if you own property, it is your problem too." Everyone, everyone except the TWU, was in the same boat, according to the company.[53]

There was some logic to this argument, and Quill understood that it had the potential to score his union deeply. He and Local 234 leaders argued to the public that they should not fall for the PTC's case. The company, they said, all but forced the workers to go on strike, which allowed the PTC to then request fare increases. The company "fostered hardships on the riding public," the TWU claimed, because the PTC was frustrated by contract negotiations with a real union and wanted to "take it out on their employes and the citizens of Philadelphia." "The public," Quill claimed, should place the blame for strikes on "the PTC, not the TWU." The company had enough money, the TWU argued, it just chose to spend it on items other than paying its workers. PTC employees thus "refused any longer to subsidize the transit system." It was bad enough that Ebert and other officials received substantial pay raises; more galling were the stock dividends. The PTC blamed fare increases on labor costs, then fought tooth and nail over pennies for workers while it paid out millions to bond and stock holders. Between 1940 when the company reorganized and 1952, the PTC paid out some $60 million, which amounted to a 5 percent return on preferred stock and an 8 percent return on common stock at par value. When city auditors used market as opposed to par values, the return skyrocketed to 12.9 and 13.8 percent respectively. Sure, there was a wealth transfer taking place, but it was from riders to stockholders, not riders to workers.[54]

Clark, Dilworth, and other elected officials, who saw the protection of the interests of ordinary Philadelphians as their primary interest, increasingly opposed the PTC. They did not necessarily always support transit workers' demands for higher wages—the frequent strikes or threats of strikes were in fact quite an irritant—but they saw through the PTC's arguments, understood the way the company spent new revenue generated by higher fares, grasped that management was adept at "making a profit out of a labor crisis," as Clark once put it. In what became a running battle between municipal officials and the PTC, city attorneys executed a number of lawsuits in court and counterarguments before the Pennsylvania Public Utility Commission to try to prevent fare increases. Established by Pennsylvania law in 1937, the Utility Commission had the authority to set fares at the PTC based on two factors: the valuation of the company and a 1944 state Superior Court ruling that the PTC could not earn more than 6.5 percent on its valuation. The value of the company was thus a vital question, and when city officials claimed it was "only" worth $50 million, they were not just debating the purchase price but were also arguing to state authorities how much money the PTC could legally pay out to stock and bond holders. Assuming it had enough income, a company worth $100 million could pay out $6.5 million, but one worth $50 million could pay only half that. If the total payout of the legally allowable dividends were lower, then the PTC did not need a fare increase after all.[55]

The city essentially made two arguments against the PTC. The first was that the PTC's primary obligation was to provide "adequate service, dependable service, and safe service at the lowest rate of fare which will meet the costs of that service." That rate was of course in dispute, but higher fares certainly did not help the company meet its obligation. In fact, they reduced the number of riders millions at a time. The second was that the PTC's finances were not so dire as management claimed. With the correct valuation, Clark argued, "PTC can operate safe and adequate transportation facilities and continue to earn a fair return on its securities at the present rates of fare." Even if it could not, "inability of a public utility company to meet wage demands without operating at a loss is not a legal defense to such wage demands," especially when the company was already receiving a subsidy of $3 million per year from the taxpayers to pay for improvements on the Broad Street Subway and Frankford elevated. The PTC countered that Clark and other officials were "short-sighted, unjust and unwise," engaging in "politically-motivated and destructive" attacks that would damage "the public interest [and] wreck this Company and its ability to provide transportation service." Clearly, the city was using the rate cases as groundwork for taking control of the PTC. That, the company argued, would be an unmitigated disaster for Philadelphia: "All the deficits and problems of transit municipal ownership and operation that have burdened taxpayers in New York City and elsewhere will have to be faced here in Philadelphia." The public had to understand the political subtext, understand the need for a "fair rate," and back a company that would "continue the fight for the right, not denied to other businesses, to combine good service for the public with a reasonable return for investors and fair treatment of employes." Despite the city's protests, the PTC tended to be more successful in its arguments before the Utility Commission. Fares varied for rapid transit, trolley, and bus service, and patrons could purchase multiple fares at a discount, but overall the trend was obvious: fares stood at eight cents from 1924 to 1946, moved to ten cents in 1947, then twelve cents in 1950, fifteen cents in 1951, eighteen cents in 1954, and twenty cents in 1955. In ten years, a PTC patron paid two and a half times as much for the same ride, and often on inferior buses or equipment that was a decade older.[56]

Liberal city officials were not always successful in opposing the PTC, but their efforts did make it apparent that overall they supported the transit workers in their running conflict. This set them apart from most people in the Philadelphia area: the city's newspaper editors, its judicial officials, and most importantly, and pervasively, its residents. By the early 1950s, Clark often mediated contract talks and made it clear that he found the TWU's arguments more compelling. Strikes, such as the one in 1953, angered him, but he understood that the PTC's intransigence forced the TWU to take drastic measures to improve the lot of "the lowest paid of any [transit] employes of any great city in the United States north of the Mason-Dixon line." Clark even vowed to reverse the city's stance and support fare increases, but only if the money went to workers' wages. Transit workers deserved this, the raise and the support, because they were valuable residents of the city, "our neighbors and fellow-citizens [who] have always been respected as some of the most responsible people in our community." If the public wanted to find a villain in the negotiations, it should look to the company with its "cheap press agentry . . . which has done so much to make normal, mature labor relations between the company and union impossible." The way forward lay in the company paying higher wages, ignoring the clash of personalities, and finally putting to rest its "old fashioned labor policy." Clark's forthright stand on behalf of Philadelphia's transit workers led Quill to thank him in a letter for his "invaluable help during the negotiations." Clark's reply praised Quill's "courageous leadership." More and more the TWU urged its members to vote the Democratic ticket, to know which politicians were on their side. In the fire of the strikes and turmoil at the transportation company, an alliance, sometimes uneasy, between the TWU and the city's liberal political leadership was born. It was an alliance, between politicians and transport workers, unlike anything in the city's history.[57]

Philadelphia's major newspapers, the *Evening Bulletin*, the *Inquirer*, and the *Daily News*, all with generally centrist or right-leaning politics, were more ambivalent about the transportation workers' cause. Contract negotiations and especially transit strikes were of course front page news, but the newspapers had a difficult time teasing out the "truth," apportioning fault for the incessant conflicts. Editors felt much like "the average citizen [who] doesn't have the remotest idea who is right and who is wrong—nor is there any completely objective mediator to whom he can turn." Ambivalence, however, did not equal objectivity. The papers repeatedly trumpeted the PTC's case, telling Philadelphians that there was an "almost-inevitable corollary" between wage increases and fare increases. The newspapers failed to give equal coverage to the TWU's argument about how revenues generated by fare increases went to dividends for stockholders. The papers' limited analysis mattered because it led them to argue that the TWU failed to take the interest of "the people" into consideration. Strikes, from their point of view, could not be in the public interest and almost by definition harmed the union's cause. The union, in this line of argument, was merely a special interest, not a representative of working people. The TWU held the PTC hostage, "forcing" it to implement fare increases that harmed the public, who ultimately had to pay the bill. "A great number of the workers might be more temperate in their demands if they knew those demands were going to increase the expenses of 2,000,000 other workers and their families," the *Daily News* wrote in one of the most direct statements of the argument. The newspapers in some ways tried to appear unbiased, but in reality helped the PTC make its case.[58]

Philadelphia-area judges and more conservative state legislators drew a harder line against the TWU. Judge Joseph Kun led this charge, warning Quill and his attorneys in 1952 that they should rein in their demands or be held in "anticipatory contempt." A bewildered reporter commented that he could find no such phrase in the state's law books and asked Kun to clarify "for the benefit of lawyers and others who might think he pulled something out of a hat." "No one is permitted to take the law into his own hands. There is such a thing as anticipatory contempt," Kun replied. It may be a new concept, he added, but it was real and necessary for "a much more serious offense than other types of contempt." "Similar things are going on behind the Iron Curtain," a chagrined O'Rourke observed.[59]

The idea of "anticipatory contempt" had too many obvious legal problems to hold—the Supreme Court had ruled in *United States v. Bryan* in 1950 that there was no such doctrine in American jurisprudence—but Kun's call for a strengthened public utility law that would ban transit strikes found greater support. Such laws gained passage in many states in the postwar period as the number of transit strikes surged. The Amalgamated despised these laws and led court challenges in 1951 that ultimately got many of them overturned because they violated the rights workers held under the Taft-Hartley Act. Pennsylvania passed one of these laws in 1947, its "Public Employe Anti-Strike Law," but the state excluded urban transit workers from its coverage. After each major transit strike in this period and into the 1960s, a drumbeat started for strengthening the law. Lawmakers such as State Senator A. Evans Kephart (R-Philadelphia) pushed the legislation, and they found backing from the press and Philadelphia-area constituents. "Perhaps the best preventive lies in State legislation outlawing transit strikes in Pennsylvania and requiring compulsory arbitration of contract disputes," wrote the *Inquirer* in the midst of the 1953 strike. "Street transit is a public utility, and no public utility should be permitted to be the instrument of injury to the public by means of unjustified strikes." The city could not be "cruelly victimized" by the TWU anymore. Letters to political officials similarly urged that "legislation should be enacted outlawing all strikes on public utilities of every nature: Gas, electric, water and transportation." The TWU and its political allies, bolstered by the questionable constitutionality of such laws, fought off the legislative efforts, but they nonetheless showed the difficult legal and political circumstances generated by judges and legislators opposed to the transit union or at least to the turmoil they believed it caused in Philadelphia.[60]

The perpetual conflict swirling about the PTC, the way fare increases were pitched as the fault of overbearing unionists, the charges that union leaders were outsiders and maybe even Communists (the last being a self-inflicted wound by the TWU as much as anything), all combined to turn much of the public against Philadelphia's transit workers and give them a vocabulary to express their discontent. Not everyone, of course, bought the company's line. Letters to the mayor on occasion charged that the PTC had a "public be damned policy," that its service and equipment showed that it "has not given any consideration to the transportation needs of the thousands of new residents along [its] lines." Most of the workers' support came from other unions, mostly in the left wing of organized labor, or the Progressive Party. The United Office and Professional Workers of America, for example, demanded that public officials make the PTC bargain in good faith. Other unions, such as the United Paperworkers and the UAW, wrote that the city's political class had an obligation to recognize the TWU's just demands. Philadelphia's Progressive Party added that Mayor Samuel had a responsibility as a public servant to demand an end to "the squeeze play of the PTC against the union and the public."[61]

Most of the limited support the workers received was more mixed, putting a plague on both houses. "Why is it possible for one small group to cause the expense and inconvenience that the P.T.C. union has caused?" wrote Alice Stratton. But, she continued, "Why is it possible for one company to set itself above the combined needs of the citizens of Philadelphia?" Any transit strike, James Waide told Mayor Clark, is a "damnable outrage," but one could not really understand the difficulty of a transit worker's position "unless you have worked on that job." "I wouldn't want it!" he concluded. The PTC did not deserve a fare increase, wrote "Mr. Happy Devotion," when streetcar crews were "very impolite" to the public and at the same time the company refused to use its resources to keep the cars in decent shape. "Why should we, the people of this Democracy, sit silently still and not give vent to our feelings on such matters?" he added.[62]

Many Philadelphians, at least those who chose to express their views to politicians, the press, the PTC, and the TWU, came down strongly against the workers. They did not necessarily support the company, but they clearly opposed the employees, often with vitriol. The distress caused by work stoppages and higher fares sparked their reaction, which developed along three major lines: that unions had too much power, that the TWU and especially Mike Quill were outsiders, and that no one took the public's interest into account.

Many Philadelphians argued that transit workers thought too highly of themselves and held their city hostage because their union, like many unions, had too much power. "It takes no particular education to drive a trolley car or make change, just a week or two to get familiar with the routine," "Car Rider" wrote to the *Evening Bulletin*. They were not worth a raise or the trouble they caused, but the TWU and the rest of organized labor had so much power that they threatened to "strangle this country." "Labor," wrote another correspondent to Governor James Duff, "has industry by the throat and sits in the saddle." "Just who is running the city," another writer asked Bernard Samuel, "you the mayor or the unions?" Sam Mathews captured the sentiment in (not very good) poetry that he sent to the TWU:[63]

> Rioting at Plants and Mills
> Is the cause of most of our Ills,
> IN AMERICA.
> Picketing is one disgrace
> Union Thugs must soon erase,
> OR ELSE
>
> . . . .
>
> Rip the STRIKE THUGS FROM THEIR THRONES,
> BAG AND BAGGAGE, SKIN AND BONES,
> WE WANT PEACE
> OUR DEAR COUNTRY WE MUST SAVE,
> FROM THE UNION THUG AND KNAVE.[64]

The TWU's strength, many writers asserted, posed particular danger to Philadelphia because of Quill. Over and over again, Philadelphians called him an "outsider," a "New Yorker," and an "Irishman." "Are we going to let an outsider, this man Quill come from New York and start another P.T. Co. strike," wrote J. H. Rummelman. "It seems a shame," wrote "Phila. Voter and Taxpayer," "that one man can come from another city and just because he is head of a racket like the C.I.O. union put so many people to such a disadvantage." Quill, others wrote, reflecting their city's difficult history with Irish immigrants, was nothing more than an "Irish windbag" who should "take a slow boat to Ireland." No one should "fall for the brogue or the shillelagh, he is a stinker and a common low down mobster."[65]

Quill's outsider status, in Philadelphia and more ominously the United States, plus his labor politics made it easy for many writers to make the leap to his being a Communist. "Red labor" was taking over in the city, wrote one correspondent to Governor Duff. If no one fought Quill and the TWU, the union would continue to "breed more and more Communists." "To hell with all Communists, and to hell with Quill," wrote another man. He was "satanic," a "son of a bitch," in need of being "dropped in the ocean." The city did not need to negotiate with such a man, according to a number of Philadelphians who forgot the mayhem of 1910 if not 1944; instead it needed to call in the state police, the militia, the national guard. As the rhetoric ratcheted up, violence frequently seemed in the offing. A man styling himself "One Shot Charlie" wrote to Quill, "Mike Quillinski or whatever your damn old red name is you bald headed bastard . . . unless this strike is called off by Saturday nite . . . a 45 slug will enter that thick bald-headed skull of yours. . . . I'm not foolin' you son of a bitch your life ain't worth a dime when I catch up with you." Two PTC strikers who were beaten up in the 1953 strike (and saved by none other than police captain, and future law-and-order mayor, Frank Rizzo) may have gotten off lucky.[66]

Given this perception, Philadelphians could not understand how the PTC and government officials could not stand up to the TWU. Who, they repeatedly asked, was looking out for the people, for the public? The TWU did not go on strike against the company, from this perspective. No, these were strikes "against the public," and it was time, as one man put it, that "the general public is given some consideration." The city had to get its priorities straight, wrote "A Voter and Taxpayer." Why do "we the citizens and taxpayers have to put up with the P.T.C. workers union and their leader?" Many correspondents, like this writer, signed their letters "Taxpayer," "Mr. Citizen," and similar. Some even recalled the recent hate strike of 1944: "Several years ago the employees of PTC went on strike in protest against the company's employment of a few Negroes. The federal government immediately stepped in, threatening to use the armed forces of the Nation if the strike were not recalled. Is the welfare of three-million persons less important than that of a few Negroes?"[67]

Such arguments suggested that many Philadelphians, although by no means all, had by the mid-1950s come to see transit workers as a separate class. They were no longer part of the people, members of the community. They were instead outsiders, or at least led by outsiders, such as Quill, who were tainted with communism. The city's transit union, for the first time since the Amalgamated came to town over a half century earlier, had become at least to some Philadelphians too powerful, dangerous to the functioning of an orderly city. Quill's efforts certainly paid off in securing higher wages for his rank and file, but the PTC's campaign proved more powerful in the long term as it played a part in setting in motion a politics that tarred public service workers and their unions as a special interest that stabbed at the vitals of the city. This politics continued in the late 1950s and 1960s, but was tempered by a different context as National City Lines took over the PTC and proved an even more difficult employer for Philadelphia's transit workers. By the late 1960s, the company would be largely gutted, the union constrained, and the PTC, what was left of it anyway, turned over to the city to operate as best it could. The decade after World War II was hard, the ensuing fifteen years far harder.

NATIONAL CITY LINES AND THE IMPERATIVES OF POSTWAR CAPITALISM

The *Evening Bulletin* warned its readers what they should expect when Douglas Pratt of National City Lines took over at the PTC in 1955. "Douglas M. Pratt, who was elected new president of PTC today, takes a dim view of private automobiles," the reporter wrote, "and thinks privately operated transit systems should make money—or else. The 'or else' in his three-year performance as head of the Baltimore Transit Co. brought increased fares, curtailed service and reductions in force. Baltimore was an unhappy city transit-wise when the 43-year-old Pratt moved in three years ago. And there were a lot of unhappy stockholders in the transit company, too. They hadn't gotten a dividend in 25 years. Baltimore is still an unhappy city transit-wise—perhaps even unhappier—but in 1953, a year after Pratt took over, preferred stockholders received a $2.50 dividend and common stockholders got 50 cents. In 1954 the preferred got another $2.50 and the common was boosted to 65 cents." Clearly, dividends mattered greatly to NCL, and Pratt knew the two methods for generating the necessary income: "Pratt took over Baltimore Transit on February 15, 1952. By April 8, he was knocking on the Public Service Commission's door with proposals for a 20 cent fare—a five cent boost—discontinuing four lines, shortening six and reduction of trips on seven others. He fired 310 maintenance workers." "If Mr. Pratt goes to Philadelphia," one Baltimore associate told the reporter, "you people are really in for a jolt."[1]

Pratt said all the right things when he first came to Philadelphia, that he was "looking forward to working with the PTC family," that he understood "the public interest inherent in an essential community enterprise," that his company was "inextricably linked with Philadelphia's general progress and prosperity." But no one, given NCL's track record just one hundred miles away in Baltimore and in many other cities across the United States, could have been surprised by the policies the new owners of PTC would implement. Mayor Clark certainly was not. He had said a year earlier that he would "welcome any strong group which had a basic understanding of successful transit operation," but when Pratt was introduced to the Philadelphia media, Clark had no substantive response, only that he and the four other city representatives on the PTC's board were "uncommitted."

That proved to be the high point of NCL's relationship with the city and, for that matter, with its transit workers.[2]

National City Lines gained a measure of fame, or infamy, in the early 1970s with the publication of Bradford Snell's *American Ground Transport*. According to Snell, NCL was a small-time company that became a front for GM, Standard Oil, and Firestone. Those three corporations used NCL to purchase controlling interest in dozens of urban transit systems, demolish serviceable if older electric trolley lines, and replace them with diesel-powered buses that ran on rubber tires. There was a kernel of truth to Snell's theory, enhanced by a criminal conspiracy conviction in 1949 (the court fined the corporations five thousand dollars apiece and individual defendants one dollar) and GM's control of 80 percent of the motor bus market in the 1950s, such that it gained credence and was picked up by popular periodicals, *60 Minutes*, and academic and more popular books such as Stephen Goddard's *Getting There*. Scholars largely laid the Snell thesis to rest after that, highlighting the flaws in the conspiracy theory, pointing out how public transportation was already on the decline by the time NCL arrived, and arguing that the culture of the automobile had a far greater impact than any conspiracy GM could concoct.[3]

National City Lines played a large role in the final chapter of PTC's life as a private enterprise, but not for the conspiracy theory that Snell propounded. Rather than taking up a mostly debunked conspiracy theory, this chapter shifts the focus to the impact NCL's policies had on the transit company, its equipment, its workers, and the city it served. Ultimately, NCL's career in Philadelphia highlights what corporate leaders such as Douglas Pratt saw as the imperatives of postwar capitalism. Their job, he believed and his actions showed, was first and foremost to earn as large an income as possible for his company and a corresponding dividend for his stockholders, not to provide a vital service at a price that secured a reasonable profit while meeting the social and economic needs of Philadelphia and its residents. When NCL took control of the PTC, the urban transit business was in obvious decline. Not only did National City Lines never figure out how to turn the business around and make it profitable while serving the needs of Philadelphians; NCL management in fact never tried to do so. Instead, the company sought to increase profits by cutting expenses, particularly on payroll and maintenance. The company thus came off as an interloper whose ultimate goal was not providing a public service so much as maximizing dividends by whatever means possible. Looking back on the last years of the PTC, SEPTA board chairman David Girard-DiCarlo captured the spirit of NCL's efforts, telling the Pennsylvania state legislature that Pratt and his staff were bent on "sucking the lifeblood out of the system." PTC management under Charles Ebert had undoubtedly committed itself to private enterprise, drove a hard bargain in contract negotiations, and employed ad hominem attacks on Mike Quill and others, but the company always seemed committed at some level, even in the eyes of its critics, to meeting its social obligations to the city. Once the PTC came under NCL control, however, few thought that was the case.[4]

NCL's policies sharpened class conflict well beyond what had taken place in the previous decade, drastically reduced the quality and quantity of the city's transit service, and finally forced Philadelphia to buy out the PTC and put transportation under the power of a regional authority. The city's transit workers combated NCL's policies by staging two authorized strikes during the National City Lines years (1961 and 1963) and so many unauthorized wildcat strikes that journalists had trouble keeping count. One student of Philadelphia transit recorded eleven such stoppages just from 1955 to 1957, leading him to call the era "the Wildcat Period." Clark, Dilworth, and the city's other liberal political leaders, who had already shown their commitments to the workers as opposed to management over the previous decade, knew that the labor strife and constant strikes that paralyzed the city could not continue. NCL, they all believed, was to blame. In response, they worked with other Pennsylvania officials to establish SEPTA, but by that point NCL had stripped what value it could out of the company. The problems SEPTA faced in the 1970s and beyond had much more to do with NCL's exploitative policies and the dwindling resources that bedeviled urban America than bad management or the ineptitude of public servants.[5]

National City Lines Comes to Philadelphia

For all the controversy it provoked, National City Lines came from humble origins. The Fitzgerald family owned, managed, or worked for small bus services in the upper Midwest, especially Minnesota and the Chicago area, in the 1920s. As they saw transportation companies consolidate around them, they formed the Rex Finance Corporation in 1928 to serve as a financing and holding company to purchase local transit concerns. By the early 1930s, Rex developed a profitable business model that worked even during the Depression (although it was cash-starved in the early years) and lasted for forty years: buy distressed local transit companies, frequently electric-powered, convert them to bus systems as possible, and operate them with an eye on dividends and eventual sale. Oshkosh (Wis.) came first in 1933, quickly followed by Galesburg, Joliet, and East Saint Louis (Ill.) and Kalamazoo (Mich.). The Depression created many opportunities for Rex, but the Fitzgeralds needed more capital. In 1936 they established National City Lines and sold thirty thousand shares of stock at $12.50 apiece. A public stock offering followed that brought in another $2 million. Still, NCL needed more money, and in 1939 the Fitzgeralds entered into an agreement with Firestone for $250,000. GM followed four years later with a $1.25 million stock purchase when it bought the Yellow Coach Manufacturing Company, with which NCL had a contractual relationship (GM sold its NCL stock in 1949). Over the next thirty years NCL acquired fifty-two transportation companies as subsidiaries, mostly in small or mid-size cities such as Jackson (Miss.), Port Arthur (Tex.), Terre Haute (Ind.), and Tulsa. NCL also obtained controlling interest in larger cities, including Philadelphia, Baltimore, Los Angeles, and Saint Louis. In its heyday NCL's transportation empire was concentrated in the Midwest but stretched from Pennsylvania to California, Michigan to Florida. The company also diversified into trucking and park-and-ride lots, often selling transit systems to bolster these ventures. By the early 1970s, NCL had all but divested itself of its transit interests, and in the early 1980s a Texas-based holding company, Contran, bought National City Lines and liquidated its assets.[6]

When NCL set its sights on the PTC, then, it was employing a well-tested business model, albeit at a much larger system than usual. NCL started buying PTC stock in 1954 and acquired 12.7 percent of the company at a value of $2 million by early 1955. This interest, roughly half of what NCL had in the Baltimore Transit Company, was sufficient for the company to take voting control of the PTC that year. In March 1955, NCL shook up the leadership of the PTC, ousting President Ebert, who had worked in Philadelphia transit for thirty-five years, removing six out of sixteen members of the board of directors (the city had five additional seats), and firing fifty-one supervisors. At the same time, NCL installed Douglas Pratt in Ebert's position and placed six loyalists—all worked for or served on the board of directors of NCL or its subsidiaries—in the vacated spots on the PTC's board of directors. Pratt, age forty-three when he took over in Philadelphia, had spent fifteen years with NCL, rising quickly through the ranks. After graduating from the University of Nebraska, he entered the transit business as a chief clerk in Omaha and then moved to Pacific City Lines as a general manager in 1941, becoming vice president and a board member for that company in 1943. He joined NCL when it bought Pacific City Lines in 1946 and asked the company to make him head of Baltimore Transit just six years later, which it did. By the time he arrived in Philadelphia, the PTC's new president, despite his youth, was known as a driven man and imperious leader who expected his associates to call him "Mr. Pratt."[7]

Everyone knew Pratt would bring a new ethos to the PTC. Mayor Clark may have been uncommitted on the subject of NCL, and Ebert, who offered a curt "No comment" to the press about Pratt, was of course chagrined by the change, but most commentators hoped for the best. Albert Greenfield voiced his approval of NCL's takeover, no doubt in part because he retained his chairmanship of the

PTC's executive committee, but also because of the company's reputation for doing anything necessary to generate profits. That included taking on organized labor and refusing to "buckle under to union threats." The *Evening Bulletin*, more circumspect about the PTC's labor issues than Greenfield in this case, well captured a common sentiment: "Now is the time for the working of a miracle. Investors in transit securities as well as riders on transit lines would welcome one."[8]

In less than two months, Pratt gave city officials and the press reason for pause. Pratt, in an offhand conversation with Mayor Clark that signaled his willingness to try to expand management's power, confirmed that NCL planned to remove the city's sole representative on the executive committee of PTC's board of directors. Pratt wanted to reduce the committee's size from eight to five and believed that the PTC, under NCL's guidance, should be able to set policy without city interference. Clark rejected the idea. Philadelphia had only five representatives on the twenty-one-member board, and the larger body, he said, "is too often either a rubber stamp for the executive committee, or is presented with an accomplished fact after it is too late to do anything about it." City council members expressed their anger at the proposed move. Victor Moore told the *Evening Bulletin*, "It is rather unfortunate that financial control of PTC is going into the control of outside interests [who] are out solely for their own benefit." Louis Schwartz went further, calling NCL "a group of industrial carpetbaggers" who wanted to turn "a public service system into a source for private gain." "When an outside group of promoters with a minimum investment can endanger the entire economy of the city through sheer selfishness then it is time to remember how stupid it is to have the tail wag the dog," he added. Perhaps, in light of the critical nature of public transit to the city's future, it was time to establish a transit authority "to take over the system and to protect the city's $200,000,000 investment in it." The *Inquirer* told its readers that Pratt had at best gotten off on the wrong foot, and at worst he regarded the public and their elected officials as "snoopers" and was staking a claim to NCL being allowed to operate in "darkest secrecy." NCL looked like an outside company with little interest in a true partnership with the city. "Pratt of National City Lines," the paper concluded, "may be a first-class transit man. Pratt of PTC has a lot to learn in public relations." Facing this criticism, Pratt realized he had overplayed his hand and backed down, leaving the city with one member on the executive board. But by this point to many city officials it was too late. The attempted removal of the city's representative, Councilman Moore said, "was ruthless and signified relations hereafter would be hostile." Moore was right. Over the course of NCL's tenure in Philadelphia, Pratt often complained of the city's adversarial stance against the PTC, especially when the city sued his company for implementing policies that it claimed demonstrated National City Lines "acted in bad faith" and caused "irreparable harm and damage" to Philadelphia's economic interests. As the city-NCL antagonism grew more apparent, the TWU and city officials grew closer together. They certainly did not share all the same interests, but to both of them NCL presented the real problem.[9]

Throughout NCL's tenure, from 1955 to 1965, the PTC lost riders and seemed, at least in the company's estimation, to move from one crisis to the next. Whether the crises were self-inflicted or even real became more and more of a question over the years of NCL's ownership. Ridership numbers, because of issues that predated NCL such as automobile usage, rising fares, and strikes, plus new fears of crime, continued their postwar slide. But revenue and profits, as city officials and TWU leaders pointed out, "curiously" held steady for the duration. Fare increases and reduced labor costs kept the bottom line firm, helping NCL to remain a profitable company in the 1950s and 1960s, PTC to periodically turn a profit itself, and the local transit company to pay dividends to its stockholders with regularity. This led NCL's critics increasingly to wonder how a company charged with providing a public service could continue generating the same revenue while hauling fewer than half as many passengers a decade after it took over the transit system. NCL did not help its case when it repeatedly complained about its "constant struggle to strengthen [its] financial stability," lamented the reduced fares for schoolchildren that put a "burden" on the company, and argued that it should no longer have to meet its contractual tax and snow-removal obligations.[10]

With ridership numbers declining, NCL pursued three strategies for maintaining its income. It increased revenue by obtaining higher fares. It controlled payroll costs by laying off workers. And it converted most of the remaining trolley lines to buses, reduced service, and deferred maintenance. Each strategy was not so much a new effort as an intensification of what PTC management had done in the past, and all brought NCL into conflict with the city's political leaders and the TWU. Most Philadelphians, who had sided with the PTC in the first decade after World War II, still found the TWU problematic, but NCL's draconian policies and outsider status brought the company greater scorn.

TABLE 2. Philadelphia Transportation Company ridership and revenues under National City Lines

YEAR	NUMBER OF RIDERS (IN MILLIONS)	GROSS REVENUE (IN MILLIONS)
1956	596	$69
1958	343	$64
1961	316	$66
1964	289	$67

Source: Philadelphia Transportation Company annual reports.

Transit fares provoked searching debates about what NCL "owed" the people of Philadelphia and how the company should use its money. The company, Pratt often argued, wanted "to hold fares at the lowest possible level," but it could not afford to do so when the TWU called for "exorbitant" pay raises and the city government backed their demands while opposing fare increases. Clark and Dilworth did not oppose fare increases in principle, but supported them only if the revenue went to purchase new equipment or provide higher pay for the PTC's workers. Instead, the city saw NCL pay dividends to PTC shareholders while claiming it had to raise the fare for schoolchildren or at least have the Philadelphia school system push its start time back to 10 a.m. so the company did not "lose money" on students' rides. NCL publicly claimed that fare increases had no direct tie to dividends, but to city officials the link was clear. Either Pratt and other officials demanded fare hikes to gain leverage over the TWU and the city, or, as city council president Paul D'Ortona put it, the "scheming," "greedy" company had a "public-be-damned" attitude. Philadelphia could no longer tolerate the "financial chicanery" and "political skullduggery" of NCL, Dilworth told the press. NCL had to recognize the city's $200 million investment in its transit system [primarily the Broad Street Subway and the Frankford elevated that it had paid to construct] and the PTC's primary responsibility to serve the common good.[11]

In claiming that higher wages necessitated higher fares, the PTC made a reasonable-sounding claim but one that did not stand up to scrutiny. Under Quill's leadership, the TWU negotiated wage increases that took PTC workers' pay from $1.95 per hour in 1955 to $2.13 in 1957, $2.23 in 1960, $2.85 in 1965, and $3.26 in 1968. The base fare rose commensurately, starting at twenty cents when NCL first took over, rising to twenty-two cents in 1960, and then hitting twenty-five cents in 1965, where it remained until SEPTA took control of the system in 1968. Those rising numbers, however, were matched by reductions in the workforce that took employment levels from 8,100 in 1956 to 7,000 in 1959, 6,500 in 1962, and 5,800 in 1967. The reductions in the labor force meant that PTC's payroll expenditures remained roughly between 57 and 60 percent of revenue despite the pay raises. Higher fares did not generate more money for the workforce as a whole; the money instead went into stock and bond payouts and compensation for management, with Pratt earning $47,500 per year.[12]

NCL's policy of slashing its workforce came as no surprise to the TWU. Quill intimated to the press as soon as NCL came to Philadelphia that he knew his union would have its first fight with the company over layoffs. In Baltimore, Pratt had slashed the workforce by 38 percent, from thirty-seven hundred to twenty-three hundred employees, in the last three years. The PTC over that same period had cut its workforce by only 8 percent. "They're even worse than the present management," Local 234 President Paul O'Rourke said. "In other cities where they've gone in, they have curtailed service, laid off enough help to pay for any wage increases, have raised fares, and have declared tremendous dividends to stockholders." NCL made a habit of talking about "surplus employees" at the PTC, and wanted, in the words of one union official, to use one-man operation and layoffs to "break the union" and thus pay its dividends "out of the blood, sweat and tears of the employees." NCL, Local 234's Charles Robinson argued, was taking labor relations at the PTC back to the old days of "Jungle Law." When the TWU did approach Pratt to discuss layoffs, he cut meetings short (one lasted eighteen minutes) and announced, "You will never get an agreement with us. You will never get anything out of me." Maybe not "Jungle Law," but Pratt certainly called to mind Charles Kruger's transit company from decades earlier.[13]

NCL policies and Pratt's intransigence left the TWU in a fighting mood. The company, wrote Joe Marks, a Local 234 section chairman, was "completely circumventing [the] protective clauses . . . of our contract." "We are prepared to fight for our jobs," he concluded. "Local 234," wrote another PTC employee in a telegram to Quill, "is now down to 8,500 members and when National City Lines is finished it will number 5,000. Action is needed now to prevent this hardship to the working man." TWU leadership assured its members that it had "taken the position they must fight National City Lines on all fronts, [including] pushing for City ownership of the Philadelphia Transportation Company which would give the schedules and working conditions to which our members are entitled." They vowed to use television, radio, newspapers, and rallies to take their case to the public in an effort to "protect our interests and fight the monied groups." Layoffs, not wages, were the source of most NCL-TWU tensions, including two authorized and myriad wildcat strikes. NCL, the TWU asserted, was treating its Philadelphia system as badly as Baltimore's, "where the needs of the public and employes were completely ignored in management's ruthless efforts to serve only stockholders." Whatever Pratt thought NCL's mission was, he had to learn that PTC "cannot operate shops, streetcars, buses and subways with stock certificates."[14]

As the TWU found greater success in securing wage increases than staving off layoffs, it did its best to bring the public to its side, arguing that pay raises and fare increases did not have to be, in fact were not, linked. Quill pushed to have negotiating sessions conducted in public so Philadelphians could hear both sides, and told Pratt that the PTC's so-called "inadequate fare," its "financial plight," was "not the business of the union." "TWU," he wrote, "will steadfastly refuse to enter into any agreement that would in any manner tie our wages to your fare." Rank-and-file workers agreed with Quill on this score. Few PTC negotiating tactics, wrote an *Evening Bulletin* reporter, engendered more "bitter resentment" among the workforce than claims that the workers were forcing fare increases on the public. "The question of a fare increase is a matter to be settled by the Public Utility Commission and City Council's transportation and utilities committee, and we refuse to become the whipping boy in this situation," Local 234 said in one statement to the press.[15]

The TWU's case made some headway with the public. The *Evening Bulletin* had captured a sentiment common in letters to the editor and public officials when it argued that fares should be at a level that generated a profit for the transit system. "Users of the street transportation service [should] pay for what they get," wrote the editors, rather than "carrying passengers at less than cost and letting the taxpayers meet resulting deficits." But now pressures for higher fares led Philadelphians to hold the PTC and its new ownership equally to blame rather than lambasting the "greedy" unionized workers. The transportation system had gone downhill "since the new P.T.C. Co. has taken over," wrote one man, "and I certainly pity the older people in this city that have to get around." Philadelphians, "the paying and squeezed public," were "being caught in a trap" between NCL and the TWU. "We the people," wrote Philadelphian Rose Levin to Governor William Scranton, "are beginning to believe that 'government for the people and by the people' means just that—certainly not government for only a chosen few with the right connections." Higher transportation fares and other levies were ruining Philadelphia, she argued, destroying "the backbone of any nation, the great middle class." "Having a fare increase foisted upon us," Levin concluded, helped the PTC and the TWU, but was tantamount to "a conspiracy to crush all but the very rich and very poor." By the late 1950s and early 1960s the vitriol and the communist charges against the TWU had faded. Philadelphians certainly did not absolve the union, but the company received a greater amount of the blame for fare increases and labor turmoil on the transit system. The fact that the company cleared an average of $3.6 million per year before taxes from 1958 to 1965 did not help NCL's public perception.[16]

The continued conversion to buses and cuts to service and maintenance also highlighted NCL's commitment to profits at what most observers felt was the city's expense. In the conversion to buses, NCL played a significant role nationally but was by no means alone. Across the United States the total rail miles of public transportation systems continued their drop from 6,251 in 1955 to 3,197 in 1960 to 2,224 in 1965 to the nadir of 1,860 miles in 1970 (and most of those miles were subways and elevateds). By the late 1960s, only a handful of cities, including Philadelphia, San Francisco, Boston, and New Orleans, still operated electric trolleys. A half decade later streetcars carried 0.9 percent of the passengers they had at their peak.[17]

In Philadelphia, NCL mirrored and furthered these trends. From the beginning, Philadelphians knew NCL was a bus, not a trolley, company. NCL's wholly owned subsidiaries, the *Evening Bulletin* told its readers in 1955, operated 1,961 buses and only 20 streetcars. Philadelphia's transit company had started the conversion process under Ebert, taking the streetcar fleet from 1,898 trolleys at the end of World War II to 1,513 in 1954. But NCL sped the process, arguing that buses offered greater flexibility, were more economical, could extend routes to new suburbs, and in the Cold War era could handle "evacuation and rescue work in event of attack." Whether or not anyone actually believed a PTC bus offered the best means of escape from a city aflame was an open question. Nonetheless, in one year NCL cut in half the number of trolleys in Philadelphia (to 741), and that number fell to 508 by 1960. Overall, NCL converted eleven trolley carbarns to bus garages within two years of taking over PTC and turned twenty-nine streetcar lines into bus routes. PTC management had only converted eight in the previous decade. A few streetcar lines survived in West Philadelphia into the twenty-first century (after SEPTA also converted several lines following a carbarn fire in 1975), serving some seventy thousand riders a day after the turn of the century, but they were a vestige of Philadelphia's once robust system.[18]

As NCL's conversion project gained momentum, city officials grew increasingly skeptical of the company's motives. Clark, Dilworth, and other city leaders understood that economics dictated the need to convert some of the more lightly used lines to buses, but the company's wholesale conversion policies rankled them for two reasons. First, replacing streetcars with buses on Market, Chestnut, and Walnut Streets, as well as other Center City avenues, was supposed to help traffic flow, but each bus hauled fewer people than the trolley it replaced, and the company "consolidated" a number of routes in Center City and beyond, which further reduced the service it provided. All told, NCL abandoned thirty-five routes during its tenure; PTC had shut down only eleven in the previous decade. NCL's policies would lead, the TWU's O'Rourke told the press, to "more people for less vehicles and a spreading out in the waiting time between buses." The company's actions actually seemed designed to push more people into their cars and make Center City even more crowded with vehicles. Second, NCL trumpeted how it had spent over $20 million on more than one thousand buses, which allowed it to "modernize the company's surface-vehicle fleet [and] retire and scrap the last of its old-type streetcars." The purchases were certainly necessary to replace some of the older buses and streetcars that PTC had neglected, but city officials wondered why NCL elected to buy the buses from GM at $22,000 per unit, when Mack had bid $1,000 less. More, NCL and the city repeatedly wrangled over who actually owned the buses if the city provided support for the purchase. If PTC owned the equipment, as it claimed, then that raised the company's valuation, making it more expensive for the city to purchase. In floating bonds to assist with the purchase of operating equipment, the city was, in effect, subsidizing the company so it could then charge Philadelphia more when it came time to buy the transit system. NCL and city officials settled on each side owning half of the newly purchased equipment.[19]

As NCL's "bus blitz," as some observers termed it, hit Philadelphia, the company drastically cut back on maintenance. NCL forcibly "retired" Albert Lyons, superintendent of the rolling stock and shops department, after thirty-nine years of service to the PTC, at age fifty-nine, when the normal retirement age, as the *Evening Bulletin* pointedly noted, was sixty-five. In his place, Pratt installed his assistant, Leroy Olson, and gave him the mandate to close carbarns and reduce the department's workforce to the point that NCL would save $1.5 million per year in maintenance expenses. Olson proceeded to fire several senior men in his department, as well as 158 rank-and file-employees. Under NCL's management, the state's Public Utility Commission heard testimony that the PTC did not have enough workers to maintain its passenger facilities, the mayor told the public that PTC policies left buses and subway cars dirty and in disrepair, and the TWU said personnel cuts made it impossible to keep the system in good condition. "The point the TWU is trying to make," a union official told the *Philadelphia Inquirer*, "is that you are not going to sell transportation without decent-looking buses, decent-looking stations, decent-looking subways." Philadelphia's press agreed with the criticism, charging NCL with "inexcusable subway neglect," "disgraceful dirt and deterioration." The transit system, especially the Broad Street Subway, needed to "be made a focal point of civic pride instead of shame."[20]

NCL's failure to maintain PTC property created conditions that city officials and the TWU claimed were dangerous for workers and the riding public. "Right now there is unsafe equipment operating on the street," the TWU charged. "PTC employes will refuse to operate unsafe equipment," the union continued, and would engage in "an all-out fight" over maintenance issues. PTC workers argued in petitions that NCL allowed buses, trolleys, and subway cars to fall into such a deplorable state that "bad steering, unsanitary conditions, inadequate heat, broken windshields and windshield wipers, broken windows, [and] worn out gears" put the safety of the workforce and the "life and limb of every Philadelphia transit rider" at risk. Some seventy thousand riders, showing their agreement with the transit workers or at least displeasure with management, signed TWU petitions protesting the state of the transit system under NCL, and Philadelphia's commissioner of public property demanded that the company explain how its layoffs impacted the transit system's equipment and the safety of its patrons.[21]

TWU members' most effective, if controversial, tactic for calling attention to how NCL's policies impacted Philadelphia's workforce and transit riders was the wildcat strike. These strikes, which took place outside the bounds of regular contract negotiations and agreements, offered workers an alternative opportunity to express grievances and try to stymie NCL's agenda. The greatest rash of these service stoppages took place in the early years of NCL control, from 1955 to 1957, when workers struck eleven times and threatened to strike eleven more. The city's creation of a Commission on Transit Labor Relations quelled much of this unrest, but threats of such low-level yet highly disruptive activism kept labor relations at the PTC simmering throughout the 1950s and 1960s.[22]

The TWU's approach to formal contract negotiations with NCL, after the ferment of the immediate postwar period, focused mostly on typical questions of wages, hours, and job security. For many rank-and-file workers, this was a starting point, but the formal contract did not adequately address the problems they confronted in their day-to-day work. Wildcat strikes flared for a host of reasons: management compelling maintenance workers to send out vehicles that they did not believe were fit for the road (often called a "safety strike"); workers fired for stealing or tardiness without receiving due process according to their contract; too few employees to do a job safely and correctly; transferring workers without their consent or even input; the threat of layoffs. Much of this came down to worker and rider safety and the right, as the TWU put it, "to be treated with respect." The PTC under NCL management, in the employees' estimation, just did not seem to care about its workforce. Workers often bought homes near their depots, so transferring them without their input forced them into long commutes when they already worked long, odd hours. Layoffs not only threatened employees' livelihoods but also meant more work for those who remained employed. And that work often became more dangerous because employees had to work on a faster schedule and equipment was not repaired promptly or adequately.[23]

Safety also emerged as an issue for PTC workers in the 1950s and after because of increasing incidents of crime on the transit system. Clifton Hood has argued that the subways in New York City were safe and perceived to be so until the 1960s. In Philadelphia, the perception and reality of crime on the transit system came earlier. A major exposé in the *Inquirer*—under the overarching heading "Corridors of Crime"—used lurid headlines to call attention to the problem and, of course, sell papers. Articles warned readers that "Fear Haunts Subway Riders," "1000 Arrests Made," and that a "Crime Peril" stalked the transit system. Anyone venturing into underground Philadelphia, into "tunnels like a tomb," had to have a "stout heart."[24]

The *Inquirer* no doubt sensationalized the issue, but there was a noticeable uptick in transit workers reporting violence on their vehicles in the 1950s. Drivers especially in North Philadelphia complained of high school kids threatening them and their riders, leading to many ejections from vehicles and arrests. In the first five months of 1961, for example, PTC officials reported that 350 people had been taken off the PTC for various types of "rowdyism." Some of these reports highlighted the worst racial stereotypes, blaming "negro youths" and raising the specter of women—drivers as well as riders—being most at risk. Women composed only about 1 percent of the PTC's drivers in the 1950s, so the company offered the "solution" of removing them from night work. To the TWU, doing so would not only violate the union's contract, but would also abnegate PTC's obligation to provide a safe workplace. This issue, the notion that black youth were responsible for endangering transit workers, could have fractured the TWU along racial lines, but for the most part PTC workers, black and white, rallied together around the problem of crime. There is little evidence that the approximately one thousand African Americans who worked for the PTC in the 1950s and 1960s, increasingly in driving jobs, differed from their white colleagues on the question of crime. Philadelphia's transit system was of course not perfect: some black workers, led by civil rights leader and NAACP president Cecil B. Moore, did complain that the TWU's all-white office staff revealed the union's willingness to practice racial discrimination, but most found that the union treated them fairly and represented their interests. Philadelphia's police department agreed with the transit workers about crime, but argued that it had neither the duty nor the manpower to provide security on PTC vehicles; doing so was the company's responsibility, and the cost should come out of corporate profits. The situation got so bad by the mid-1960s that the TWU offered to have military veterans from its ranks provide security details at Center City stations. Still, NCL would not attend to drivers' safety, refusing for years even to implement a policy that required exact change only, removing from drivers the job of handling significant amounts of cash. By the late 1960s thirty-nine cities had such a policy, but in Philadelphia it took a record ninety-seven holdups in the first nine months of 1968 for the system, by then under public management, finally to require riders to have exact fare. Crime on the PTC, of course, was certainly not NCL's fault. It took place before NCL arrived and remained a major problem after SEPTA took over. But NCL's refusal to deal with it heightened tensions between the company and its workforce and highlighted for the workers how little their employer seemed to care about them, especially when the solution would eat into corporate profits.[25]

Whatever the cause of the wildcats, PTC officials charged that they created a nuisance to which the company and the public should not be subjected. Few of these wildcat strikes involved many people or lasted long: most were waged by fewer than one hundred employees and lasted a handful of hours. Often only ten thousand to twenty thousand riders lost their transportation. One of the largest, in 1956, lasted from 8 p.m. on May 8 to 2:30 p.m. on May 9, as overhead linemen walked off the job to protest layoffs, and many trolley and subway lines lost power. Overall, the strike impacted half a million riders. To PTC management, all these stoppages, whatever their size, stemmed from issues that should have gone through the contractually bargained grievance procedure. In this view, the company was not at fault for its draconian policies; the union was at fault for not controlling its workers. Quill argued that it was not the union's job to "control" the company's workforce, plus, as one PTC employee put it, the grievance procedure was "hopelessly clogged and it sometimes took months to obtain satisfaction." To many observers, this seemed purposeful. In response, the PTC repeatedly sued the TWU for wildcat strikes that harmed the company's interest (the company dropped the suits), proposed that the union agree in advance to pay a $20,000 fee for each wildcat (the TWU rejected the idea), and wrote stipulations into the contract that allowed management to fire workers who led wildcat stoppages (the TWU reluctantly agreed to this).[26]

Although TWU leadership fought off most of the PTC's reaction to wildcat strikes, union leaders had mixed feelings about the grassroots job actions. On the one hand, they knew that the incessant work stoppages angered many Philadelphians because they brought "grave inconvenience to the working people." More, they aided "conniving management" who wanted to stir up public antagonism and use it to "bring on punitive labor legislation and weaken the union in . . . contract talks." The TWU went so far as to pass a resolution stating that it would "refuse union protection to any member or members who willfully creates a work stoppage, sit-down or any other job action without authorization and approval [of the union]." Some PTC employees agreed that the wildcats had to stop because they antagonized the public. Quill had to crack down on "irresponsible and unauthorized 'Wildcat' strikes, which gained for Labor's cause nothing but the ill-will of the riding public," wrote one man. But most correspondents argued that the company had put them in a bind, that all the cutbacks, the safety issues, and other concerns left them with little recourse when the TWU could not solve the problems in contract negotiations. Quill and local union leaders understood these concerns and despite their formal resolution tacitly accepted the wildcat strikes as a way of pressuring management. The company's policies, they argued, were wrecking workers' lives, and they had to fight back any way they could. The TWU fought at every turn PTC efforts to fire wildcat strikers. While Pratt labeled the wildcats "guerrilla warfare," the TWU retorted that the workers had no other recourse. "These wildcat strikes and sitdowns unquestionably were provoked by the crude insolence and tyranny of inept supervisors," the union told the *Evening Bulletin*. At times, the TWU disavowed wildcats because they antagonized the public, but more often it fought to protect those who walked out and damned the company for giving them no better options.[27]

To the press, wildcat strikes were unconscionable. Editors at the *Inquirer* and the *Evening Bulletin* understood the way NCL's policies impacted the city's transit workers, but rather than holding the company accountable they pointed their fingers at the employees. "Whatever justness there may be in underlying griev-ances," wrote the *Evening Bulletin*, any wildcat strike was an "ornery display of contempt for everybody." "Wildcat strikers seem to have their own philosophy," the editors wrote another time, "which boils down to a system of punishing the public for challenged procedures of the company." It was, the editors argued, up to TWU officials to discipline their workforce, make them abide by their con-tract. "If they cannot deliver what they promise, what validity will future con-tracts have?"[28]

The newspapers had a point if the wildcat strikes occurred in a vacuum, but Philadelphia's political officials saw the larger picture. To Clark, Dilworth, and others, transit workers were indeed too ready to walk off the job, but labor ten-sions were largely NCL's fault. "In the pre-NCL days," said one government report, "unauthorized stoppages which affected the public were almost unknown. . . . Since the assumption of control by the present management, the same grievance procedure has not worked successfully." NCL had created "an environment in which strike action appears to the employes as preferable to more peaceful meth-ods." Philadelphia investigators repeatedly rapped PTC for its "persistent refusal to make concessions which . . . were minor, not matters of principle," and wrote that they "were astounded at the arbitrary and unyielding attitude of PTC nego-tiators." Mayor Dilworth finally felt compelled to ask labor arbitrator Eli Rock to help with PTC-TWU relations, but Rock found the company and union locked into hardened positions. Dilworth knew the city could not accept constant tur-moil in its transit industry and in August 1957 proposed a three-member Com-mission on Transit Labor Relations. The PTC readily accepted the idea, which made the TWU skeptical at first, at one point calling the commission "window dressing" that would not deal with the root of labor's problems. Dilworth con-vinced the union that the commission would be a serious body and named three highly qualified people to it: Rev. Dennis J. Comey, director of the Department of Industrial Relations at St. Joseph's College; John Patterson, vice president of the Penn Fruit Company; and Nochem Winnett, a former Municipal Court judge. With Dilworth's full backing, the commission used its public platform to soothe some of the tensions at the PTC, reducing although not eliminating the wildcat strikes. It had no legal standing, though, and could not sway management to ameliorate its most draconian policies.[29]

NCL's policies—on labor that triggered strikes and wildcats, on relations with the city that led to an attempt at disenfranchising Philadelphia's elected leaders, on maintenance and service that reduced the PTC's utility—confirmed for polit-ical officials and the TWU what they had suspected all along. NCL just wanted to "milk the taxpayers," as James Tate, the Democratic mayor from 1962 to 1972, put it, raising fares while reducing ridership and laying off workers. Pratt, the mayor claimed, was "substituting self interest for the public interest." Everything, the TWU charged, came down to cutting expenses "in order to pay the biggest dividends." Philadelphians had to beware the fate of New York City, Quill argued, where "private companies bled the subways white, from 1905 to 1940. When the subways were no longer good milking cows they abandoned the subways to the City of New York. . . . Let's be sure that none of the people's money in the City of Philadelphia will go down the drain." Pratt, hearing the frequent charges, periodically told the press that NCL's "aim is not to 'milk' the company, but to improve its investment by providing better transportation for Philadelphia." Dilworth heard Pratt's remonstrations, reviewed the evidence, and hired the highly regarded transit expert Edson Tennyson as Philadelphia's chief transit engineer and deputy commissioner of transportation, with the charge to "keep a watch on NCL's 'gutting' of the system." Rumors circulated throughout Philadelphia, especially when city officials considered purchasing the transit company, that NCL wanted to extract what profits it could and then "unload the system on the City of Philadelphia." Perhaps NCL, if it so chose, would then stay on to manage the gutted system, for a sizable fee, of course. Local 234 member Ernest Mozer put it more colloquially: "Mr. Pratt has been nicknamed 'Skippy,'" he wrote, "for his ability to wreck transit operations and then skip."[30]

A scandal on the Market Street elevated finally proved Pratt's undoing. For some time in the late 1950s and early 1960s, the TWU complained about structural problems on the Market Street line. The elevated structure had sinking pillars, poor concrete work, shoddy welds and expansion joints, and even places where support beams were repaired rather than replaced as engineers required. NCL first claimed that the TWU was just trying to scare the public in an effort to damage the PTC, but city inspectors soon backed the union's claims. Investigators found that the company had overcharged the city of Philadelphia hundreds of thousands of dollars for repair work not performed or done shoddily. Former PTC officials who served under Pratt came forward to implicate their boss directly in the fraud, alleging that he directed them to overcharge the city by as much as $800,000. Pratt first told the press that he "vehemently and vigorously denied it" but subsequently admitted to overcharging the city $200,000. He continued to claim, however, that the overcharges were not company policy, that his mysterious granting of a pension for George Smith (one of his accusers) had nothing to do with the scandal, and that he never told his staffers to use the inflated charges to "hit city hall" as they claimed. Albert Greenfield, the PTC's board of directors, and city officials found Pratt notably unconvincing. Greenfield vowed to have an open investigation of the PTC's books, and the board held a closed-door meeting, with Pratt absent, to discuss his fate. At the same time, Philadelphia officials prepared to file a $2 million suit against Pratt and his accomplices for their scheme to defraud the city. NCL ultimately settled the city's suit in 1962 for just over $1 million, and a few months later Pratt left PTC. He assumed the presidency of NCL. Albert Greenfield, growing older and winding down his business commitments, also chose to step down from the PTC's board after twenty-six years of service to the company.[31]

PTC's new president, Robert Stier, tried to calm tensions among the company, the city, and the TWU. Stier had started with the PTC as a mechanic at the age of seventeen and rose to become vice president of operations before Pratt removed him in 1955. One of Stier's first actions was to send a letter to all employees offering his greetings to them and their families, and asking that they work with him on a "friendly service" campaign. He promised a new beginning for the company, one that would come from drivers committing themselves to better service and the company providing cleaner, safer vehicles and stations. Under his leadership, management and labor would put forward "combined efforts to do our jobs as they should be done. To do them better." "Can we brighten the face of PTC?" he exhorted in a not-so-subtle slap at NCL. "Can we make it a company we can brag we work for? In the past I've never found you unwilling to make an extra effort when it was needed. It's needed now." In letters to middle management, Stier vowed that he would do all he could to make them aware of his plans, and he solicited their support in rebuilding the PTC with a focus on increasing ridership rather than cutting costs. Labor relations with the TWU continued to be difficult, but Stier sounded a note of conciliation, reminding the workers that he had hammered out previous contracts with Quill and that in tough times management and labor had to "live and work as Siamese Twins—neither one of us getting along without the other."[32]

Despite Stier's best efforts, the transit strike of 1963 revealed the degree of control Pratt and NCL continued to wield behind the scenes and finally pushed the city to move toward taking over the system. When the TWU walked out in mid-January 1963, Stier negotiated a contract that would have ended the strike after ten days. Pratt, however, flew in from Chicago to meet with the NCL-dominated executive committee of PTC's board of directors and persuaded the members to reject the deal. Stier immediately resigned, telling the press that he was just a "hired hand" who could not "overrule his bosses." The strike dragged on for another grueling week and a half, and the press, although annoyed by Quill as always, grew furious with Pratt for the impasse, calling his move a "Pratt Fall" that the city could not laugh off. "Mayor Tate and City Council," wrote the editors of the *Philadelphia Daily News*, had to "take a good hard look at PTC and its operation—particularly the long-distance masterminding by Pratt from his Chicago office." Private ownership, especially the variety practiced by NCL, more and more looked like a sure path to service cuts, shoddy maintenance, and labor strife. The press, politicians, and the TWU all increasingly asked, in the words of the *Evening Bulletin*, "Is Public Ownership the Answer?" By 1963, after nearly a decade of NCL, Philadelphia was finally willing to say "Yes."[33]

End of the Line for the PTC

Philadelphians had discussed the possibility of public ownership of the city's transportation system, and the pitfalls of private control of this vital enterprise, for decades. The move to public ownership, spurred by NCL policies and the 1963 strike, finally pushed the city into action. The great irony, of course, was that it took NCL implementing the most nakedly extractive capitalist policies to finally move Philadelphia to end private ownership. It was the combination of NCL's policies, working-class activism that highlighted and heightened the problems of the PTC, and the rise of liberal political leadership in the city that made public ownership possible. Making the transit system into a public operation was ultimately a political project, one that required conceptualizing, funding, and organizing the system. Although Philadelphia had the obvious transportation problems and will for change, it took five years to finally establish SEPTA, purchase the PTC, and run a regional system under public auspices.

Philadelphia's political leaders, especially Mayors Clark, Dilworth, and Tate, were nationally known for championing the cause of urban public transportation. As early as the mid-1950s, they had commissioned studies of southeastern Pennsylvania's transportation problems and called for regional solutions to Philadelphia's challenges. They all argued that a portion of federal money, chiefly assigned to highway construction, should be spent on urban transportation infrastructure and pushed Congress to make a greater financial commitment to America's cities. "Washington's darlings have long been highways, airways and waterways," Tate once said. Federal policy had led to every major city being "strangled with traffic and jammed with cars." Now it was time for a more "balanced transportation program" predicated on federal support. But in the Eisenhower era of highway building, their position found no large interest groups to back it, and until the 1960s Congress regarded urban public transportation as a local concern. A few officials in the Commerce Department recognized the need for adequate public transportation and despite President Eisenhower's neglect of the issue worked with mayors of the nation's largest cities to keep the idea of investment in urban transportation alive until John F. Kennedy and especially Lyndon Johnson assumed the presidency. As part of President Johnson's Great Society, which as much as anything was an unprecedented commitment to urban America, Congress passed the Urban Mass Transportation Act of 1964 that established the Urban Mass Transportation Administration and enabled it to provide matching funds to cities that wanted to develop regional systems and improve facilities and equipment. The Philadelphia City Council immediately threw its wholehearted support behind the law. It argued that "mass transportation systems are vital to the health, safety and general welfare and continued economic growth and development of both large and small cities." Moreover, the city council claimed that Philadelphians had paid all they could afford over the decades to build the PTC and that it was now time for the federal government "both because of its wider jurisdiction and because of its broader financial base [to] aid in the solution of [transit] problems."³⁴

Federal assistance to mass transportation at first seemed a lifeline to American cities. Declining ridership in the post–World War II decades greatly reduced or eliminated profits in the industry, exacerbating trends that led smaller cities to end their service and larger ones to take their systems public. As a few examples, New York City's IRT and BMT went public just before World War II, Cleveland followed suit in 1942, and Chicago and Boston in 1947. Between 1959 and 1970, 235 private transportation systems disappeared from the United States, while public transportation authorities grew up in the Pittsburgh, Miami, Los Angeles, Chicago, and San Francisco regions, as well as many others. By 1980, 90 percent of North Americans rode on publicly owned and operated systems, all of which faced similar financial pressures. Analysts increasingly noted that with revenues and thus profits dropping and private operators leaving the business, cities had to meet the social goal of maximizing service without the income necessary to do so. Philadelphia's transportation system on this score was one among many. As SEPTA chairman James McConnon once put it, his company could not "produce sufficient revenue from the fare-box to meet operating expenses and depreciation charges. . . . It is unmistakably clear, therefore, that money to supplement payments by riders must be forthcoming from public sources." Out of a sense of social fairness, McConnon continued, "riders, who by and large are members of lower-income groups, must be relieved of burdens that should be borne by the public as a whole." Such burdens included everything from maintaining vehicles and repairing vandalized stations to proposed expansions of the Broad Street Subway and subsidies for the region's commuter rail lines. Over time, the 1964 mass transportation law and its successor legislation put substantial resources— some $50 billion by 1989—into urban transit systems, but never enough. The federal government, for example, spent $375 million on urban transportation in the first half of the 1960s; it spent $24 billion on highways, airways, and waterways over that same period. Federal funds did help Washington, D.C., Baltimore, San Francisco, and Atlanta construct rapid transit systems, but the money was insufficient to pull New York City, Boston, Chicago, Philadelphia, and other large systems out of their financial predicaments. Philadelphia's political leaders, then, confronted a transit problem that was also an economic and social problem that beset all of urban America.[35]

City and state reports in the early 1960s emphasized to Philadelphia's mayors that southeastern Pennsylvania had to be treated as a region, and that adequate transportation was vital to the economic life of the state. Across the United States, wrote the Mayor's Transit Study Task Force, cities had taken over their transportation systems, usually with "great reluctance, in the face of stern facts that often allowed no other reasonable solution." Philadelphia had to follow suit; it could no longer allow NCL or any other private entity to control this essential industry. The chief challenge facing the city, according to the reports, was that no agency had control of regional policy, no group of professional planners coordinated rail, highway, subway, and bus transportation. A regional authority could, for two examples, ensure that commuter railroads, especially the Pennsylvania Railroad, received the subsidies necessary to continue passenger service and also plan the extension of public transit lines such as the Broad Street Subway. Doing so would tie roads, rails, and bus lines together into a system and help the public understand that investing in transit of all types would be cheaper and more efficient than building ever more highways and parking lots that swamped the city in traffic. Such a system should strive to "operate on a self-sustaining basis so far as possible," by raising fares when necessary and attracting more customers with better equipment and service, but the reports all noted that some government aid—local, state, and/or federal—would be necessary and that a regional authority could secure and manage those resources in ways that a private company could not.[36]

The Mayor's Transit Study Task Force also made it clear that NCL policies at the PTC, toward labor and its service commitments, had forced the city's hand. "The public," wrote the mayor's task force, "has lost confidence in the principal transit company because the latter has not provided the high level of service which should be in keeping with its repeated demands for fare increases." All the labor turmoil, too, had soured the public, and a regional authority would implement "sound machinery for contract bargaining and grievance settlement" and develop "excellent personnel merit systems covering hiring, discharge, promotion, retirement, fringe benefits, etc." Transit workers' activism had played a notable role in pushing the city finally to take over its transportation system.[37]

The question confronting Philadelphia's political leaders was how best to develop a system that served the entire region. A report to the city council in 1961 laid out the key issues. Developing a Philadelphia-specific system, essentially replacing the PTC with a city-owned and city-operated company, would solve the immediate problems with NCL but not "recognize [the] importance of

Delaware Valley as a metropolitan area." A regional authority, on the other hand, would give planners the opportunity to knit the entire region together, hopefully fostering "cooperation between political subdivisions." "A broader area for operation," the report continued, would also "contribute to the financial stability of the operation." The chief obstacle to such an approach, the report concluded, lay in "obtaining cooperation of neighboring counties."[38]

That cooperation, Philadelphia officials knew, would be difficult to obtain, yet it was vital to the city's transportation system and its future. Mayor Tate and others repeatedly argued that people had to understand that "Philadelphia is not an island or entity unto itself, but is now conceived of and is, in actuality, the hub of a growing metropolitan area sometimes referred to as 'the Delaware Valley.'" The region had to coordinate its transportation services, but Philadelphia repeatedly found its efforts blocked by suburban counties such as Bucks, Delaware, and Montgomery that had "an ingrained reluctance . . . to become associated with the big city in any kind of joint venture." "This reluctance," a city report continued, "is an outgrowth of political differences, a fear of domination by the big city and a certain rural attitude. In order to overcome this natural reluctance, any plan proposed must be sold to the suburban counties and they must be assured of adequate representation in the planning and management of such a system." Philadelphia's political leaders understood what this meant in practical terms: to create a regional authority the city would have to surrender much of the power over transportation policy to smaller suburban counties. They did just that in establishing the Southeastern Pennsylvania Transportation Authority pursuant to the Metropolitan Transportation Authorities Act of 1963. That law, which emphasized SEPTA's responsibility to set fares, operate facilities, and determine the services the system would provide, also made it clear that SEPTA would act as a state agency and not "be an instrumentality of any city or county or other municipality."[39]

Philadelphia, by law, could not see SEPTA as its province, and the makeup of the authority's board of directors made the point concrete. Five counties composed SEPTA (Philadelphia, Bucks, Chester, Delaware, and Montgomery), with each getting two representatives, plus an eleventh member appointed by the governor. Although the city of Philadelphia had by far the greatest interest in decisions made by SEPTA, the suburban counties wielded the great bulk of the power. The city did have limited veto power over SEPTA decisions, but the distribution of authority was so clearly disproportionate that the state added one Democrat and one Republican from each chamber of the state legislature to SEPTA in 1991. That formulation created a fifteen-member board that is still in place today, but it did little to address SEPTA's imbalance of power. Despite SEPTA's limits, Mayor Tate and the Philadelphia City Council pushed for and gratefully accepted passage of the Metropolitan Transportation Authorities Act, seeing it as "a giant step forward in the struggle of the people of Southeastern Pennsylvania to overcome the massive and ever-growing problem of congestion and inadequate mass transportation in the metropolitan area."[40]

A "giant step forward" perhaps, but in the context of the Philadelphia region's difficult racial politics, what the city gave up to create SEPTA hamstrung the transportation system. Philadelphia well represented the emerging political alignment in the urban North with a largely minority Democratic core and a white suburban Republican periphery. In the mid-twentieth century, Bucks, Delaware, and Montgomery Counties overwhelmingly voted Republican, while having white populations above 90 percent. The Republican Party made no apologies for its efforts to block passage of state fair-employment legislation and keep public housing out of Philadelphia's suburbs, in essence to defeat initiatives that would have eased segregation in the region. SEPTA board members from the suburban counties—all Republican—followed suit, refusing to authorize studies that would show the economic justification of new routes in their communities and opting out of programs that would have helped the transit system, especially in Philadelphia, meet its financial needs. New routes would have especially assisted the city's black population, who found themselves caught in a deindustrializing economy while locked into deteriorating, segregated communities mostly in North and West Philadelphia. When SEPTA proposed bus routes to take black Philadelphians from the city to suburban industrial parks, however, collar counties defeated the idea. Republicans did not necessarily oppose a regional transit authority on principle. Republican governor William Scranton signed the legislation, after all, calling it a solution to a "tremendous problem." And the politics of suburbia were not just about race: many people did move to the suburbs to buy a home, get more space, raise their families away from urban problems (often, although not always, a euphemism for racial conflict). Nonetheless, few suburban GOP officials wanted SEPTA used to integrate the region. Philadelphia's Democratic political leaders may have had to surrender authority to get SEPTA created, but doing so placed long-term limitations on how public transit could shape the development of the metropolitan area.[41]

Regardless of its constraints, SEPTA always envisioned itself as a "truly regional operating agency." Its master plan entailed purchasing the PTC, acquiring all of the metropolitan area's bus lines (some twenty-eight companies) that carried anywhere from 876 to 95,000 passengers per day, and integrating the commuter rail lines of the Pennsylvania and Reading Railroads. Having a regional authority in charge allowed transit planners to think more deeply about how to coordinate the system, including how radial highways competed with commuter rail while circumferential highways could feed those same train lines; how the city could expand its subway system to provide better service and attract more customers; how to improve transit connections between southern New Jersey and Philadelphia. Acquiring and integrating existing companies and routes, however, proved much easier than expanding the system. Much of the latter planning work, which built on earlier reports such as the Penn Jersey Transportation Study, proved fanciful in the more difficult economic times in which SEPTA operated. Plans to expand the Broad Street Subway in North Philadelphia or to turn the rapid transit system into a grid like those in Paris and London, to give two examples, foundered. The Lindenwold line to New Jersey did get built and opened in 1969 (operated by the Port Authority Transit Corporation—PATCO—rather than

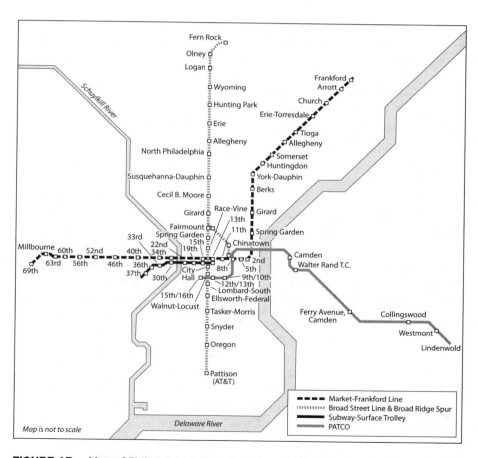

FIGURE 17. Map of Philadelphia's transit system. Philadelphia's transit network, as managed by SEPTA, never attained the size or capacity that planners at various times in the twentieth century had hoped for. Map drawn by Bill Nelson.

SEPTA), the Broad Street Subway extended its service in 1973 from Snyder Avenue to Pattison Avenue (now AT&T Station), and the long-discussed Airport Line finally opened, but not until 1985.[42]

Much of SEPTA's most effective system-building in the 1960s involved connecting the suburbs to the city, chiefly through incorporation of the Philadelphia Suburban Transportation Company (PSTC—commonly known as Red Arrow) and through coordination with the commuter railroads. Red Arrow, which began operation in 1936, and its predecessors, especially the Philadelphia and West Chester Traction Company, served Delaware County plus portions of Philadelphia, Montgomery, and Chester Counties. It was by far the largest of the suburban transit companies, and critical to SEPTA's plan of forming a regional system. From the terminus of the Market Street elevated at Sixty-Ninth Street, Red Arrow's trolleys and buses fanned out across the region west of Philadelphia and played a vital role in the area's development. As the population of those counties expanded, Red Arrow grew profitable under the direction of A. Merritt Taylor and his sons and grandsons who ran the company from 1899 to 1970. Red Arrow continued to turn a profit into the 1960s, but its margins grew ever smaller. Declining ridership and rising wages led the PSTC to try to solidify its finances by buying car washes and shifting from streetcars to buses and Hy-Rail vehicles (buses that could run on streets or railroad tracks). None of the efforts helped much, and in 1963 Red Arrow's drivers, represented by the Brotherhood of Railroad Trainmen, staged the first strike in the company's history. Red Arrow lost nearly $62,000 in the thirty-four day strike, and from there the company's fate was sealed. In the same annual report that announced the lost revenue, PSTC president Merritt H. Taylor Jr. told investors that with the rise of SEPTA "it is reasonable to assume that [Red Arrow's] facilities will be a component of the proposed coordinated transit system." After several years of negotiations, SEPTA purchased Red Arrow for some $13.5 million and began operating that part of the transit system in January 1970. With this purchase, SEPTA took control of 97 percent of bus and street railway traffic in the region. President Taylor, whose family had played a prominent role in the transit business for a century, accepted Red Arrow's fate with equanimity. He and his investors cleared $9.1 million in the deal, renamed their company the Bryn Mawr Group, and became real estate investors in Florida. "The urban transit business," Taylor assured his shareholders, "is not a promising one for private industry, and we are fortunate to now have the opportunity of directing our full resources to areas of greater growth." Red Arrow management, unlike NCL, never pillaged its property or had running difficulty in labor relations, but even a transit stalwart like Taylor knew the industry no longer offered returns attractive to private capital. It was time to let the public have the business.[43]

The Pennsylvania and Reading Railroads, like passenger railway companies across the country, would have agreed with Taylor about the financial prospects of providing transportation service. But unlike the PSTC, the railroads were bound by state and federal regulations that made abandoning lines, even unprofitable ones, difficult. Philadelphia area railroads, like many large cities across the United States, developed their service networks in the nineteenth century, when commuter lines generated significant profits. The industry had its dips, but private companies generally found the business profitable until the post–World War II period, when the automobile and highways siphoned thousands of riders out of the system. The Pennsylvania and Reading, unable to end their commuter service, scaled back their schedules and deferred maintenance in the 1950s and 1960s, which drove away commuters. Mayor Dilworth and other political officials knew that such policies put thousands more people on the roads, worsening traffic congestion in the city, so they developed programs such as Operation Northwest, which connected the PTC to the Chestnut Hill line and subsidized commuters' travel. Such efforts set the foundation for the creation in January 1960 of the Passenger Service Improvement Corporation (PSIC), which used public money to improve railroad service. PSIC did entice more riders to the rail lines (for example, an increase of 15 percent on the Chestnut Hill line, and 250 percent on the Fox Chase line, which had limited service beforehand), but it was an expensive endeavor that subsidized the travel of railroad commuters over people who rode PTC vehicles. The newspapers applauded the effort, calling it "a constructive and dynamic approach," but Albert Greenfield argued it was no solution. The city of Philadelphia subsidized the railroads with $1,775,000 in 1962, a "discriminatory transportation policy" that transferred wealth—twenty-six cents for each one-way ride—from those living closer to the city center to those in more outlying districts. "PTC," Greenfield said, "could carry its passengers absolutely free—and throw in a morning cup of coffee—if it received the same per-passenger subsidy as do PSIC railroads." Dilworth knew the PSIC approach was a stopgap until SEPTA could gain control of the rail lines, which it began to do in 1973 when it secured long-term leases from the Reading and the Penn Central (created by a merger of the Pennsylvania and the New York Central Railroads in 1968) along with guarantees that public money would cover any deficits in operating or capital expenses. Despite public assistance, bad economic times continued for the Penn Central and Reading, leading to Congress merging them into the Consolidated Rail Corporation (Conrail) in 1976. Seven years of difficult finances and public criticism for "perform[ing] very poorly" led Conrail to turn control of the commuter rail lines over to SEPTA in 1983. Much as with the PSTC, SEPTA had little choice but to take over a service no longer attractive to private capital.[44]

As SEPTA coordinated service with and then took over suburban companies, it also worked its way through negotiations to purchase the PTC. To head all its efforts, SEPTA hired Brigadier General Michael Reichel as general manager in 1966. Reichel, who had served as director of transportation for the U.S. Army, brought a reputation as a first-class administrator who could build the entire network. Just as important to SEPTA's board of directors, however, was his ability to inspire "the loyalty of his work force." "That," the board wrote, "is at least half the battle." At the PTC, Albert Lyons returned from his exile to serve as president and negotiate the sale of the company. NCL meanwhile sold its stock in the PTC. Shortly afterward, in 1965, the PTC's board of directors ousted Pratt and other NCL officials from the company's executive committee. The removal of NCL helped thaw tensions between the PTC and the city, and Lyons even admitted that public ownership might improve the transit system's service. He just had to ensure that the company safeguarded the rights of its stockholders in any sale.[45]

Safeguarding those rights entailed years of negotiation and a number of trips to court. The PTC argued that its property was worth $75 million to $80 million; anything less, Lyons wrote, bordered on "confiscatory," "a takeover of this company against the will of its owners at a price far below its fair value." SEPTA, to which the city granted its right to buy the PTC under the 1907 agreement, argued that such a valuation was far too high, one last rip-off to be foisted on the public. The two sides finally negotiated a purchase price of $59 million in November 1965, but the stockholders voted against the deal the following April. In subsequent court hearings, a three-judge panel ruled that SEPTA had the right to purchase PTC and commanded the company to reach an agreement and sell its property. PTC appealed the decision to the Pennsylvania Supreme Court in December 1966 and, in one last effort to extract as many assets as possible, declared a one-dollar per share dividend. SEPTA obtained an injunction to halt the dividend, and the state Supreme Court upheld SEPTA's rights to purchase and stop the payments to shareholders. This "disastrous" decision, in Lyons's words, forced the PTC's hand. The company agreed to sell its business to SEPTA for $48 million, plus another $17 million in pension liability. Lyons advised his shareholders to accept the deal. The political, economic, and legal climate had become too difficult for a private company to provide public transportation, he argued. SEPTA, he continued, would "take over an efficient, competently staffed and well run transit system [that] has served the community well." "Responsibilities of corporate good citizenship," he concluded, "have been met." Lyons's interpretation of the company's "good citizenship" was, of course, up for debate, but SEPTA nonetheless consummated the deal and took control of Philadelphia's transportation system on September 30, 1968. After a century of private companies providing this vital public service, an era came to an end.[46]

SEPTA, an Epilogue

Public ownership, of course, was no panacea. SEPTA inherited from PTC a welter of interlocking problems that were too often beyond solution, given the authority's limited financial resources. Some of these problems were long-term and structural, related to political, economic, and social trends that were reshaping the country: the declining economies of urban areas, especially in the North and Midwest; politicians' neglect of cities; suburbanization and governmental policies that supported the continued use of and infatuation with the automobile; inflation fueled by the Vietnam War; disillusionment with the government and its ability to solve social issues. Other problems were particular to Philadelphia's transportation system, especially as NCL left it: old and shoddy equipment; deferred maintenance; skepticism that the company did its best for the people of Philadelphia; deep mistrust between workers and management. The story of SEPTA goes beyond the scope of this book, but sketching the connections between the eras of private ownership and public ownership into the 1970s highlights the difficulties the public authority faced from the start and how the perpetual problem of financing the system kept workers and management at odds for decades to come. The greatest legacy of private ownership of public transportation in Philadelphia was an adequate, if not great, system of rail, trolley, and bus lines, but one that had such limited financial resources that workers felt exploited, management felt pressed into an impossible bargaining situation, and Philadelphians received declining service that left many believing SEPTA was uniquely incompetent or even corrupt. The last view made sense, but only to people who did not know the history of how private ownership had developed and managed, or mismanaged, the system for the last century.

Finances remained SEPTA's most salient problem throughout the 1970s. The authority's costs and expenses (fuel, maintenance, wages, etc.) exceeded revenues by $7.3 million in 1970, $12.1 million in 1971, $14.7 million in 1972, and the trend continued. SEPTA's financial situation was common across the United States: one federal report found that between 1960 and 1970 the industry as a whole went from a deficit of $400,000 to $80 million, most of that because of rising expenses rather than falling passenger revenues. But Philadelphia's situation was especially troubled because of how the metropolitan area funded the system. Over the years fares covered between 40 and 50 percent of SEPTA's costs (a high proportion by North American standards), leaving the rest of the budget to come from public sources. The mix could vary, but 1978 offers a representative picture of SEPTA's public funding: 40 percent from the federal government, 40 percent from the state, 15 percent from the city, and less than 5 percent from the four suburban counties. Of some thirty cities surveyed in 1979, Philadelphia's system was the only one that did not have taxes and/or tolls specifically earmarked for public transportation. "SEPTA is proof," McConnon testified to the Pennsylvania State House of Representatives, "that it is impossible to run a good transit system without predictability of funding." The *Evening Bulletin* agreed with McConnon about the need to rethink SEPTA's funding. The transportation system needed resources because it had "to speed the rehabilitation of Philadelphia's rundown subway system, to form a transit police force, to restore service that had been slashed under the system's private owners and, now, with the acquisition of the Red Arrow Lines, to improve the quality of its suburban operations." "Public subsidies to mass transit were once unthinkable," the paper concluded, but "today, with the transit system under public ownership, they can be as readily justified as maintaining the city's streets and highways, or the water works. Efficient mass transit is no less a public utility than these other services." The *Evening Bulletin*'s stance represented a sea change in thinking about transit funding within the city, but such views could not overcome what one journalist called "the lack of taxing authority [and] the subordination of the needs of the city system to suburban financial opt-outs." The deal that Philadelphia's politicians struck to get SEPTA created hamstrung the authority in its efforts to provide the service that the city needed.[47]

Recognizing these limits, SEPTA's board members made what they thought was a rational choice to attempt to secure the necessary financial resources. In 1970, the board brought in William Eaton, a senior manager at General Electric, to apply his business expertise after General Reichel stepped down. The city estimated that it would need some $1.5 billion for capital improvements and system expansion in the 1970s, and SEPTA's board believed that Eaton, who had no experience in transit, had the expertise to obtain the money through grants. Assisted by federal and state legislation such as the National Mass Transportation Assistance Act, the Public Transportation Act, and the Pennsylvania Urban Mass Transportation Act, which together provided up to 66 percent of the cost of capital improvements, covered up to 50 percent of transit agency deficits, and allowed the transfer of interstate highway funds to aid mass transit, Eaton successfully secured resources that kept the system afloat, if not out of the red. Under his watch, SEPTA purchased the most needed new equipment, continued subsidies to the commuter railroads, and refurbished some of the most deteriorated stations.[48]

Although Eaton was successful in obtaining grants, his focus on securing public funding created a sense among some observers that SEPTA cared more about the pursuit of government dollars than executing its core mission of providing an essential public service for its users. Eaton's emphasis fostered a turbulent time—one federal official called it a "disastrous decade"—and unhappy relations between SEPTA and the city until he was fired in 1978. Government reports repeatedly found "a lack of confidence in SEPTA on the part of the general public and its elected representatives," a view that management squandered public money and always asked for more than it needed, and a common belief that the authority needed to pay closer attention to administering the service it provided. To be sure, Philadelphians understood that SEPTA did valuable if thankless work in coordinating public transit for the city, that its continued use of light rail served as an exemplar for other urban areas in the United States, and that the authority did in some ways try to pay greater attention to citizens' needs with reduced fares for senior citizens and Braille guides for blind passengers, to give two examples. But too often, government analysts argued, SEPTA neglected its primary mission "to insure that service to the public was maintained." "Private ownership, with an eye on profits, could cut back unprofitable service lines, regardless of the effect that would have on riders," the analysts wrote in one State House of Representatives report, but "state subsidies to transit are based upon the need to permit service which may be socially desirable but economically unjustified. . . . When the current SEPTA system [1980] is compared with that of twelve years ago . . . it is seen that the level of service has not been maintained." SEPTA, the report concluded, had to refocus its energies on service, and it had to hire a general manager with "transit management competency and experience as a prerequisite." SEPTA heeded the state's advice, hiring lifelong transportation men David Gunn and Joseph Mack to run the system from 1979 to 1987, and they were widely credited with stabilizing the system and improving SEPTA's performance, although the authority never did solve its financial woes, and Gunn was particularly known for being a difficult opponent for organized labor.[49]

McConnon understood Eaton's weaknesses but backed him by emphasizing how the years of neglect under PTC had played a large role in the authority's problems. "It was a constant struggle for money in which we were always [asking] for more," McConnon said. "We were so far behind on maintenance, the need for which goes back 40 years. . . . In a desert, it isn't a choice of luxuries, it was a matter of survival. . . . Long run considerations were long out of date, and nothing had been done about them." McConnon's views may have been self-serving, but were in many ways accurate. In its last year of controlling the PTC, NCL generated a surplus of nearly $3 million, while at the same time the company deferred maintenance and allowed the rolling stock to become so old that SEPTA estimated it would need to spend $600 million on capital improvements in the first half of the 1970s just "to keep the system alive." Renovation, SEPTA officials argued, was a stopgap approach at best; the goal had to be "the development of a substantially new system to replace one which, through several years of neglect, was in poor condition." Among many projects, SEPTA had the immediate need to buy 144 new railroad cars and one hundred buses, refurbish six subway stations, and modernize all its maintenance shops and power stations. Equipment surveys found that as much as 75 percent of the city's subway cars were inoperable, buses on average broke down after only 1,398 miles on the road (in comparison, the number was 2,700 miles in Baltimore, 3,334 in Minneapolis–Saint Paul), and even the bus washers had been used beyond their expected service life. NCL had so drastically cut the maintenance workforce that SEPTA had to hire 450 more mechanics in the 1970s, an increase of 29.4 percent, which added greatly to the payroll. SEPTA found it a constant struggle to train the new employees and keep the antiquated stock rolling, which at times led drivers and TWU officials to complain that they were asked to operate equipment unsafe for them and their passengers. All of SEPTA's efforts, one state official said, were just keeping the system one step ahead of "chaos and total disintegration."[50]

For Philadelphia's transit workers, who had struggled so greatly with NCL, SEPTA offered hope but also concern. Public takeover, they feared in the early 1960s, might abrogate the contract they had worked so hard to negotiate with PTC. But if the city would honor their rights, then they could support SEPTA. Once the transportation authority agreed to accept their contract, the TWU hailed SEPTA as "a step forward in our public transportation system." Union president Dominic DiClerico, a World War II veteran and trolley driver who headed the local from 1965 to 1974, added that finally "our program that we have advocated for years may be put into action. Too long has our transportation system been ill-managed, badly equipped and inadequately serviced." Perhaps SEPTA offered a new beginning. Federal officials certainly thought it would be a good idea. "Real improvement in system performance," one report stated, "must entail the involvement of SEPTA's employees; their own performance and knowledge of SEPTA's operations are both too important to be ignored." Workers should get greater control over their work and participate in setting policies, and union leadership should work with management to obtain that participation and thus boost employee performance. "Such a commitment need not be based on idealism alone," the report concluded; "it should result in clear gains for both sides."[51]

To the transit system's workers, such a recommendation made sense, but SEPTA management proved to have mixed and sometimes even conflicting views about its workforce. On the one hand, SEPTA reports held that improving labor relations was the first order of business for the new authority because it could not afford the conflict that plagued New York City or the PTC. Another SEPTA-authorized report found that the employees composed an "efficient, trained operating work force" that merited the $2 million that had been invested in their preparation. Other studies, however, argued that SEPTA, like other transit authorities, had a problem with "the quality of manpower," that the system needed "competent, skillful and dedicated personnel, interested in their work and enthusiastic about potential improvements." Philadelphia's transit workers "expressed a lack of identity with SEPTA as an organization and had low morale.... They did not feel financially rewarded [and] viewed their physical surroundings negatively, as well as their work and supervision." That same report concluded, however, that "the blame for SEPTA's mismanagement cannot rest at the feet of its employees.... The Board has the final responsibility for the personnel within SEPTA and for the actions of those personnel."[52]

Whatever dissatisfaction some SEPTA employees may have had, pay and benefit increases over the postwar years made transportation work a decent job in the 1970s, especially as the Northeast and Midwest lost better-paying industrial jobs. A study commissioned by the U.S. Department of Transportation reported that the transit industry "has long been noted as one providing highly stable employment," in part because 95 percent of transportation workers had a union contract. Their hourly pay, on average, had grown from $1.44 in 1949 to $3.71 in 1969, a 158 percent increase. Their wages fell short of building-trades journeymen, who made $5.87 an hour (a 151 percent increase over the same period) or truck drivers, who made $4.01 (a 159 percent increase), but compared to the composite pay for all manufacturing employees in 1969, which stood at $3.19 per hour (a 131 percent increase), transit workers had done well. Nationally, the transportation workforce had dwindled from 240,000 employees in 1950 to 138,000 in 1970, but transit systems still offered a significant source of good semiskilled jobs. In large cities such as Philadelphia, Chicago, New York, and Detroit, transit was an especially vital employer for growing black populations. Only 3 percent of the industry's employees in 1940 were African American. That number grew to 11 percent in 1960 and 30 percent in 1970. In Philadelphia, African Americans composed over 40 percent of SEPTA's workforce in 1970, and at the end of the decade the authority created an Office of Civil Rights staffed with twelve employees to make sure the company adhered to civil rights law.[53]

Studies found that the strength of transit unions, much more than the protections provided by labor law, helped ensure that a transportation job remained a good one. With almost all transportation companies going public by the 1970s, management's emphasis shifted from maximizing profits to maximizing service. Despite the new priority, transit authorities still had to safeguard their budgets so they could spend money on equipment and routes rather than wages. This created an atmosphere ripe for labor conflict, although it was much more peaceful than in earlier eras. The United States had 375 transit strikes in the 1940s, 482 in the 1950s, and 532 in the 1960s. With transit companies going public, their labor relations fell under state rather than federal labor law, and most states first passed compulsory arbitration laws and then laws that made strikes by public employees illegal. Management and labor both disliked the arbitration laws for hamstringing the collective-bargaining process and were relieved when the U.S. Supreme Court struck them down as unconstitutional. The laws banning public-employee strikes, on the other hand, had permanence—as of 2010, thirty-nine states prohibited such labor actions—and management generally backed the laws because they allowed public employers to dictate terms to their workers. Employees had the option of accepting what management offered or engaging in an illegal strike, a tack that the TWU was especially known for being willing to take. Pennsylvania was an exception to the national scene, with the state's Public Employe Relations Act of 1970 allowing public transit employees to strike if doing so did not create "a clear and present danger or threat to the health, safety or welfare of the public." SEPTA's employees, then, had a more hospitable atmosphere for negotiating with their employer than did most transit workers.[54]

Contract negotiations between SEPTA and the TWU focused primarily on wages and layoffs. The authority, pressed by all its financial burdens, cried poor in every negotiation, repeatedly claiming it could not "afford to give a wage increase to union employes," especially when TWU demands could add as much as $100 million to SEPTA's bills. To the union, however, SEPTA's financial problems did not by default have to mean stagnant wages for workers. DiClerico understood that raising transit workers' wages could hurt "the little people who ride the vehicles [and] can't afford it," but argued that was true only if the money actually came out of the pockets of poorer Philadelphians. In the new era of government funding, he continued, it was SEPTA's "job to find the money. All we're saying is that we want SEPTA workers to be treated fairly and decently." SEPTA operators earned $4.68 per hour in 1973, 12.5 percent less than the average hourly wage of $5.35 paid in the six largest cities in the Northeast. "The company," DiClerico told a boisterous crowd of SEPTA workers, "is attempting to take money out of your pocket and pay for their subsidization." The question, as an *Inquirer* article put it in a bit of unlikely sympathy, was "Does SEPTA's precarious financial condition justify putting a lid on its workers' pay raises?" Inflation, which averaged approximately 7 percent per year in the 1970s, impacted transit workers as much as anyone, and they too needed to keep up with the cost of living. More often than not, management and the TWU were able to reach agreements on wages that helped workers keep up with rising costs, but raises often necessitated greater financial support from the state or federal government—since SEPTA and city officials feared raising fares would reduce ridership—and on occasion nearly caused the authority to fail to meet payroll.[55]

On a number of occasions in the 1970s and after, the TWU could not reach an agreement with SEPTA and engaged in strikes that conjured for Philadelphians memories of the NCL years, but not the more tumultuous clashes of earlier eras. Wildcats disappeared from the system, but official TWU walkouts shut down SEPTA in 1968, 1971, 1975, 1977, 1981, 1983, and 1986. Most of these strikes were relatively brief, lasting a week or two, although the ones in 1977 and 1983 lasted 44 and 108 days. The press, more balanced in its assessment of the transportation system's labor issues than in previous decades, understood transit workers' need for a raise, but regularly admonished TWU leaders to keep the public interest in mind and remember that their members were relatively lucky, living in an era where 10 percent unemployment meant every SEPTA opening had many applicants. "A job with SEPTA is a valuable property," as Eric Schmertz, a state-appointed fact-finder, put it in 1975.[56]

SEPTA's workers of course understood the larger point that they had to take care not to alienate the rest of Philadelphia with their demands, but they also knew that they had never gotten job security or better pay without a fight. In negotiations and strikes they not only won higher pay but also fought off service cuts and prevented layoffs of hundreds of workers, especially furloughs of maintenance workers that would "endanger the welfare and safety of the public." The TWU also drew the line at SEPTA implementing cost-saving measures that transportation companies across the United States were beginning to use in the 1970s: subcontracting out maintenance work and hiring part-time workers for cashier and driving positions. Such overtures, the TWU's DiClerico held, were dangerous if honest, "extortion" if a negotiating ploy. Philadelphia's political leaders, especially Mayor Tate and Mayor Frank Rizzo, mostly sided with the TWU in its conflicts with SEPTA. "The TWU," Tate said in 1968, "has been treated shabbily" by management, and a strike would fall on the authority's head. TWU International president Matthew Guinan singled out Rizzo for praise in 1975, citing his "perceptive evaluation of the priorities among the conflicting interests: your city's need for an efficient mass transit operation; the transit workers' need for costly wage and fringe improvements; and the public's resistance to any increased tax load." "Your prompt response to our request for your help, your firm insistence on a settlement fair to the employees and to the riding public and your effective leadership in difficult days—all are a matter of record to which I am happy to attest," Guinan concluded. Impartial external observers agreed with the mayors' stance. They may not have used Tate's words—he called the authority's board, hyperbolically and probably unfairly, "anti-labor," "amateur," and dedicated to "scab operation"—but as one study of SEPTA management put it, negotiators had to become more professional, had to "incorporate in [their] preparation and deliberation efforts a capability to realistically evaluate and appraise the long-term as well as short-term financial implications of demands." The workers deserved nothing less than honest, good-faith, competent negotiations with management, especially now that the public owned the system and authority board members worked for them.[57]

In the end, after so many decades of fighting to improve their work lives, Philadelphia's transit workers knew that public ownership had shifted management's emphasis but that labor conflict would remain. They had to continue to make demands, negotiate forcefully, and hold SEPTA accountable. "If we are state employes," TWU member Richard Harper said, "we want state benefits. . . . We can't live on promises." Around the city, when negotiations grew heated, when strikes loomed, the chant "No contract, no work!" echoed through the depots. The phrases may have changed, but the city's transit workers in the 1970s did not sound all that different from how they had generations before. They would not get what they needed without a fight.[58]

ADVANCES HARD WON
AND WELL DESERVED

Labor all over the United States has been feeling for a long time
these crushing attacks of capital and has been awaiting this
opportune time to strike a blow for freedom. It is in your power to put
the Government in the hands of all the people, not a privileged few.

—C. O. Pratt, Amalgamated organizer in Philadelphia, 1910

In this survey of nearly a century of history at Philadelphia's transportation system, the key theme that emerges is the shifting but enduring nature of capital's
quest to control its workforce, and workers' efforts in turn to contest management's agenda. Company officers carried out their quest with talents that drew on
the cultural tools and socially acceptable ideas and practices available in different
eras of American history. From raw violence and company unionism to race-
mongering and ideas about the imperatives of capitalism, management shifted
its tactics sometimes sharply, sometimes adeptly, sometimes both, in its effort to
control its workforce. Philadelphia's transit workers responded with decades of
protests, strikes, and organizing campaigns that helped them earn higher wages,
better hours, and union recognition. Their efforts were often hesitant, uneven,
riven with internal dissent, but they nonetheless helped the city's transit workers
pose an alternative answer to the one that management gave to the labor question.

This theme of the shifting nature of capitalist initiatives and workers' response,
central as it is, cannot stand alone. Because Philadelphia from the earliest days of
the industry relied on private capital to provide public transportation, a number
of other themes came to the fore throughout the late nineteenth and twentieth centuries, including the impact of financial pressures on the transit system's
routes, its workers and its riders, the debate about public versus private provision
of a social good, and the importance of technology to the shape of the transportation system itself as well as Philadelphia's growth. These themes cannot be
neatly teased apart; all were intertwined and impacted each other. Moreover,
because of the size and longevity of the city's system, these themes not only illu-
minate the history of public transportation in Philadelphia but gesture toward
the experience of other American cities. Pitched more broadly, they also deepen

our understanding of twentieth-century American history, especially the shifting nature of capitalist efforts at social control, working-class experience and organizing, and urban development.

In the early years of the industry, when horses supplied the motive power on the lines, urban transportation suggested great opportunity to Philadelphia's leaders, its entrepreneurs, and its riders. Philadelphians knew how vital transportation was to their city's geographic and economic growth. Horsecars essentially doubled the size of the city, bringing to market thousands of acres for homes and industry. Economic prosperity followed geographic growth, as the transit system enabled goods and people to move across the city more easily, particularly along numerous north–south routes and east–west through Center City. As a new industry, albeit employing older technologies, horsecar transportation showed the propensity for generating significant returns on investment and thus attracted many entrepreneurs. From the beginning with James Boxall, small-time entrepreneurs with enough capital for horses and a coach, and later rail lines, developed routes that crisscrossed the city. This frothy early capitalism created many routes, but with no coordination the industry was inefficient—as both an economic venture and a transportation system, although the former dominated entrepreneurs' concerns—and by the late nineteenth century, wealthier capitalists, especially Peter Widener, consolidated Philadelphia's transit lines.

The end goal of unifying the system, especially rationalizing its routes and timetables, made sense to most Philadelphians, but the means used called the project into question. In a city famed for its corruption, Widener and his ilk manipulated the political system not only to consolidate Philadelphia's transportation network but also to water corporate stock and keep out competitors so that millions of dollars flowed into a few owners' pockets. Some Philadelphians, especially those on the political Left such as members of the Socialist Party, abhorred the idea of the wealthy using public streets for private gain. They also disliked how the business of transit further perverted the political system, once again aligning the interests of the rich and the politically powerful. Yet the notion of public ownership, although afloat in the universe of thought about public services, found little purchase in a city with a long and fraught history of private interest shaping the public domain. Such an orientation toward private ownership in an age of laissez-faire dominance ordered much of America's, not just Philadelphia's, political, economic, and social discourse and policy.

The dominance of private capital in the realm of public services played a key role in shaping transit workers' lives in the horsecar era and beyond. Horsecar drivers before the turn of the twentieth century struggled to secure living wages and humane working conditions. Their incomes were often too meager to support their families, their hours kept them on the job for as many as eighteen hours a day, and their working conditions frequently resulted in illness, sometimes death. The shift to electric power made the job somewhat more skilled, but did not lead to a commensurate improvement of wages and working conditions. In fact, the increasing traffic and chaos of the early twentieth-century city in some ways made their jobs more difficult. In an era without minimum wage laws, restrictions on working hours, or any state-backed safety net, the larger political and social context around the labor question stacked the deck against workers, if it did not foreordain a negative answer. The first capitalist strategy for controlling its workforce, then, was in some ways constructing and naturalizing an ideological order that made unions, even the very act of working-class organizing, a threat to the state, its political system, its economy.

In response to their difficult working conditions and low wages, Philadelphia's transit workers engaged in a running class conflict that deployed protests, strikes, and union organizing. Particularly in the pre–New Deal era, none of this activism came easily, and it often entailed real danger. Joining the Amalgamated and fighting for working-class rights in the years surrounding the turn of the twentieth century brought a fierce managerial response that involved busting unions, firing workers, importing strikebreakers, and wielding the violent power of the state. The trolley strike of 1910 represented the most notable example of this working-class activism and the lengths to which management would go to put it down. Such class conflict galvanized working-class Philadelphians, their solidarity fired not only by the PRT's treatment of its workforce, but also by the company's disregard of their communities as it "modernized" its system and the city. Large, fast, heavy trolleys on the narrow streets of working-class Kensington, South Philadelphia, and other communities infuriated crowds of Philadelphians who watched their children die under a streetcar's wheels. Working-class Philadelphians understood the necessity of the transit system but could not abide the violence it brought to their communities—both to their families and to their fellow workers who labored on its lines—nor could they quietly accept corporate dominance over their lives. The solidarity that working-class Philadelphians displayed in the general strike of 1910 brought class conflict and the distrust, even hatred, of corporations to the fore. The clash caught the transit company's management and its political allies off guard, leading them to respond with great force, and the conflict became one of the largest, most violent general strikes in American history.

After the turmoil of 1910, leaders of Philadelphia's business and political class swore their city could not tolerate such conflict again. They brought in Thomas Mitten, who over nearly twenty years in power brought mostly peaceful labor relations and a boosterish hope that the PRT was leading the way in answering the labor question. To Mitten, his combination of a company union and a suite of benefits including the cooperative plan and stock purchases heralded a new order where labor would become capital. At least so he claimed. Over time it became apparent to many transit workers and an increasing number of commentators that Mitten offered no salvation. As the Mitten Plan matured, it led transit employees to mix up their interests as workers and petit investors and implicated them in their own workplace subordination. They banished the ineffectual Amalgamated and voted to forgo raises in favor of maintaining their stock purchases. It took Mitten's death and the near simultaneous onset of the Great Depression to expose the Mitten Plan as the quasi–pyramid scheme that it was. Mitten's methods of control—using "softer" power to eliminate real unionism, sanding down the rougher edges of capitalist exploitation, and drawing workers into supporting their own subordination—felt different, less combative, than the previous era. But the resulting power of management over its workforce, the Mitten era's answer to the labor question, was notably similar.

The Mitten Plan's successful implementation highlights the great impact that finances had on the history of the transit system. Mitten held power in Philadelphia in an era of high urban population density and general economic prosperity before the automobile posed an existential crisis to public transportation. With the PRT at the height of its market dominance, Mitten could afford the raises that kept workers mostly docile, especially when the company deducted money from the employees' wages to pay for the stock purchases, welfare plan, and other fringe benefits. Labor's quiescence on the transit system went hand in hand with a rare moment of financial strength in the company's history. This financial strength also enabled Mitten to put resources into new services and technologies (including one of the nation's earliest airlines) that made the PRT a national leader in the provision of transportation.

Yet the PRT's market dominance also raised questions, particularly about whether a privately held company could best meet the city's needs. By consolidating subway, elevated, trolley, bus, and taxi service, Mitten cornered the market on transportation in Philadelphia, making his company one of the most unified and economically efficient in the nation. A financially beneficial situation for Mitten and his stockholders, however, did not translate into a system that provided optimal transportation for all Philadelphians. The city's planners called for a great expansion of the rapid transit system to help the city grow and bring more land for housing and industry into the orbit of Center City, thus expanding the tax base. This required Mitten's agreement and the PRT's investment, a challenge because new subways and elevateds would siphon traffic off the streetcars, lines that the PRT had spent significant money to construct. How, Mitten wanted to know, could he justify to his stockholders the expenditure of funds (that could otherwise go to dividends) on rapid transit lines that might help riders but reduce the company's income? Mitten, for all his entrepreneurialism, his funeral cars, and his airlines to Washington, D.C., purposely retarded the growth of Philadelphia's transit system at one of the few moments it had the resources to do more. As was often the case, private capital first had to look after its own interests, not those of the public the company was charged to serve.

The Great Depression and World War II challenged management's ability to control its workforce. The onset of the Depression caused Mitten's plan, his empire, to crumble. Financial difficulties hit the PRT with a vengeance, halting dividends, eliminating grander plans for the system, stopping most development under way, suspending much basic investment. Difficult times and the continuation of the company union in the guise of the PRTEU kept transit workers, fearful of losing their jobs, restrained at least in the Depression's early years. But management could not keep the activist ferment of the era, the widespread and searching questions about the verities of capitalism, at bay. Working-class activism drew upon many sources, chief among them the rise of the CIO and its brand of industrial unionism, the growth of a liberal Democratic Party that finally challenged the Pennsylvania GOP, and the establishment of a New Deal labor law regime that outlawed company unions and opened space for Philadelphia's transit workers to organize a real union for the first time since 1910. It took years for the city's transit workers to fully embrace and exercise their rights, but in 1944 the PTC's workforce elected the TWU to replace the PRTEU. The new union quickly demonstrated that it would be a champion for its members, a difficult and expensive foe for management.

With the PTC's finances in a precarious state and the company union banned, management groped for a new strategy and ultimately found race to be the issue that could cleave the workforce. But much like the atomic bomb that symbolized the ultimate destructiveness of the era, racism proved to be too powerful, too dangerous, of a weapon. PTC managers first experimented in limited ways with red-baiting tactics but learned that they found little purchase with the workforce. Workers who just months before had joined a CIO union because it would champion their interests had little time for charges that it advanced a Communist agenda. Pivoting to racist innuendo, employing a strategy that built on decades of managerial practices that had created a racially segmented workforce, enabled management to gain far greater traction. PTC managers certainly did not invent racism among their workers. In fact, the naked racism displayed by many white PTC employees marks one of the lowest points in their history and one of the lowest for the transit system more generally. Still, managers fanned the flames, created an environment accepting of racist talk and actions, and were thus complicit in fomenting one of the nation's worst hate strikes during World War II.

With the city on the edge of mass racial conflict, and war production endangered just weeks after D-day, the federal government had no choice but to put down the disturbance. President Roosevelt authorized the deployment of five thousand armed troops, and they quickly ended the strike. Although the 1944 hate strike highlighted the racism of many white employees, the fact that race faded so quickly as an issue emphasizes the role management played in fomenting the strike as a tool for undermining the TWU.

After the war, the TWU, and organized labor more generally, were at high tide. Unions had their highest membership rate in American history, political clout, and unprecedented financial resources (although they of course did not rival corporations and their trade groups). At the PTC, transit workers and TWU leaders challenged the company over traditional matters of wages and working conditions, but also over layoffs of employees, adoption of equipment, maintenance of vehicles and track, and many other issues. The relative, and rare, balance of power left management casting about once again for new tactics. Open violence was no longer acceptable, company unions were illegal, racism was too inflammatory. Surprisingly, PTC management never tried blatant red-baiting. Instead, it launched a vigorous, and rather successful, public relations campaign that emphasized the TWU as an outsider attempting to hold the company and its riders hostage. The PTC, management argued with language that had notes drawn from prevailing anticommunist discourse, wanted to provide proper service while earning a reasonable return on investment, but the "greedy," "alien" union leaders were forcing the company to raise wages, and thus fares, or go bankrupt. Transit employees, in this portrayal, were a special interest looking out for their own pay, the public be damned.

The TWU knew how damaging this attack was and tried to respond with its own claims on the public's allegiance. TWU leader Michael Quill, local leaders, and rank-and-file workers repeatedly argued that transportation employees labored for long, hard hours, that they deserved pay commensurate with the work they performed, and that rising fares had more to do with the PTC's commitment to bond payments, stock dividends, and high managerial wages than with workers' efforts to keep up with postwar inflation. The TWU's arguments helped build closer ties with the city's new liberal political leadership, which was sympathetic to the workers' cause and recognized the importance of public transportation to Philadelphia's growth and economic vitality. But the press and many, although by no means all, Philadelphians bought the PTC's case. This led them to concur with the company and voice the argument that the system's union and its workers were separate from, and dangerous to, ordinary Philadelphians.

The TWU no doubt secured higher wages for its members, but management's success in its public relations campaign gave the company greater latitude in controlling the other issues at stake. Workers had little say about layoffs, fares, equipment, and routes. In the first decade after World War II, Philadelphia's transit employees got the most satisfying answer to the labor question that they ever received, but that positive response was limited to union acceptance, wages, and hours. Management's success in its public relations campaign enabled it to take other issues of vital interest to the workforce off the table, allowing the PTC to offset higher wage costs with layoffs and fare increases. Higher wages actually thus cost the company little. At the same time, fare increases pitted much of the public against the transit workers and helped set the context for the politics surrounding public service employees for decades to come.

The final chapter of private ownership of public transportation in Philadelphia saw NCL obtain control of the system and take to new levels the priorities of capital over those of workers, riders, or political leaders. NCL officials, especially Douglas Pratt, used their notions of the imperatives of capitalism to speed the trolley-to-bus conversion process, defer maintenance, and siphon as much money as possible out of the city's transit system. These policies troubled Philadelphians from all backgrounds who understood the investment they had made in their transit system, an investment that far outstripped what NCL had paid for its shares of stock. Moreover, NCL's policies drove riders off the system, exacerbating the gridlock that paralyzed the city every workday. For the PTC's workers, layoffs were as problematic as any NCL effort, and with the company's offensive putting the TWU on its heels, rank-and-file workers felt they had no choice but to use wildcat strikes to protest management's actions. Mayors Clark, Dilworth, and other liberal politicians saw the chaos that NCL's policies brought to the transit system and its workforce and sided ever more strongly with the employees. NCL's strident stance even moderated the anti–transit worker politics that had grown in the previous decade. As the labor turmoil continued, Clark, Dilworth, and other political leaders recognized that the city's only hope was a consolidated transit system operated under public authority. A multitude of factors—management's desire to maximize its profits and assert its dominance over the TWU, labor's running effort to combat management's agenda, the battle over technology (the conversion to buses) and its implications for working-class jobs, and the financial pressures instilled by capitalism—all came to a head in the 1960s as Philadelphia finally converted the world's largest privately owned public transportation company into a publicly held service. By that point, there were no profits left to be had in the industry, and investors wanted to withdraw their capital for more lucrative pursuits.

SEPTA took over a system gutted by decades of extractive managerial policies that pulled resources out of the company while deferring maintenance, paring back service, and laying off workers. Philadelphia's political officials, liberal or not, did not have a great desire to purchase the PTC. But they recognized the essential nature of public transportation to an urban center and knew by the mid-1960s, as ridership declined and resources became scarce in American cities, that they had no choice but to take over the private company. In the Reagan era and afterward to this day, people who champion privatization have forgotten or ignored this history: American cities, Philadelphia in this case, have a long track record of providing for public needs through private companies, and as the profits dissipated, so did the service. Reagan and his followers did not understand, or at least did not want to admit this point, a fact that Clark, Dilworth, and other Philadelphia politicians knew. By the time SEPTA took over, urban public transportation was a vital industry, but no longer a profitable one. The lack of investment for so many years, combined with the declining economic condition of the industry and scanty support from public sources, condemned SEPTA to decades of tight budgets and financial crises. For workers on the system, SEPTA's economic condition meant that the need to fight for recognition of their interests, for decent wages and hours, continued unabated, albeit in yet another form.

For nearly a century, management at Philadelphia's various transit companies engaged in a quest to control its workforce. For transit workers, the fact that their company could not pick up and move seemed at first blush to give them an advantage over time that their compatriots in the city's garment factories and textile mills did not have. Yet in fact management's inability over the generations to haul stakes for what looked to be more hospitable climes forced it to become shrewdly adaptive, to develop a repertoire of tools to try to control its workforce. In doing so, management's actions highlighted the intrinsically supple nature of capital's efforts at social control. Managers proved themselves adept at reading the political, economic, and social cues and developing their strategies accordingly. All along the line, they found their agenda contested by Philadelphia's transit workers, who staged an array of protests, strikes, and organizing campaigns. The workforce seldom had equal footing, never the upper hand, but they too proved themselves to be shrewd and tenacious in defending their interests. Capital and labor's contest, the key theme in this history of Philadelphia's transit system, did not play out in a vacuum. The financial condition of the industry, as well as the nation's political and social ideologies, always loomed as backdrop to their dance, helping to set constraints on what each side could do or even consider as possible. Given all these constraints, capital's great power in American society, and management's enduring quest to control its workforce, it is remarkable to see by the time SEPTA arrived how far Philadelphia's transit workers had come, how much they had achieved.

Notes

CAPITAL AND THE SHIFTING NATURE OF SOCIAL CONTROL

1. Easton, *Practical Treatise*, 7; Morley, *Travels in Philadelphia*, 60.

2. Speirs, *Street Railway System*, 97.

3. Sklansky, "Labor, Money, and the Financial Turn," 34. The literature on the links among capitalism, labor, and the working class is vast. For a primer see ibid., 23–46. Beckert, "History of American Capitalism," and Sklansky, "Elusive Sovereign," 233–48, are also useful. Two studies of particular value on the way capital has tried to control its workforce, especially by moving to new locations when workers organize, are Cowie, *Capital Moves*, and Sidorick, *Condensed Capitalism*. For recent works on the "labor question" see Currarino, *Labor Question*; Stromquist, *Reinventing "The People."* Informative analyses of the connections among capital, class conflict, and urban geography may be found in Stowell, *Streets, Railroads, and the Great Strike of 1877*; Welke, *Recasting American Liberty*; and Herod, "From a Geography of Labor," 1–31. On the politics of public service work, with a focus on Philadelphia, see Ryan, *AFSCME's Philadelphia Story*.

4. Skelsey, "Streetcar Named Endure," 11. Works examining transit systems in other major U.S. cities include Hood, *722 Miles*; Freeman, *In Transit*; Warner, *Streetcar Suburbs*; and Barrett, *Automobile and Urban Transit*. Even one of the nation's newest transit systems, the Metro in Washington, D.C., has found its historian in Schrag, *Great Society Subway*. Charles Cheape provided the most sustained analysis of Philadelphia's system in his comparative study *Moving the Masses*.

5. Recent scholarship on workers and globalization includes Cowie, *Capital Moves*; Sidorick, *Condensed Capitalism*; and Fink, *Sweatshops at Sea*.

6. To sample the literature on the connections among public transportation, technology, and urban growth see Cheape, *Moving the Masses*; McShane, *Technology and Reform*; McShane, *Down the Asphalt Path*; Post, *Urban Mass Transit*; Warner, *Streetcar Suburbs*; Cudahy, *Cash, Tokens, and Transfers*; Barrett, *Automobile and Urban Transit*; Hood, *722 Miles*; and Schrag, *Great Society Subway*. A handful of studies do focus on workers in transportation. See Freeman, *In Transit*, and Molloy, *Trolley Wars*. For a non–United States history see Rosenthal, "Arrival of the Electric Streetcar," and Rosenthal, "Streetcar Workers."

7. Warner, *Private City*. Debates over "gas and water" socialism—the desire to wrest control of public services from private interests—shook many cities in the United States, especially in the Progressive Era. For examples see Leidenberger, *Chicago's Progressive Alliance*, and Righter, *Battle over Hetch Hetchy*.

8. On the labor question, in addition to Currarino, *Labor Question*, and Stromquist, *Reinventing "The People,"* see Schneirov, *Labor and Urban Politics*, and Leidenberger, *Chicago's Progressive Alliance*.

9. Recent influential works on twentieth-century Philadelphia include Countryman, *Up South*; McKee, *Problem of Jobs*; Wolfinger, *Philadelphia Divided*; Ryan, *AFSCME's Philadelphia Story*; Cole, *Wobblies on the Waterfront*; Perkiss, *Making Good Neighbors*; Arnold, *Building the Beloved Community*; Hepp, *Middle-Class City*; and Delmont, *Nicest Kids in Town*.

1. BEGINNINGS

1. Hallowell and Creighton, *Frankford*, 5.

2. Federal Writers' Project, *Philadelphia*, 142; Andrews, *Short History*, 1–2; Vance, *Capturing the Horizon*, 174–75; UTTB, *History of Public Transportation*, I-1.

3. Andrews, *Short History*, 1.

4. Vance, *Capturing the Horizon*, chaps. 3–5, provides illustrative comparisons between the United States and Europe.

5. Speirs, *Street Railway System*, 9–10.

6. Ibid.

7. Andrews, *Short History*, 3; UTTB, *History of Public Transportation*, III-1; Speirs, *Street Railway System*, 10–11.

8. McShane and Tarr, *Horse in the City*, on urban transit see especially chap. 3; Cudahy, *Cash, Tokens, and Transfers*, 7; Vance, *Capturing the Horizon*, 358–59, 363; Moffat, "*L*," 8; McChord, *Report of C. C. McChord*, 6–7; Miller, "History of the Transit System," 3; UTTB, *History of Public Transportation*, VIII-3–4; Ford, Bacon & Davis, "Pennsylvania State Railroad Commission," 120. On Philadelphia's growth in this period, in large part because of improved transportation, see Hershberg et al., "'Journey-to-Work.'"

9. Vance, *Capturing the Horizon*, 129; McChord, *Report of C. C. McChord*, 7; *Facts Respecting Street Railways*, 22, 24, 27–28; Easton, *Practical Treatise*, 4–7.

10. Edmund Stirling, "Hardships of Car Riding in Early Days of System," file 15, box 3, John F. Tucker Papers, Hagley Museum and Library, Wilmington, Del.; Meyers and Spivak, *Philadelphia Trolleys*, 30; Cheape, *Moving the Masses*, 5; McShane, *Down the Asphalt Path*, 18; Vance, *Capturing the Horizon*, 377; Post, *Urban Mass Transit*, 22, 32; McShane and Tarr, *Horse in the City*, chap. 3.

11. UTTB, *History of Public Transportation*, II-1, unpaginated; Simon, *Philadelphia*, 33; McShane, *Down the Asphalt Path*, 1; Stowell, *Streets, Railroads, and the Great Strike of 1877*.

12. Andrews, *Short History*, 15; Hallowell and Thomas, *Frankford*, 35, 39; Meyers and Spivak, *Philadelphia Trolleys*, 20; Schmidt, *Industrial Relations*, 9 (Martin quoted on p. 10); McShane, *Technology and Reform*, 8.

13. Meyers and Spivak, *Philadelphia Trolleys*, 21; McShane, *Technology and Reform*, 8–9; Post, *Urban Mass Transit*, 32; Cudahy, *Cash, Tokens, and Transfers*, 22, 27, 32–33; Vance, *Capturing the Horizon*, 376–77; SEPTA, *SEPTA History*, unpaginated; Skelsey, "Streetcar Named Endure," unpaginated; Cheape, *Moving the Masses*, 6; McLain, "Street Railways of Philadelphia," 237.

14. Schmidt, *Industrial Relations*, 12–15; Vance, *Capturing the Horizon*, 378–85.

15. Schmidt, *Industrial Relations*, 14, 16; Moffat, "*L*," 37–38; Edmund Stirling, "New Controversy Attended the Switch to Electricity," file 15, box 3, Tucker Papers; Higgins, *Street Railway Investments*, 8, 12; Cudahy, *Cash, Tokens, and Transfers*, 49; Barger, *Transportation Industries*, 216.

16. Andrews, *Short History*, 22–28; "Street Railways in the United States," *Motorman and Conductor* 2 (March 1896): 3; Chandler Brothers & Co., *Philadelphia Rapid Transit Company*, 3, 7; Cheape, *Moving the Masses*, 169.

17. Steffens, *Shame of the Cities*, 193.

18. Andrews, *Short History*, 12; Speirs, *Street Railway System*, 28–30; Edmund Stirling, "Fifty Companies Make Up Present Transit System," file 15, box 3, Tucker Papers; Roberts, "History and Analysis," 2; UTTB, *History of Public Transportation*, IX-1–3; Higgins, *Street Railway Investments*, 87.

19. Higgins, *Street Railway Investments*, 98; Schmidt, *Industrial Relations*, 28–32.

20. Cudahy, *Cash, Tokens, and Transfers*, 91; Moffat, "*L*," 68; Burton J. Hendrick, "Great American Fortunes and Their Making," *McClure's Magazine*, December 1907, 34, 236;

Cheape, *Moving the Masses*, 162–63, 174, see part 2 on Boston; "Southeastern Pennsylvania Transportation Authority, History," 2, finding aid, SEPTA Collection, Hagley Museum and Library, Wilmington, Del.; McCaffery, *When Bosses Ruled Philadelphia*, 146–47.

21. McChord, *Report of C. C. McChord*, 9–11; Cox and Meyers, "Philadelphia Traction Monopoly," 415–16; *Report of Transit Advisory Committee*, 17–18; Cheape, *Moving the Masses, 179.*

22. McShane, *Technology and Reform*, 83, 87, 102; McShane, *Down the Asphalt Path*, 115. On urban growth and the politics of transportation see, for example, Barrett, *Automobile and Urban Transit*. On urban growth boosterism, often with a critical eye, see Warner, *Streetcar Suburbs*; Hood, *722 Miles*; Pushkarev and Zupan, *Urban Rail in America*, 1–4; Goddard, *Getting There*, 68.

23. Simon, *Philadelphia*, 134; Ford, Bacon & Davis, "Pennsylvania State Railroad Commission," 115; Cheape, *Moving the Masses*, 199; McShane, *Down the Asphalt Path*, 26; Metropolitan Rail Road Company, *Rapid Transit*, 9, 16, 21; *Philadelphia Rapid Transit*, unpaginated; *The Overhead Electric Trolley Ordinances*, 23; Parry, *History of Rapid Transit*, 1.

24. Speirs, *Street Railway System*, 94–95; McCaffery, *When Bosses Ruled Philadelphia*, 17–22; Schmidt, *Industrial Relations*, 34; Barrett, *Automobile and Urban Transit*, 40; Goddard, *Getting There*, 70–71; Painter, *Standing at Armageddon*, 33; McLain, "Street Railways of Philadelphia," 250–51; Edmund Stirling, "Financial Pyramid Set Up by Street-Railway System," file 15, box 3, Tucker Papers.

25. McShane, *Technology and Reform*, 25; McLain, "Street Railways of Philadelphia," 253–55; Caskie, "Philadelphia Rapid Transit Plan," 189; UTTB, *History of Public Transportation*, X-1; Feustel, *Report on Behalf of the City of Philadelphia*, 92–95; UTTB, *Survey of Investigations*, 44; McChord, *Report of C. C. McChord*, 14–15; *Philadelphia Transit System*, 5. The clearest listing of Philadelphia's many transit companies and the way they were consolidated may be found in McCaffery, *When Bosses Ruled Philadelphia*, 154.

26. Cox and Meyers, "Philadelphia Traction Monopoly," 407–10; Speirs, *Street Railway System*, 76, 80, 117; Edmund Stirling, "Tricks Invented to Evade Car Lines' Tax Obligations," and Edmund Stirling, "New Controversy Attended the Switch to Electricity," file 15, box 3, Tucker Papers.

27. Cox and Meyers, "Philadelphia Traction Monopoly," 409–17.

28. Woodruff, "Philadelphia Street-Railway Franchises," 216, 221; Conway, "Street Railways," 73.

29. Conway, "Decreasing Financial Returns," 15; Post, *Urban Mass Transit*, 92–93; Edmund Stirling, "Today's Problems Rooted in Beginnings of Car Service," file 15, box 3, Tucker Papers; Shaw, *Mitten*, 227.

30. Cheape, *Moving the Masses*, 159; Edmund Stirling, "Fifty Companies Make Up Present Transit System," file 15, box 3, Tucker Papers; *Rapid Transit Talks* (1910): 5; "Says God Raised Fares," book 11, box 115, "City to Make Up Transit Deficit," "Transit Paving Deal Robs City $125,000 a Year, Audit Shows," "A Stockholder on Transit Service," "Mayor Censors Vaudeville Song," book 5, box 113, Harold Cox Papers, Historical Society of Pennsylvania, Philadelphia. The Cox Papers, as these endnotes make clear, are an indispensable source base for this book. The papers include correspondence and reports, and, even more valuably, dozens of scrapbooks prepared in the late nineteenth and early twentieth century with what appears to be every newspaper article related to public transportation published in Philadelphia's newspapers. Although vital to this project, the scrapbooks have their limitations: the numbering system is somewhat haphazard, stopping and starting over again; the pages are not always numbered; and a number of articles lack dates or the name of the newspaper in which they were published. In these

endnotes, I chose to use uniform citations, giving the title of the article, the number of the book, and the number of the box that contains the pertinent scrapbook.

31. Ford, Bacon & Davis, "Pennsylvania State Railroad Commission," 5–7, 18; "Complain of City Trolleys," "Demand for Heated Cars," book 15, box 117, political cartoons in book 5, box 113, Cox Papers; Cox, *Early Electric Cars*, 117; *Overhead Electric Trolley Ordinances*, 13, 18. The lack of heat in transit cars was a complaint common to other cities too. Cudahy cites the example of Saint Louis in *Cash, Tokens, and Transfers*, 25.

32. *Overhead Electric Trolley Ordinances*, 14–15, 20, 27, 30; Cheape, *Moving the Masses*, 172–73.

33. Speirs, *Street Railway System*, 99, 105–9.

34. Welke explores how steam and street railroads made danger a more central element in most Americans' lives in *Recasting American Liberty*.

35. Speirs, *Street Railway System*, 148; Post, *Urban Mass Transit*, 46; *Overhead Electric Trolley Ordinances*, 8–10, 35–36; *Annie E. Chittick and John F. Chittick, Her Husband vs. Philadelphia Rapid Transit Co.*, *Pennsylvania State Reports*, vol. 224, 9. For more on lawsuits against the PRT see cases in the *Pennsylvania State Reports* more generally.

36. Second Vice President to Mr. T. W. Wilson, April 22, 1907, file JR-62–9, box 62, "Man Killed under Loaded Ash Car," file JR-52–1, box 52, Cox Papers; Harring, "Car Wars," 868; Amalgamated Transit Union Staff, *ATU*, 25; *Rachel M. Saunders vs. Philadelphia Rapid Transit Company*, *Pennsylvania State Reports*, vol. 240, 12–13; Ford, Bacon & Davis, "Pennsylvania State Railroad Commission," 77–78. For more on urban transportation changing the nature of public streets see McShane, *Down the Asphalt Path*, 62; Barrett, *Automobile and Urban Transit*, 58.

37. "The Slaughter of Children," "Car Kills Baby of 3; Is Eighteenth Child Victim since First of Year," book 27, box 123, "Newsboy Killed by Car," book 15, box 117, "Girl Electrocuted under Car; Throng Attacks the Crew," book 11, box 115, Cox Papers.

38. "Attack Motorman When Girl Meets Death under Car," book 11, box 115, "Streetcar Accidents to Children," book 15, box 117, Cox Papers.

39. McShane, *Technology and Reform*, 16; Ford, Bacon & Davis, "Pennsylvania State Railroad Commission," 35; "Fender Kills Girl of 7; Is Victim 32," "Call Mass Meeting to Compel P.R.T. to Adopt New Fenders," "Fender Inventors Asked for Models," "Woman, Stirred by Child Deaths, Invents Fender," book 27, box 123, "Get on the Right Side of Trolleys," "Board Cars on Right Side," book 5, box 113, PRT press release, June 26, 1909, file JR-52–1, box 52, Cox Papers; Cox, *Early Electric Cars*, 18.

40. Leidenberger, *Chicago's Progressive Alliance*; Righter, *Battle over Hetch Hetchy*; Molloy, *Trolley Wars*; Painter, *Standing at Armageddon*; Hood, *722 Miles*, 125.

41. Easton quoted in Cudahy, *Cash, Tokens, and Transfers*, 27; *Nation* quoted in McShane, *Down the Asphalt Path*, 43; Easton, *Practical Treatise*, 60. Such tensions were common to many American cities. For Chicago see Barrett, *Automobile and Urban Transit*, 3; for New York City see Cheape, *Moving the Masses*, 42, and Hood, *722 Miles*, 125.

42. Schmidt, *Industrial Relations*, 51–52, 63–64; Cheape, *Moving the Masses*, 31, 63; Barrett, *Automobile and Urban Transit*, 4, 83. For an in-depth contemporary study of regulation and municipal ownership see Johnson, "Public Regulation," 31–47.

43. Cheape, *Moving the Masses*, 177, 182.

44. Speirs, *Street Railway System*, 33, 50–51, 87, 97, 104; Conway, "Street Railways," 354; Roberts, "History and Analysis," 11.

45. Speirs, *Street Railway System*, 51; Cheape, *Moving the Masses*, 188–89; Meyers and Spivak, *Philadelphia Trolleys*, 124.

46. *New York Evening Call*, March 5, 1910, p. 2, February 28, 1910, p. 2, March 9, 1910, p. 2; Taylor, "Philadelphia's Transit Problem," 30.

2. WORKING ON THE LINE

1. Toynbee Society, *Philadelphia Trolley Companies*, 13.

2. Fones-Wolf, *Trade Union Gospel*, 159; Amalgamated's call found in its journal, *Motorman and Conductor*, as quoted in Schmidt, *Industrial Relations*, 78.

3. Harris, *Bloodless Victories*, 29; Scranton, "Large Firms," 419, 423, 430; Simon, *Philadelphia*, 64–67; Burt and Davies, "Iron Age," 479–83.

4. Harris, *Bloodless Victories*, 29.

5. Currarino, *Labor Question*, 1, Gompers quoted on p. 87; Montgomery, *Citizen Worker*; Montgomery, *Workers' Control*; Trachtenberg, *Incorporation of America*; Wiebe, *Businessmen and Reform*, vii; Toynbee Society, *Philadelphia Trolley Companies*, 8; Painter, *Standing at Armageddon*, 14.

6. Stromquist, *Generation of Boomers*, 4; Adamic, *Dynamite*, 111; Painter, *Standing at Armageddon*, xxiv–xvii, 176; Bruce, *1877*, 19, 68–69; Higbie, *Indispensable Outcasts*; Montgomery, *Workers' Control*, 159; Bonsall, *Handbook of Social Laws*.

7. Federal Writers' Project, *Philadelphia*, 149–50, 153; Fones-Wolf, "Industrial Giant," 53.

8. Adamic, *Dynamite*, 18–19, 50; Brecher, *Strike!*, 28, 30; Fink, *Workingmen's Democracy*; Blatz, "Titanic Struggles," 104–5; Greene, *Pure and Simple Politics*.

9. Green, *Death in the Haymarket*; Brenner, Day, and Ness, *Encyclopedia of Strikes*; Montgomery, *Workers' Control*, 20, 159–60; Bruce, *1877*, 90–92; Gilje, *Rioting*, 3.

10. Stowell, *Streets, Railroads, and the Great Strike of 1877*, introduction; Bruce, *1877*, 90–91; Stromquist, *Generation of Boomers*, 3; White, *Railroaded*; Adamic, *Dynamite*, 28–29 and chap. 3 more generally; Gilje, *Rioting*, 118.

11. Stromquist, *Generation of Boomers*, 18; Scott quoted in Painter, *Standing at Armageddon*, 21; Adamic, *Dynamite*, 85; Norwood, *Strikebreaking and Intimidation*, 120–25; Bruce, *1877*, 304; Fogelson, *America's Armories*, 40, 57–62, 140–42, 208, 213, 217.

12. Bruce, *1877*, 56; Painter, *Standing at Armageddon*, 95; Adamic, *Dynamite*, 101–2; Arnesen, *Waterfront Workers*, 130–35; Montgomery, *Workers' Control*, 24, 59.

13. Greene, *Pure and Simple Politics*, 91; Wiebe, *Businessmen and Reform*, 25–33, 169, 191, Philadelphia Board of Trade quoted on p. 197; Wike, "Pennsylvania Manufacturers' Association."

14. Wiebe, *Businessmen and Reform*, 111; Greene, *Pure and Simple Politics*, 104; Drayer, "J. Hampton Moore"; Wolfinger, *Philadelphia Divided*, 29; Simon, *Philadelphia*, 70.

15. Department of Commerce and Labor, *Bulletin of the Bureau of Labor*, 550, 559, 563, 578–79; Schmidt, *Industrial Relations*, 27.

16. Department of Commerce and Labor, *Bulletin of the Bureau of Labor*, 564; Fairchild, *Training for the Electric Railway Business*, 52, 55; Schmidt, *Industrial Relations*, 85–86; Toynbee Society, *Philadelphia Trolley Companies*, 13.

17. Department of Commerce and Labor, *Bulletin of the Bureau of Labor*, 551, 568; "All Sorts of Men Run Street Cars," book 6, box 113, Cox Papers; Bayor and Meagher, *New York Irish*, 229; Clark, *Irish Relations*, 166; Golab, "Immigrant and the City," 204–5; Simon, *Philadelphia*, 134; Arnold, "Building the Beloved Community," 207; Lane, *Roots of Violence*, 39; Schmidt, *Industrial Relations*, 166. On Philadelphia's immigrant history see also Golab, *Immigrant Destinations*. The PRT did not keep records on the ethnicity of its workforce, so I pieced together the demographics based on sources in the Albert Greenfield Papers, corporate journals, and the United States census. I discuss this methodology in *Philadelphia Divided*, p. 264.

18. Goddard, *Getting There*, 72; "All Sorts of Men Run Street Cars," book 6, box 113, Cox Papers; Mahon, "History of Organization," *Motorman and Conductor* 16 (April 1908): 7; Amalgamated Transit Union Staff, *ATU*, 3; Toynbee Society, *Philadelphia*

Trolley Companies, 11–13; Department of Commerce and Labor, *Bulletin of the Bureau of Labor*, 638, 642; Speirs, *Street Railway System*, 100–101.

19. "Car Man's Leg Burst from Long Standing," book 6, box 113, "Shirtwaist Motorman Arrives," book 12, box 112, Cox Papers; Charles Kruger to John B. Parsons, June 13, 1906, file JR-52–10, box 52, Cox Papers; Mahon, "History of Organization," *Motorman and Conductor* 13 (November 1905): 18; Schmidt, *Industrial Relations*, 142; Toynbee Society, *Philadelphia Trolley Companies*, 12–13; Department of Commerce and Labor, *Bulletin of the Bureau of Labor*, 644; Cudahy, *Cash, Tokens, and Transfers*, 145; Edmund Stirling, "Fare Disputes with Public and Trouble with Employes," file 15, box 3, Tucker Papers.

20. Amalgamated Transit Union Staff, *ATU*, 25; "Carman Electrocuted on His First Day at Work," unnumbered book, box 123, F. W. Johnson to Charles O. Kruger, September 9, 1909, file JR-53–7, box 53, Cox Papers; *Yearbook of the Amalgamated* (1910): 6–7; Department of Commerce and Labor, *Bulletin of the Bureau of Labor*, 610, 635.

21. Barrett, *Automobile and Urban Transit*, 17; Skelsey, "Streetcar Named Endure," unpaginated; Cox, "Tram Subways," 206; accident reports of S. Money, October 26, 1909, W. McCoach, October 26, 1909, William Sweigart, October 27, 1909, James M. Lets, October 27, 1909, B. Monaghan, October 27, 1909, file JR-54–2, box 54, W. A. Flounders to James Bricker, April 23, 1912, "Ridge Avenue Depot," April 15, 1912, report of John Burns, June 19, 1912, file JR-50–8, box 50, accident reports of S. W. Sedgling, March 14, 1910, and T. Pilson, March 19, 1910, [illegible name] to M. F. Ryan, February 12, 1910, file 47–2, box 47, Cox Papers.

22. C. E. Calkins, "The Motorman," *Motorman and Conductor* 12 (July 1904): 1.

23. Schmidt, *Industrial Relations*, 211–12; Speirs, *Street Railway System*, 100–101; Department of Commerce and Labor, *Bulletin of the Bureau of Labor*, 603, 608–9; Toynbee Society, *Philadelphia Trolley Companies*, 7; "Why Aaron Scott Is a 365-Days-in-the-Year Man," book 16, box 117, Cox Papers.

24. Schmidt, *Industrial Relations*, 71, 78–79, 93; Easton, *Practical Treatise*, 103–11; Department of Commerce and Labor, *Bulletin of the Bureau of Labor*, 568, 581, 584; Cudahy, *Cash, Tokens, and Transfers*, 15; *Philadelphia Public Ledger*, January 8, 1910, p. 1, January 23, 1910, p. 2; [illegible name] to Jas. J. Springer, August 29, 1911, unnumbered file, box 47, Cox Papers.

25. Schmidt, *Industrial Relations*, 81–82; Emmons, "Relations of the Electric Railway Company," 88; Edmund Stirling, "Hardships of Car Riding in Early Days of System," file 15, box 3, Tucker Papers; James Gay to C. O. Kruger, June 23, 1910, untitled file, box 66, Cox Papers.

26. Department of Commerce and Labor, *Bulletin of the Bureau of Labor*, 564, 579; Toynbee Society, *Philadelphia Trolley Companies*, 7; Schmidt, *Industrial Relations*, 82; Emmons, "Relations of the Electric Railway Company," 89, 92; L. E. Summers, "Operating Delays Reduced," book 32, box 120, Cox Papers.

27. Pierce, "Strike Problem," 93; Schmidt, *Industrial Relations*, 83–84; "Statement Relative to the Length of Service of the Conductors and Motormen of the Electric Lines as of June 30, 1909," file JR-63–1, box 63, Cox Papers; Person, "Mitten Plan," 1–2.

28. Schmidt, *Industrial Relations*, 78.

29. Mahon, "History of Organization," *Motorman and Conductor* 12 (July 1904): 6; Amalgamated Transit Union Staff, *ATU*, 4.

30. Mahon, "History of Organization," *Motorman and Conductor* 12 (July 1904): 7; Amalgamated Transit Union Staff, *ATU*, 5.

31. Mahon, "History of Organization, Chapter Two," *Motorman and Conductor* 12 (August 1904): 6; Roberts, "History and Analysis," 2, 5, 8–10, Weidler quote on p. 6; Department of Commerce and Labor, *Bulletin of the Bureau of Labor*, 611; Speirs, *Street Railway System*, 102–3.

32. Speirs, *Street Railway System*, 102–3; Department of Commerce and Labor, *Bulletin of the Bureau of Labor*, 610; Mahon, "History of Organization, Chapter Two," *Motorman and Conductor* 12 (August 1904): 6.

33. Mahon, "History of Organization, Chapter Ten," *Motorman and Conductor* 13 (September 1905): 6; Amalgamated Transit Union Staff, *ATU*, 13.

34. Mahon, "History of Organization, Chapter Thirteen," *Motorman and Conductor* 18 (March 1910): 8; "Objects of the Amalgamated Association," *Motorman and Conductor* 1 (April 1895): 1; Schmidt, *Industrial Relations*, 113, 155–58; Roberts, "History and Analysis," 11; Amalgamated Transit Union Staff, *ATU*, 47–51; Mahon, "History of Organization, Chapter Eight," *Motorman and Conductor* 13 (April 1905): 16; Mahon, "What Has Organization Done for the Street Railway Workers of America?" *Motorman and Conductor* 15 (July 1907): 7.

35. "Objects of the Amalgamated Association," *Motorman and Conductor* 1 (April 1895): 1; Mahon, "History of Organization," *Motorman and Conductor* 12 (July 1904): 1.

36. "The Railway Man's Commandments," *Motorman and Conductor* 1 (April 1895): 16.

37. Mahon, "History of Organization, Chapter Two," *Motorman and Conductor* 12 (August 1904): 6; Harring, "Car Wars," 865; Molloy, "Trolley Wars," 523.

38. Mahon, "History of Organization," *Motorman and Conductor* 18 (February 1910): 18; Mahon, "History of Organization, Chapter Fourteen," *Motorman and Conductor* 14 (May 1906): 15–17; Mahon, "History of Organization, Chapter Twelve," *Motorman and Conductor* 14 (December 1905): 8; Roberts, "History and Analysis," 12; Mahon, "History of Organization, Chapter Seventeen," *Motorman and Conductor* 14 (August 1906): 31; Mahon, "History of Organization, Chapter Nineteen," *Motorman and Conductor* 18 (November 1910): 35.

39. UTTB, *History of Public Transportation*, XI-1; Speirs, *Street Railway System*, 106.

40. Albert and Miller, *American Federation of Labor Records*, reel 142; articles from file Clips Bulletin, Strikes—Transit—Phila.—Hist. and Lists, box 227A, Urban Archives, Temple University, Philadelphia; Mahon, "History of Organization, Chapter 14," *Motorman and Conductor* 18 (April 1910): 36; Roberts, "History and Analysis," 11–12; Schmidt, *Industrial Relations*, 134–35; Speirs, *Street Railway System*, 104–6; "The Philadelphia Strike," *Motorman and Conductor* 3 (May 1897): 2; Toynbee Society, *Philadelphia Trolley Companies*, 8; untitled article, *Motorman and Conductor* 1 (April 1895): 1; Hepp, *Middle-Class City*, 45–47.

41. Mahon, "History of Organization, Chapter Fourteen," *Motorman and Conductor* 14 (May 1906): 16; Speirs, *Street Railway System*, 108–9; *New York Times*, December 18, 1895, pp. 6, 8, December 19, 1895, p. 6, December 20, 1895, p. 1, December 21, 1895, p. 1, December 22, 1895, p. 1, December 26, 1895, p. 1; Albert and Miller, *American Federation of Labor Records*, reel 142; Mahon, "History, Chapter 14," *Motorman and Conductor* 18 (April 1910): 37; Barrett, *William Z. Foster*, 19–20; articles from file Clips Bulletin, Strikes—Transit—Phila.—Hist. and Lists, box 227A, Urban Archives; Johanningsmeier, "Philadelphia 'Skittereen,'" 300–301.

42. Roberts, "History and Analysis," 13; articles from file Clips Bulletin, Strikes—Transit—Phila.—Hist. and Lists, box 227A, Urban Archives; Albert and Miller, *American Federation of Labor Records*, reel 142; Connelly, *History of the Archdiocese*, 295–96; Roberts, "History and Analysis," 14–16; Mahon, "History of Organization, Chapter Fifteen," *Motorman and Conductor* 14 (June 1906): 30; *New York Times*, December 24, 1895, p. 3, December 22, 1895, p. 1; Speirs, *Street Railway System*, 109.

43. Schmidt, *Industrial Relations*, 134; untitled article, *Motorman and Conductor* 2 (January 1896): 2; untitled article, *Motorman and Conductor* 2 (February 1896): 5–6; Mahon, "History of Organization, Chapter Fourteen," *Motorman and Conductor*

18 (April 1910): 36; Mahon, "History of Organization, Chapter Fifteen," *Motorman and Conductor* 14 (June 1906): 30; Roberts, "History and Analysis," 15–16; Speirs, *Street Railway System*, 112; Mahon, "History of Organization, Chapter Sixteen," *Motorman and Conductor* 18 (June 1910): 36; "The Philadelphia Strike," *Motorman and Conductor* 3 (May 1897): 2; Albert and Miller, *American Federation of Labor Records*, reel 142.

44. Speirs, *Street Railway System*, 114.

3. TIME OF TROUBLES

1. Levinson, *I Break Strikes!*, 97–98; articles from file Clips Bulletin, Strikes—Transit—Phila.—Hist. and Lists, file Clips Bulletin, Strikes—Transit—1946 Phila., box 227A, Urban Archives.

2. Higgins, *Street Railway Investments*, 98; Chandler Brothers, *Philadelphia Rapid Transit Company*, 4; Schmidt, *Industrial Relations*, 43, 52.

3. Cheape, *Moving the Masses*, 214; Cunningham, "Electric Railway Stocks," 657–59, 665; Conway, "Decreasing Financial Returns," 15, 27; Schmidt, *Industrial Relations*, 48.

4. Schmidt, *Industrial Relations*, 40, Street Railway trade association quoted on p. 43; Vance, *Capturing the Horizon*, 386, 388; Daniel T. Pierce, "Street Car Problems," file JR-52–1, box 52, Cox Papers.

5. Cunningham, "Electric Railway Stocks," 663–64; Hood, *722 Miles*, 181, 221; Conway, "Decreasing Financial Returns," 30.

6. Chandler Brothers, *Philadelphia Rapid Transit Company*, 9–12; Cheape, *Moving the Masses*, 183–84, 196.

7. Metropolitan Rail Road Company, *Rapid Transit*, 18–22.

8. Cox, *Road from Upper Darby*, 3–5; Edmund Stirling, "Beginnings of the P.R.T. Came in Difficult Times," file 15, box 3, Tucker Papers; Shaw, *Mitten*, 236; PRT, *Annual Report*, 1926, unpaginated; PRT, *Annual Report*, 1929, 7.

9. Cox, *Road from Upper Darby*, 6–9, 16; UTTB, *History of Public Transportation*, XIII-3; Edmund Stirling, "Beginnings of the P.R.T.," file 15, box 3, Tucker Papers.

10. Parry, *History of Rapid Transit Development*, 275; Edmund Stirling, "New Controversy Attended the Switch to Electricity," file 15, box 3, Tucker Papers; *Report of Transit Advisory Committee*, 7, 28–29; *Agreement of July 1, 1907*.

11. *Bulletin Almanac*, 1944, 374; Cheape, *Moving the Masses*, 204; McChord, *Report of C. C. McChord*, 16–17.

12. Lewis, "Philadelphia's Relation," 606, 608, 610; untitled article, *Motorman and Conductor* 18 (March 1910): 11; Lewis, *Street Railway Situation*, 20–21, 26–27.

13. "Philadelphia Strike and Settlement," *Motorman and Conductor* 17 (June 1909): 4; "'Square Deal for All,' Motto of 477," *Motorman and Conductor* 16 (March 1908): 21; "Labor Leaders Make Demands on Parsons," "Trolley Employes Make Demands," "P.R.T. Throws Down Gauntlet to Union," "Transit Employes Meet in Secret," "Car Men Meet and Listen to Leaders," "Street Car Men Declare Strike Is Imminent," "Trolley Men Plan Big Meeting Today," book 5, box 113, Cox Papers; "The Car," "John B. Parsons Dies at Seashore," in file John B. Parsons, *Evening Bulletin* Morgue, Urban Archives, Temple University, Philadelphia (hereafter EBM).

14. "Sketch of Charles O. Kruger," file JR-52–1, box 52, "Violence Is Imminent," "Mass Meeting of Trolley Men Called," "Carmen, Corralled by Company, Vote to Drop Demands," "To the Officials of the Phila. Rapid Transit Co.," "Rapid Transit Runs a Rival Meeting for Men," "Trolley Meeting Fiasco," book 5, box 113, Cox Papers.

15. "P.R.T. Takes Precautions," "Transit Co. Ready for Strike; Men in Doubt," "1800 Men Here as Strike Breakers," "Willow Grove a Fort," untitled article, "Strike Breakers Become Unruly," "James Farley Praises His Strike Breakers," book 5, box 113, Cox Papers; Smith, "King of the Strikebreakers," 21–37.

16. "Clay Orders on Duty Men of Two Bureaus," "Police on War Footing," "Union Leaders Laugh," "Gen. Clay, Retired, Gives Army a Rest," "General Clay, Self-Made Hero, Directs Mighty Army All Night in His Pet War," "Violence Is Imminent," book 5, box 113, Cox Papers.

17. "Transit Company's Stand," "Car Men's Union Accused of Sham," "Union's Reply to Mayor," "Mayor Will Urge Street Car Peace," "Union Leaders Drop Plan of Trolley Strike," "Gompers Not to Interfere in Car Men's Troubles," "Mahon Here for Trolley Labor War," "Mahon Here for Trolley Labor War," book 5, box 113, "Notice to Motormen and Conductors of the Philadelphia Rapid Transit Co.," December 21, 1907, file JR-52–10, box 52, Cox Papers.

18. "Parsons Agrees to Meet Trolley Men," "Carmen to Vote on Company's Offer," "Union Men Accept Parsons' Peace Move," "Street Car Men Rescind Decision," book 6, box 113, Cox Papers.

19. Greene, *Pure and Simple Politics*, 88, 142; Douglas, "Analysis of Strike Statistics," 866–77; Montgomery, *Workers' Control*, 93–98; Tomlins, *State and the Unions*, 14–15; Wiebe, *Businessmen and Reform*, 168, 177. For more on the key workplace issues of the era see Roediger and Foner, *Our Own Time*; Rosenzweig, *Eight Hours*.

20. Greene, *Pure and Simple Politics*, 142, 223; Adamic, *Dynamite*, 157–64; Painter, *Standing at Armageddon*, 259, 262–63; Cole, *Wobblies*.

21. Cole, *Wobblies*; Albert and Miller, *American Federation of Labor Records*, reel 142; Sidorick, "'Girl Army,'" 333–35, 347–50. Many examples of labor activism may be found in Wm. J. Boyle, "Philadelphia," *American Federationist* 17 (February 1910): 159; Wm. J. Boyle, "Philadelphia," *American Federationist* 17 (April 1910): 351.

22. Molloy, *Trolley Wars*, 175; Molloy, "Trolley Wars," 519–33; Young, "St. Louis Streetcar Strike"; Norwood, *Strikebreaking and Intimidation*, 34–45. Streetcar strikes were by no means a phenomenon strictly found in the United States. See, for example, Rosenthal, "Streetcar Workers"; Kaplan, *Red City, Blue Period*, 61–67.

23. Molloy, "Trolley Wars," 525; Lefebvre, *Writings on Cities*, 19–20, 154–59; Herod, "From a Geography of Labor." Molloy, *Trolley Wars*, well captures the stakes in these clashes in the streetcar industry.

24. "C.O. Pratt, Leader in Car Strike Dies," "Clarence O. Pratt, Professional Strike Organizer," "Manager of Car Strike Has Won Many Victories," in file Pratt, EBM.

25. "In Solid Phalanx Workingmen Walk to Their Homes to Express Indignation over Fare Extortion" and other articles in scrapbook 11, box 115, "Car Men Organize to Get More Pay," scrapbook 5, box 113, Cox Papers; "Report of Transit Advisory Committee," 28; *New York Evening Call*, June 4, 1909, p. 1, June 5, 1909, p. 2.

26. Roberts, "History and Analysis," 23–24; Cameron, *Radicals of the Worst Sort*, 3; "Municipal Ownership," AFL-CIO, *Philadelphia Council (Pa.) Records*, reel 1; untitled article, *Motorman and Conductor* 18 (May 1910): 15; "How It Came About," *Motorman and Conductor* 17 (June 1909): 21; "Strikes in the Public Service," *Outlook* 92 (June 1909): 339; *Philadelphia Public Ledger*, January 18, 1910, p. 2; "Philadelphia Car Strike—Its Lessons," *American Federationist* 16 (July 1909): 605. Influential examples of works connecting gender and labor history include Kessler-Harris, *Gendering Labor History*, and Milkman, *Women, Work and Protest*.

27. "Strike-Breaker Dies after Attack on Car," multiple other articles in scrapbook 13, box 117, F. W. Johnson to Charles Kruger, June 26, 1909, file JR-53–7, box 53, Cox Papers; *New York Times*, May 30, 1909, p. 1, June 3, 1909, p. 1, June 5, 1909, p. 2; Norwood, *Strikebreaking and Intimidation*, 39.

28. Smith, *From Blackjacks to Briefcases*, 39; Norwood, *Strikebreaking and Intimidation*, 66; correspondence between Pearl Bergoff and Charles Kruger, file 5173, box 6, multiple articles in scrapbook 11, box 115, Cox Papers; *New York Times*, June 2, 1909, p. 2, June 1,

1909, p. 1; Kahn, *High Treason*, 133–34; Samuel Gompers, "Decadent Philadelphia," *American Federationist* 17 (May 1910): 403; Levinson, *I Break Strikes!*, 49–50.

29. *New York Times*, June 1, 1909, p. 1, June 2, 1909, p. 2, June 4, 1909, p. 2; multiple articles in scrapbook 13, box 117, Cox Papers; Foner, *History of the Labor Movement*, 144; Fones-Wolf, "Mass Strikes, Corporate Strategies," 448; "The General Strike in Philadelphia," *Current Literature* 48 (April 1910): 362; Roberts, "History and Analysis," 30–31.

30. "How It Came About," 21, "Philadelphia Strike and Settlement," 7, "Quaker City News," 22, "Would Classify Membership," 32, 36, all in *Motorman and Conductor* 17 (September 1909); multiple articles in scrapbook 13, box 117, Cox Papers; Harold Howland, "The War in Philadelphia," *Outlook* 5 (March 1910): 524; Pratt, "Sympathetic Strike," 142.

31. Levinson, *I Break Strikes!*, 89–94; articles from file Clips Bulletin, Strikes—Transit—1910 Phila., Urban Archives; M. H. Carpenter to Charles Kruger, September 18, 1909, file JR-55–15, box 55, "Constitution and Bylaws of the United Carmen's Association of America," p. 10, Charles Kruger to Conrad Albright, June 21, 1911, file JR-64–3, box 64, Cox Papers; "Philadelphia Lockout Culmination of Long-standing Contention," *Motorman and Conductor* 18 (March 1910): 18; Foner, *History of the Labor Movement*, 146–47; Harold Howland, "The War in Philadelphia," *Outlook* 5 (March 1910): 524.

32. *Philadelphia Public Ledger*, January 22, 1910, p. 2; AFL-CIO, *Philadelphia Council (Pa.) Records*, reel 1; *New York Times*, February 23, 1910, p. 2; articles from file Clips Bulletin, Strikes—Transit—1910 Phila., Strikes—Transit—Phila.— Hist. and Lists, box 227A, Urban Archives; Roberts, "History and Analysis," 40–44.

33. Harold Howland, "The War in Philadelphia," *Outlook* 5 (March 1910): 522; *New York Times*, February 21, 1910, pp. 1–2, February 22, 1910, p. 1, February 23, 1910, p. 2, February 24, 1910, p. 2; *Philadelphia Public Ledger*, February 20, 1910, p. 2; police reports in file 47–2, box 47, John How memo, April 20, 1910, file JR-57–11, box 57, Cox Papers; articles from file Clips Bulletin, Strikes—Transit—1910 Phila., Urban Archives; *New York Evening Call*, March 12, 1910, p. 2, March 11, 1910, p. 1, March 9, 1910, p. 1; Norwood, *Strikebreaking and Intimidation*, 38.

34. "The General Strike in Philadelphia," *Current Literature* 48 (April 1910): 364; *Philadelphia Public Ledger*, January 24, 1910, p. 2; *New York Times*, February 24, 1910, p. 2, February 22, 1910, p. 2; Sidorick, "'Girl Army,'" 323–69; Roberts, "History and Analysis," 33. For analyses of women's engagement in public conflicts see Ryan, *Women in Public*, and Cameron, *Radicals of the Worst Sort*.

35. "Public Rights Are of Primary Consideration," *Motorman and Conductor* 18 (March 1910): 6; *New York Times*, February 20, 1910, p. 1, February 21, 1910, pp. 1, 2, February 23, 1910, p. 1; Foner, *History of the Labor Movement*, 147–48; articles from file Clips Bulletin, Strikes—Transit—1910 Phila., Strikes—Transit—Phila.— Hist. and Lists, box 227A, Urban Archives; Norwood, *Strikebreaking and Intimidation*, 38.

36. *New York Times*, February 25, 1910, p. 2, March 4, 1910, p. 3, February 22, 1910, p. 1; "Philadelphia Lockout Culmination of Longstanding Contention," *Motorman and Conductor* 18 (March 1910): 19; *Philadelphia Public Ledger*, February 27, 1910, p. 1; AFL-CIO, *Philadelphia Council (Pa.) Records*, reel 1.

37. *New York Evening Call*, March 15, 1910, p. 2, March 19, 1901, p. 1, March 12, 1910, p. 2, March 9, 1910, p. 2; Roberts, "History and Analysis," 46.

38. *New York Times*, March 7, 1910, p. 3, February 26, 1910, p. 2, March 9, 1910, p. 3, March 5, 1910, p. 2, March 12, 1910, p. 3; *New York Evening Call*, March 8, 1910, p. 1, March 3, 1910, 1; Foner, *History of the Labor Movement*, 150; untitled articles, *Motorman and Conductor* 18 (April 1910): 25–32.

39. *New York Evening Call*, March 3, 1910, p. 1, March 12, 1910, p. 2, March 8, 1910, p. 1, March 9, 1910, p. 1.

40. W. D. Allen to Gentlemen, March 5, 1910, Citizens Alliance to President Kruger, March 4, 1910, Worcester Builders Exchange to Pres. Charles O. Kruger, file JR-60–8, box 60, President of Youngstown Sheet and Tube Company to Mr. C. O. Kruger, March 11, 1910, file JR-60–14, Cox Papers.

41. National Sewing Machine Co. to Philadelphia Rapid Transit Co., March 10, 1910, Fuller & Warren Company to General Manager of the Philadelphia Rapid Transit Co., March 10, 1910, file JR-60–14, J. Kirby Jr. to John B. Parsons, March 2, 1910, Chas Learned to the Rapid Transit Company, February 28, 1910, file JR-60–8, box 60, Cox Papers.

42. B. P. Howell to Chas. O. Kruger, February 26, 1910, C.L.C. to P.R.T., undated, J. E. Revis to Pres. Kruger, March 4, 1910, Frederick M. Grant to Dear Sir, March 3, 1910, W. Shoe to Mr. Chas. O. Kruger, March 2, 1910, file JR-60–8, box 60, James R. Turner to Gentlemen, undated, Gram Curtis to President Kruger, March 13, 1910, file JR-60–14, Cox Papers.

43. *New York Times*, March 6, 1910, pt. 2, p. 10; "The Philadelphia Strikes," *Outlook* 94 (March 1910): 593; "A General Strike in Philadelphia," *Outlook* 94 (March 1910): 560; "The General Strike in Philadelphia," *Current Literature* 48 (April 1910): 361.

44. C. A. Carskadon to Charles Kruger, March 1, 1910, file JR-60–8, box 60, Cox Papers.

45. Charles Franklin to Clarence Wolf, March 19, 1910, unsigned to George Earle Jr., March 2, 1910, unsigned to John Reyburn, March 2, 1910, file JR-60–14, box 60, article in scrapbook 16, box 117, Cox Papers; *New York Times*, March 13, 1910, p. 5; *Philadelphia Public Ledger*, January 21, 1910, p. 2, January 25, 1910, p. 1, January 30, 1910, p. 1, February 22, 1910, p. 3; Roberts, "History and Analysis," 54.

46. AFL-CIO, *Philadelphia Council (Pa.) Records*, reel 1; Levinson, *I Break Strikes!*, 101; "Philadelphia Strike an Advance Agent," *Motorman and Conductor* 18 (May 1910): 17; Gompers, "Decadent Philadelphia," *American Federationist* 17 (May 1910): 404; Roberts, "History and Analysis," 56–57; multiple PRT Daily Reports, file 47–2, box 47, Cox Papers.

47. Foner, *History of the Labor Movement*, 162; *New York Times*, October 2, 1910, pt. 2, p. 10; Pratt, "Sympathetic Strike," 394; "Philadelphia Strike an Advance Agent," *Motorman and Conductor* 18 (May 1910): 19; "Future Brighter Than Ever," *Motorman and Conductor* 18 (July 1910): 29; Samuel Gompers, "Decadent Philadelphia," *American Federationist* 17 (May 1910): 401; "The End of a Futile Strike," *Outlook* 94 (April 1910): 963–64.

48. Schmidt, *Industrial Relations*, 189; *New York Times*, April 8, 1910, p. 1; Roberts, "History and Analysis," 58–59, 64; PRT, *Annual Report*, 1913, 5; articles from file Clips Bulletin, Strikes—Transit—1910 Phila., box 227A, Urban Archives.

4. THE AGE OF THOMAS MITTEN

1. Shaw, *Mitten*, 250–51; "Rapid Transit to Reorganize," "Stotesbury Rules, and Mitten Now Manages P.R.T. Co.," book 27, box 123, Cox Papers.

2. Shaw, *Mitten*, 86; *Service Talks* 18 (February 13, 1922): 1; Lauck, *Political and Industrial Democracy*, 130, 281.

3. Shaw, *Mitten*, 1–6, 13–17, 24–28, 49–52, 57–58.

4. Shaw, *Mitten*, 71–74, 79–86, 95–96, 104–6; Roberts, "History and Analysis," 295; Ida Tarbell, "Ida Tarbell Tells an Amazing Story of a Father and Son," *American Magazine*, March 1930, p. 13.

5. Shaw, *Mitten*, 105, 111–16, 132–34, 143, 153, 206–9, 218–19.

6. Edmund Stirling, "Steps Which Led Up to the Market St. Subway-Elevated," file 15, box 3, Tucker Papers; *Philadelphia Rapid Transit: Stotesbury-Mitten Agreement*, 2; PRT, *Annual Report*, 1916, 5; PRT, *Annual Report*, 1918, 18; UTTB, *History of Public Transportation*, XI-6; PRT, *Annual Report*, 1911, 8–9; "Stotesbury to Quit If Loan Is Amended," book 27, box 123, Cox Papers.

7. Shaw, *Mitten*, 260; Roberts, "History and Analysis," 81; local editorials reprinted in "Good Service—Not Lower Fares," *Service Talks* 38 (April 9, 1923): 2; *Evening Star* editorial reprinted in "Food for Thought," *Service Talks* 39 (April 16, 1923): 2; *Co-Operative Bulletin* 25 (December 18, 1913): 12; "City-Company Co-Operation," *Service Talks* 64 (December 25, 1923): 1; J. M. Shaw to Mr. A. M. Greenfield, November 24, 1928, file 13, box 38, Albert Greenfield Papers, Historical Society of Pennsylvania, Philadelphia; "Save as You Ride!," *Service Talks* 85 (February 23, 1925): 1; "A Million a Day for Ten Days," Hagley Museum and Library, Wilmington, Del.; Persion, "Mitten Plan," 34; Lauck, *Political and Industrial Democracy*, 304–5.

8. "Labor's Own Story," *Service Talks* 18 (February 13, 1922): 1; "Past, Present and Future," *Service Talks* 93 (June 22, 1925): 1; *Co-Operative Bulletin No. 6* (July 17, 1912): 2; *Co-Operative Bulletin No. 7* (July 29, 1912): 2; *Co-Operative Bulletin No. 11* (January 29, 1913): 2–3; "Don't Pass This By," *Service Talks* 12 (September 1, 1921): 3.

9. Meyers and Spivak, *Philadelphia Trolleys*, 123; Borgnis, *Legacy of Thomas E. Mitten*, 61; "Real Lovefeast Is P.R.T. Dinner," book 13, box 112, "P.R.T. Employes Give Huge Picnic," book 25, box 116, Cox Papers; "Summary of Courses," *Service Talks* 9, no. 5 (March 13, 1928): 3; Lovitt, *Educating for Industry*, unpaginated, 8, 20; "Saving and Spending," *Service Talks* 7, no. 18 (September 6, 1926): 4; "As the Women See It," *Service Talks* 27 (September 18, 1922): 4.

10. Edmund Stirling, "Setbacks Caused by War Delayed Transit Progress," file 15, box 3, Tucker Papers; Roberts, "History and Analysis," 94; PRT, *Annual Report*, 1911, 13–14; "Pratt Thinks Mitten's Statement Fair to Men," "Carmen Seek Talk on Wage Question with T. E. Mitten," "P.R.T. Head Makes Peace with Union," "Carmen Present Claims to Mitten," "Mitten and Car Men Meet; Good Feeling Prevails," book 27, box 123, Cox Papers.

11. U.S. Commission on Industrial Relations, *Industrial Relations*, 2734, 2747; Lauck, *Political and Industrial Democracy*, 177; Persion, "Mitten Plan," 5; *Co-Operative Plan*, 2.

12. "Mitten Plan Forward Step, Say Carmen," "Cooperative Plan to Be Discussed by Carmen's Committee," "3000 Carmen Cheer Organizer Pratt on Return from Europe," unnumbered book, box 123, Cox Papers; Ida Tarbell, "Ida Tarbell Tells an Amazing Story of a Father and Son," *American Magazine*, March 1930, pp. 110, 113.

13. Roberts, "History and Analysis," 67, 94; *Philadelphia Public Ledger*, December 13, 1911, p. 2; Persion, "Mitten Plan," 6–7; Shaw, *Mitten*, 266–67; PRT, *Annual Report*, 1927, 5.

14. "Ovation to Blankenburg by P.R.T. Employes," unnumbered book, box 123, Cox Papers; U.S. Commission on Industrial Relations, *Industrial Relations*, 2810–11.

15. "Carmen Threaten Another Uprising," "Pratt Answers Mitten," unnumbered book, "P.R.T. Forbids Carmen Wearing Buttons Which Show Union Membership," "Dispute Result of Carmen's Vote," book 29, box 123, Cox Papers; Roberts, "History and Analysis," 92–93; Shaw, *Mitten*, 275; *Philadelphia Public Ledger*, August 3, 1912, pp. 4, 7, August 2, 1912, p. 1, August 3, 1912, p. 4.

16. Roberts, "History and Analysis," 75, 294; Shaw, *Mitten*, 271; "The Philadelphia Controversy," *Motorman and Conductor* 20 (January 1912): 14; "Carmen Are Urged to Repudiate Pratt," unnumbered book, box 123, Cox Papers; "Story of Secession in Philadelphia," *Motorman and Conductor* 21 (March 1913): 26; "Affairs in Philadelphia," *Motorman and Conductor* 20 (August 1912): 15–16.

17. U.S. Commission on Industrial Relations, *Industrial Relations*, 2749; "The Proposition for Settlement in Philadelphia," *Motorman and Conductor* 19 (September 1911): 14.

18. Lauck, *Political and Industrial Democracy*, 71, 77; Douglas, "Shop Committees," 89; Brandes, *American Welfare Capitalism*, 121–22, 126; Gemmill, "Literature of Employee Representation," 479; Kaufman, *Origin and Evolution*, 29.

19. McCartin, *Labor's Great War*, 164; Douglas, "Shop Committees," 92–94; Gompers quoted in Brandes, *American Welfare Capitalism*, 120; Lauck quoted in Gemmill, "Literature

of Employee Representation," 492; Jacoby, *Employing Bureaucracy*, 140. For a review of the literature on company unions and an alternative analysis that finds them to be "largely independent, worker-led organizations" see Rose, *Duquesne*. Quote drawn from p. 7.

20. Brandes, *American Welfare Capitalism*, 10–11, 22–24, 28; Simon, *Philadelphia*, 72; McCartin, *Labor's Great War*, 223.

21. Brandes, *American Welfare Capitalism*, 4, 11, 16, 52, 63, 75–84, 95–96; Rose, *Duquesne*, 10; Lauck, *Political and Industrial Democracy*, 141.

22. Perlman and Taft, *History of Labor*, 580–81; Brandes, *American Welfare Capitalism*, 51, 147.

23. Brandes, *American Welfare Capitalism*, 141; Cohen, *Making a New Deal*, 238.

24. "117 P.C. Increase in Nine Years," book 13, box 112, Cox Papers; PRT, *Annual Report*, 1919, 10; Mitten, "Results of Collective Bargaining," 58.

25. Painter, *Standing at Armageddon*, 338–39; Gilje, *Rioting*, 132–33; Kaufman, *Origins and Evolution*, 27; Lippmann quoted in McCartin, *Labor's Great War*, 27. McCartin's book offers the definitive account of the struggle over industrial democracy in the United States in this period.

26. Painter, *Standing at Armageddon*, 341–43; Jacoby, *Employing Bureaucracy*, 102; McCartin, *Labor's Great War*, 43–45, 88; Perlman and Taft, *History of Labor*, 489; Brecher, *Strike!*, 103; Gottlieb, "Shaping a New Labor Movement," 168–69; Simon, *Philadelphia*, 89; Jacoby, *Employing Bureaucracy*, 101, 105.

27. *Philadelphia Public Ledger*, May 10, 1918, p. 1, May 17, 1918, p. 1, May 16, 1918, p. 1, May 13, 1918, p. 4; "Mitten Stands Pat on Lockout Threat," "1000 P.R.T. Men Vote to Strike; Delay Refused," book 12, box 112, Cox Papers.

28. "Car Strike Broken, Declares Mills," "Trolley Strike Hampers 80 Routes of Cars," "Strike Fizzles Says Mitten," "Walkout Probe Held in Secret," "War Labor Body to Hear P. R. T. Strike," "War Board Will Not Fix Minimum Wage," book 12, box 112, Cox Papers; *Philadelphia Public Ledger*, August 2, 1918, p. 1; Roberts, "History and Analysis," 103.

29. "Two-Cent Increase in Carmen's Wages," book 12, box 112, Cox Papers; Roberts, "History and Analysis," 103–6; *Philadelphia Public Ledger*, August 5, 1918, pp. 1, 11, August 16, 1918, p. 5.

30. *Philadelphia Public Ledger*, May 14, 1918, p. 4, May 13, 1918, p. 1, May 10, 1918, p. 1; Post, *Urban Mass Transit*, 67–68; Conway, "Decreasing Financial Returns," 17; Cudahy, *Cash, Tokens, and Transfers*, 151–54; Vance, *Capturing the Horizon*, 499.

31. *Philadelphia Public Ledger*, May 16, 1918, p. 1, May 17, 1918, p. 8; Roberts, "History and Analysis," 118.

32. Painter, *Standing at Armageddon*, 346–47, 360, 368–71, 379; McCartin, *Labor's Great War*, 39; Brecher, *Strike!*, 110–25.

33. McCartin, *Labor's Great War*, 177, Wilson quoted on pp. 189–90; *Los Angeles Times* quoted in Painter, *Standing at Armageddon*, 347; Murray, *Red Scare*; Perlman and Taft, *History of Labor*, 491–94.

34. Painter, *Standing at Armageddon*, 384–85; Cohen, *Making a New Deal*, 38–43; Gottlieb, "Shaping a New Labor Movement," 168, 170, 181; Perlman and Taft, *History of Labor*, 500–501, 589; McCartin, *Labor's Great War*, 40, 210; Simon, *Philadelphia*, 91–92.

35. Roberts, "History and Analysis," 106–14; Mitten, "Results of Collective Bargaining," 57–58; Caskie, "Philadelphia Rapid Transit Plan," 189–204.

36. "Another Forward Step in Wage Adjustment," *Service Talks* 91 (April 20, 1925): 1; PRT, *Annual Report*, 1925, 12; "What Industrial Democracy Has Done for P.R.T. Employes," *Service Talks* 7 (February 27, 1926): 12; Shaw, *Mitten*, 591–93; PRT, *Annual Report*, 1929, 7; PRT, *Annual Report*, 1930, 7. The employees' stock ownership may be traced in *Service Talks*. See, for example, "The Midwinter Meeting," *Service Talks* 33 (February 3, 1923): 1; "President Dunbar's Talk to P.R.T. Stockholders," *Service Talks* 36 (March 21, 1923): 1.

37. PRT, *Annual Report*, 1927, 5; "The Midwinter Meeting," *Service Talks* 33 (February 3, 1923): 1; PRT, *Annual Report*, 1922, 2; "Labor Hiring Capital," "Heading for Industrial Utopia," *Service Talks* 28 (September 26, 1922): 1, 4; Mitten, "Maintenance of Contact," 114; "Our Solution of the World's Greatest Industrial Problem," *Service Talks* 52 (July 26, 1923): 3; "Chairman Mitten's Answer to Dr. Eliot," *Service Talks* 82 (January 29, 1925): 3.

38. "Mr. Mitten's Picnic Talks—Highlights," *Service Talks* 98 (September 7, 1925): 1; "Motorman-Director Likes His Job," *Service Talks* 37 (March 28, 1923): 2; "Co-Operative Policy of P.R.T.," *Service Talks* 50 (July 11, 1923): 4; "Now for the Picnic and the Helping Hand," "Motorman Kellogg's Thoughts on Buffalo," *Service Talks* 24 (August 23, 1922): 1, 2; Roberts, "History and Analysis," 125.

39. "Mitten-Mahon Understanding—1928," *Service Talks* 9 (March 27, 1928): 1; "Daily Press and Labor Press Comment on Mitten-Mahon Agreement," *Service Talks* 9 (April 30, 1928): 2; Perlman and Taft, *History of Labor*, 586; *Trade Union News*, May 3, 1928, p. 7; Muste quoted in Roberts, "History and Analysis," 125.

40. "Reflections from the Radio," *Service Talks* 87 (March 11, 1925): 2; "The Co-Operative Wage Dividend Is Justified," *Service Talks* 76 (November 15, 1924): 1; "Partners in Railroading," *Service Talks* 22 (June 2, 1922): 2; Mitten, "Maintenance of Contact," 112–13; "The Enabling Resolution," *Service Talks* 50 (June 11, 1923): 2; *Philadelphia Inquirer*, July 27, 1923, p. 3; *President Mitten's Talk*, 13; Dennison, "Employee Investor," 383; "Excerpts from Talk of President Nyman," *Service Talks* 55 (September 1, 1923): 2; Weckler and Weaver, *Negro Platform Workers*, 4.

41. Lauck, *Political and Industrial Democracy*, 276; McChord, *Report of C. C. McChord*, 27; "Table Showing $7,051,275 Fees Collected by Mittens since 1911," in file Arthur A. Mitten, "Would Ban Probe of Mitten's Fees," "Mitten Management Fees Based on 'Fair Return,'" "The Mitten Management Fee," in file Thomas Mitten, EBM.

42. PRT, *Mitten Men and Management*, 3, 12; "Economists Approve Mitten Plan—1926," *Service Talks* 7 (February 22, 1926): 4; "P.R.T. Plan Praised in 'Labor's Money,'" in file Thomas Mitten, EBM; Philadelphia press articles reprinted in *Service Talks* 7 (March 13, 1926): 3; Johnson, *Case for Co-Operation*, 38; *The Spirit of PRT*, 6; Shaw, *Mitten*, 560; Mitten, *Philadelphia's Answer*; *Fundamentals of Industrial Prosperity*; "Men and Business," T. E. Mitten to Dr. Eliot, August 12, 1925, file Correspondence Mitten, Mitten, Shaw, Eliot, box 1, John MacKay Shaw Papers, Urban Archives, Temple University, Philadelphia.

43. U.S. Commission on Industrial Relations, *Industrial Relations*, 2742–43, 2753–54, 2764, 2776, 2784–85, 2792, 2811; Mitten, "Maintenance of Contact," 111; Roberts, "History and Analysis," 96, 99–100; Cox, *Surface Cars*, 14.

44. Feustel, *Report on Behalf of the City of Philadelphia*, 140; U.S. Commission on Industrial Relations, *Industrial Relations*, 2784, 2793; Hague, "Outline of Transit Service Development," unpaginated; "The Stotesbury, Mitten, Caskie 'Coercive Plan,'" book 14, box 112, Cox Papers; "In re: Co-operative Wage Fund," p. 15, file 9, box 123, Greenfield Papers.

45. Shaw, *Mitten*, 74; PRT, *Annual Report*, 1926, 4–5; "'Rotten,' Says P.R.T., but War Cuts Supplies," book 11, box 112, "Mayor Asks P.R.T. to Annul Dividend; No, Says Dr. Mitten," book 31, box 120, Cox Papers.

46. Cox, *Surface Cars*, 7–8; Ford, Bacon & Davis, *Pennsylvania State Railroad Commission*, 9, 117–18; Cox, *Utility Cars*, 27–28, 64–75; Kramer and James, *PTC Rails*, 7; PRT, *Proposed Garage*; PRT, *Annual Report*, 1927, 12; Cudahy, *Cash, Tokens, and Transfers*, 95–96, 158.

47. Shaw, *Mitten*, 45–46, 73, 647; Goddard, *Getting There*, 122; Post, *Urban Mass Transit*, 79–92; PRT, *Annual Report*, 1923, 2; Cudahy, *Cash, Tokens, and Transfers*, 99–105; PRT, *Annual Report*, 1927, 9; PRT, *Annual Report*, 1928, 7, 12; *Report of Transit Advisory*

Committee, 18; Andrews, *Short History*, 36; "Mitten Management, Incorporated Organization," book 34, box 120, Cox Papers; McChord, *Report of C. C. McChord*, 36.

48. Shaw, *Mitten*, 74–80. A primer on early civil aviation in the United States may be found in Vance, *Capturing the Horizon*, 542–70.

49. Ida Tarbell, "Ida Tarbell Tells an Amazing Story of a Father and Son," *American Magazine*, March 1930, 117; David N. Phillips, "Wings over the Sesqui," file 15, box 3, Tucker Papers; "The Taxicab Question," "New Taxi Service Put Up to P.S.C.," book 31, box 120, Cox Papers; McChord, *Report of C. C. McChord*, 37.

50. Parry, *History of Rapid Transit Development*, 20; Shaw, *Mitten*, 238; Abrams, "Story of Rapid Transit," 7; Sechler, *Speed Lines*, 17–23; Taylor, "Philadelphia's Transit Problem," 30; Taylor, *Rapid Transit Problem*, 1; Twining, *Report to the City Council*, 9.

51. Twining, *Report to the City Council*, 7; *Philadelphia Public Ledger*, January 19, 1912, p. 3; *Frankford Elevated*, 8.

52. Hague, "Outline," unpaginated; Shaw, *Mitten*, 655; PRT, *Annual Report*, 1925, 8–9; *Annual Report of the Department of City Transit*, 1915, 844; Sechler, *Speed Lines*, 36. For a scholarly study of the impact of elevateds on Philadelphia see Marsh, "Impact of the Market Street 'El.'"

53. PRT, *Annual Report*, 1924, 11–13; PRT, *Annual Report*, 1919, 19; *Annual Report of the Department of City Transit*, 1923, 9.

54. UTTB, *Survey of Investigations*, 3, 19, 27–28; U.S. Commission on Industrial Relations, *Industrial Relations*, 2703; UTTB, *History of Public Transportation*, XI-10–XI-11.

55. *Philadelphia Inquirer*, February 28, 1922, pp. 1, 4.

56. Perlman and Taft, *History of Labor*, 575; Thomas Mitten to Albert M. Greenfield, undated, file 49, Albert M. Greenfield to T. E. Mitten, June 14, 1927, file 53, box 28, Greenfield Papers; untitled speech, undated, file Local 234 August–December 1943, Transport Workers Union Papers (hereafter TWU Papers), Tamiment Library, New York University, New York; "Aid for the Labor Bank," *Service Talks* 7 (January 21, 1926): 4.

57. Roberts, "History and Analysis," 139–45; Shaw, *Mitten*, 15, 72; "P.R.T. Men Are Developing into Successful Bankers," *Service Talks* 8 (May 21, 1927): 2; "Mitten Bank—Industrial Section—Opened August 1," *Service Talks* 8 (July 13, 1927): 1; "Labor Banking of the Future," *Service Talks* 9 (March 27, 1928): 4; "Rap Mitten's Bank Plan," in file Thomas Mitten, EBM; "Mitten Bank Directors Elect Several New Officers," *Service Talks* 10 (January 26, 1929): 1.

58. Articles reprinted as appendix to Shaw, *Mitten*, unpaginated; Ida Tarbell, "Ida Tarbell Tells an Amazing Story of a Father and Son," *American Magazine*, March 1930, p. 118; "Dr. Mitten's Talk at 1929 Midwinter Meeting," *Service Talks* 10 (December 24, 1929): 1; "President Hoover Extends His Sympathy in Letter to Dr. Mitten," *Service Talks* 10 (October 12, 1929): 3; "Coroner Scouts Idea Mitten May Have Been Suicide," "'Suppressed Romantics' Bring Mitten Myth," "Thomas Mitten Death Is Still a Mystery," in file Thomas Mitten, EBM; Borgnis, *Near-Side Car*, 59; "Estate of Mitten Put at $3,527,184," book 33, box 120, "Mitten's Estate Placed Back of Investors' Money," book 32, box 120, Cox Papers; untitled document in file 2–26–1937–3–13–1937, box 10, PRT Bankruptcy Case, Records of the District Courts of the United States, National Archives, Philadelphia, Record Group 21 (hereafter RG 21).

59. Untitled speech, undated, untitled speech, 1943[?], file Local 234 August–December 1943, TWU Papers; "The Philadelphia Traction Company Strike," November 19, 1944, file Local 234 September–December 1944, TWU Papers; "P.R.T. Wage Fund Loss Described," in file Charles Ebert, EBM; "Employe Charges Juggling of Funds in P.R.T. Dividends," book 33, box 120, Cox Papers; Philleo Nash to Mr. Jonathan Daniels, August 25, 1944, file OF 4245g, box 8, Franklin D. Roosevelt Papers, Franklin D. Roosevelt Library, Hyde Park, N.Y.

60. "$2,000,000 Estate of Mitten Settles $15,000,000 Suits," book 35, box 122, "Test Case Looms on P.R.T. Stock," book 33, "'Mittenism' Hit as City Renews Transit Battle," "Wilson to Force M.B.S.C. Return of P.R.T. Stock to Men," "City Owned Transit to Free Philadelphia of Mitten Management," book 32, box 120, Cox Papers; *Edward L. Mitten v. Philadelphia Rapid Transit Co., Jan. 16, 1941*, file 2–26–1937–3–13–1937, box 10, PRT Bankruptcy Case, RG 21; "P.R.T. Takes Title to Mitten Estate," in file Arthur A. Mitten, "McDevitt Accepts P.R.T. Offer; Puts Three Trustees in Control," in file John Gribbel, EBM.

5. HARD TIMES AND A HATE STRIKE

1. PRT, *Annual Report*, 1938, 1.
2. Simon, *Philadelphia*, 94. For the context of the Depression Era, I draw liberally on Wolfinger, "Philadelphia, PA, 1929–1941," 436–42.
3. Dudden, "The City Embraces 'Normalcy,'" 596; Tinkcom, "Depression and War," 613; Miller, Vogel, and Davis, *Philadelphia Stories*, 44, 59, 62; Miller, Vogel, and Davis, *Still Philadelphia*, 73–75, 255, 261, 263.
4. Simon, *Philadelphia*, 94; Miller, Vogel, and Davis, *Philadelphia Stories*, 45; Tinkcom, "Depression and War," 606.
5. Simon, *Philadelphia*, 92–95; Miller, Vogel, and Davis, *Philadelphia Stories*, 44–45, 51; Tinkcom, "Depression and War," 609–12.
6. Amalgamated Transit Union Staff, *ATU*, 55–57; Cudahy, *Cash, Tokens, and Transfers*, 176; Hood, *722 Miles*, 16; PRT, *Annual Report*, 1938, 1, 16; PRT, *Annual Report*, 1926, 4; PRT, *Annual Report*, 1931, 3; PRT, *Annual Report*, 1933, 3; PRT, *Annual Report*, 1934, 4.
7. "Recent Cases," 421–23; *Philadelphia Transit System*, 5; "The Underlier Hog Decides to Diet," March 16, 1937, file 21, box 84, Greenfield Papers; "Future of the P.R.T.," "Pact for Rent Slash Drafted by Underliers," book 37, box 122, Cox Papers; untitled document in file 10–4–1934–11–28–1934, box 1, PRT Bankruptcy Case, RG 21; PRT, *Annual Report*, 1933, 7.
8. *Report of Transit Advisory Committee*, 8; PRT, *Annual Report*, 1935, 5; PRT, *Annual Report*, 1932, 6; "No Tears When Old Buses Disappear," file 4, box 99, Greenfield Papers; Borgnis, *Inside Story*, 13; PRT, *Annual Report*, 1936, 5; UTTB, *History of Public Transportation*, XIII-11–12; Cox, *Tram Subways*, 212; PRT, *Annual Report*, 1933, 5; PRT, *Annual Report*, 1938, 4.
9. "P.T.C. Takes Up Transit Facilities," file 7, box 99, "6 P.R.T. Trustees Named in Surprise Move," file 46, box 79, Greenfield Papers; *Philadelphia Transit System*; "Review of the Proposal," 36; *Philadelphia Evening Bulletin*, February 3, 1936, p. 16, November 21, 1935, p. 1, December 21, 1935, p. 22.
10. PRT, *Annual Report*, 1931, 1; "Dr. Mitten Is Out of P.T.C. on Leave," in file Arthur A. Mitten, EBM; PRT, *Annual Report*, 1934, 7; Miller, "History of the Transit System," 38–54; Hague, *Outline of Transit Service*, unpaginated; "Plan of Reorganization of Philadelphia Rapid Transit Company System," file 12–4–1934–12–12–1934, box 1, PRT Bankruptcy Case, RG 21; PRT, *Annual Report*, 1938, 2.
11. *Philadelphia Transit System*, 8–9; "Welcome to the P.T.C.," file 7, box 99, "Reorganization: The Road to Modern Transit and a Greater Philadelphia!" file 3, box 98, "P.R.T. Employes Support New Plan on Reorganization," "P.R.T. Men Back Court Trustees," file 20, box 84, Greenfield Papers.
12. Testimony before Judge George A. Welsh, pp. 6, 10, file 1–3–1938–1–7–1938, box 13, Testimony before Judge George A. Welsh, p. 39, file 10–25–1938, box 15, PRT Bankruptcy Case, RG 21; "1250 Jam Drexel Offices to Collect PRT Wage Payoff," file 8, box 134, Greenfield Papers; *Philadelphia Evening Bulletin*, March 18, 1936, p. 1.

13. PRT, *Annual Report*, 1932, 3; "P.R.T. Cuts 11,000 Employes 7.7 P.C.," book 36, box 122, "Mr. Mitten's Talk—Midwinter Meeting—12–16–31," December 16, 1931, file 408–102–0, box 93, Cox Papers; Welch, *Employment Trends*, 48.

14. Wolfinger, *Philadelphia Divided*, chap. 2; Wolfinger, "Philadelphia, PA, 1929–1941"; Miller, Vogel, and Davis, *Philadelphia Stories*, 85–96.

15. Cohen, *Making a New Deal*; Simon, *Philadelphia*, 99–101; Tinkcom, "Depression and War," 612, 618–19. The literature on the working class during the Great Depression is vast. For examples, see Lichtenstein, *State of the Union*, chap. 1; Bernstein, *Turbulent Years*; Zieger, *CIO*, chaps. 1–5.

16. Gottlieb, "Shaping a New Labor Movement," 189–91; Jacoby, *Employing Bureaucracy*, 168–69; Harris, *Right to Manage*, 21–22.

17. National Labor Board Hearing, December 12, 1933, pp. 7, 15, 18, 39, file Philadelphia Taxi Strike, box 14, Richard M. Neustadt to National Labor Board, December 2, 1933, file Case No. 150 Philadelphia Taxi Strike, box 22, Records of National Labor Relations Board, National Archives, Philadelphia, Record Group 25 (hereafter RG 25); "P.R.T. Taxi Men Free to Join Union," "Closed-Shop Plea Denied by P.R.T.," book 37, box 122, Cox Papers.

18. *Philadelphia Evening Bulletin*, November 29, 1933, p. 28, November 30, 1933, p. 26, November 27, 1933, p. 34, November 30, 1933, p. 1, December 1, 1933, p. 4, December 2, 1933, p. 30, December 6, 1933, p. 2, December 15, 1933, p. 1; "Low Tide," file Strikes—Phila—Taxicabs, *Philadelphia Record* Photos, Historical Society of Pennsylvania, Philadelphia.

19. *Philadelphia Evening Bulletin*, November 28, 1933, pp. 1, 32, December 16, 1933, p. 1, December 22, 1933, p. 1; National Labor Board Hearing, December 12, 1933, pp. 19, 25, file Philadelphia Taxi Strike, box 14, H. F. Galbraith to Robert Wagner, December 17, 1933, file Case No. 150 Philadelphia Taxi Strike, box 22, RG 25.

20. National Labor Board Hearing, December 12, 1933, p. 22, file Philadelphia Taxi Strike, box 14, Albert Williams to Senator Wagner, undated, Francis Ralston Welsh to Hugh S. Johnson, December 16, 1933, file Case No. 150 Philadelphia Taxi Strike, box 22, RG 25; "The Taxicab Hearing," book 37, box 122, Cox Papers.

21. *Philadelphia Evening Bulletin*, December 27, 1933, p. 2, December 11, 1933, p. 3, December 20, 1933, p. 1, December 26, 1933, p. 1, December 27, 1933, p. 2, December 28, 1933, p. 26; "Johnson Hears Striking Taxi Drivers Today," "P.R.T. Defends Its Taxi Policy at P.S.C. Hearing," book 37, box 122, Cox Papers.

22. Unsigned letter to Robert Wagner, undated, H. F. Galbraith to M. H. McIntyre, January 4, 1934, file Case No. 150 Philadelphia Taxi Strike, box 22, National Labor Board Hearing, December 12, 1933, pp. 53–54, file Philadelphia Taxi Strike, box 14, RG 25; *Philadelphia Evening Bulletin*, December 4, 1933, p. 1, December 19, 1933, pp. 2, 32.

23. S. Davis Wilson to Robert Wagner, December 30, 1933, Maynard C. Krueger to Franklin D. Roosevelt, December 29, 1933, Emma G. Gruner to *Philadelphia Record*, undated, file Case No. 150 Philadelphia Taxi Strike, box 22, RG 25.

24. "Memorandum of Agreement," January 8, 1934, file Case No. 150 Philadelphia Taxi Strike, box 22, RG 25; *Philadelphia Evening Bulletin*, January 9, 1934, pp. 1, 30.

25. *Philadelphia Evening Bulletin*, October 8, 1935, p. 1, November 18, 1935, p. 3, November 22, 1935, p. 2, October 10, 1935, p. 1; PRT, *Annual Report*, 1935, 1; "In the Matter of the Sale of Taxicab Properties of the Debtor," undated, file 11–27–1935, box 4, RG 21.

26. Rose, *Duquesne*, 79; PRT, *Annual Report*, 1933, 1; "To the Employes of the Philadelphia Rapid Transit Company," September 26, 1933, file 408–102–0, box 93, Cox Papers.

27. Kennedy, *Freedom from Fear*, 290; Roberts, "History and Analysis," 171; "To the Employes of P.R.T.," June 7, 1937, file 408–102–0, box 93, Cox Papers; PRT, *Annual*

Report, 1937, 5. The definitive study of Pennsylvania's state-level version of New Deal policies and programs remains Keller, *Pennsylvania's Little New Deal*.

28. "To the Employes of P.R.T.," June 7, 1937, file 408–102–0, box 93, Cox Papers; PRT, *Annual Report*, 1937, 6; Louis Hicks to Albert M. Greenfield, April 22, 1941, file 14, box 107, Mildred Morris to Albert M. Greenfield, May 7, 1943, file 7, box 117, Greenfield Papers; *Philadelphia Afro American*, August 12, 1944, p. 12; untitled speech, undated, file August–December 1943, box 81, TWU Papers.

29. "Philadelphia Rapid Transit Company Election," Executive Board of the Transport Workers Union of America to Ralph Senter, file 11–1–1937–11–3–1937, box 12, RG 21; Roberts, "History and Analysis," 172–73, 296, 315.

30. PRT, *Annual Report*, 1937, 5; Roberts, "History and Analysis," 166, 170, 175–76; Wolfinger, *Philadelphia Divided*, 114–15.

31. Simon, *Philadelphia*, 94, 102; Tinkcom, "Depression and War," 641; Miller, Vogel, and Davis, *Philadelphia Stories*, 51, 113; Scranton and Licht, *Work Sights*, 234–40; Palmer, *War Labor Supply*, 2; *Bulletin Almanac*, 1945, 148. For the context of World War II Philadelphia I draw liberally on Wolfinger, "Philadelphia, PA, 1941–1952," 501–6. For a national perspective on how the war roiled American society a good starting point is Kennedy, *Freedom from Fear*, chap. 21.

32. Simon, *Philadelphia*, 134–35.

33. Wolfinger, *Philadelphia Divided*, chap. 4; Simon, *Philadelphia*, 104, 117; Tinkcom, "Depression and War," 640; Miller, Vogel, and Davis, *Philadelphia Stories*, 114, 123, 148, 161, 193, 218; Scranton and Licht, *Work Sights*, 244; Licht, *Getting Work*, 46, 195; Palmer, *Manpower Outlook*, 6–7; Palmer, *Philadelphia Labor Market*, 6–7.

34. Cudahy, *Cash, Tokens, and Transfers*, 178; Hood, *722 Miles*, 241; PTC, *Annual Report*, 1943, 1, 13. PTC statistics drawn from war-era annual reports, particularly PTC, *Annual Report*, 1940, and *Annual Report*, 1945.

35. PTC, *Annual Report*, 1943, 2, 12–13; PTC, *Annual Report*, 1944, 2–3; PTC, *Annual Report*, 1942, 13; Roberts, "History and Analysis," 180; *PTC Traveler*, file 1, box 134, Greenfield Papers; "Come Take a Ride in Subway—There's Glamor in the Cab," George Forman Scrapbook, Historical Society of Pennsylvania, Philadelphia.

36. Borgnis, *Inside Story*, 31; Roberts, "History and Analysis," 180–81; "Come Take a Ride in Subway—There's Glamor in the Cab," Forman Scrapbook; "Protect Your Seniority!," file January–July 1943, box 81, TWU Papers. For more on women during World War II, the literature includes Milkman, *Gender at Work*; Hartmann, *Home Front and Beyond*; Anderson, *Wartime Women*; and Chafe, *American Woman*. Gabin, "Hand That Rocks," particularly explores male workers' reaction to women getting "nontraditional" jobs. David Montgomery details the development and importance of seniority in *Workers' Control*, 140–43.

37. Theodore Spaulding, "Philadelphia's Hate Strike," *Crisis* 51 (September 1944): 281; *PM*, August 6, 1944, p. 3; "Happy New Year to You," *Service Talks* 15 (December 31, 1921): 4; Roediger and Esch, *Production of Difference*, 7; Licht, *Getting Work*, 45; Wolfinger, *Philadelphia Divided*, 116; "United & Alert—for Victory!," file August–December 1943, box 81, TWU Papers; George M. Johnson, "Equal Employment Opportunity for Negroes in Local Transportation Systems," May 23, 1943, file Phil., PA., Rapid Transit Strike 1943, box A467, National Association for the Advancement of Colored People Papers, Library of Congress, Washington, D.C.

38. Wolfinger, "'We Are in the Front Lines,'" 1–23; Miller, Vogel, and Davis, *Philadelphia Stories*, 128–29; *Philadelphia Afro American*, November 13, 1943, p. 2. I explore black Philadelphians' politics more fully in *Philadelphia Divided*. Chapter 5 especially focuses on the PTC.

39. Kelley, *Race Rebels*, 56; Kelley, *Right to Ride*; Wolfinger, *Philadelphia Divided*, 121–22.

40. Complaints in file Philadelphia Rapid Transit Co., box 25, Records of the Committee on Fair Employment Practice, National Archives, Philadelphia, Record Group 228 (hereafter RG 228); Wolfinger, *Philadelphia Divided*, 125–33. For more on black Philadelphia and the FEPC see Wolfinger, "Equal Opportunity," 65–94.

41. Carolyn M. Davenport to Mr. Edward Hopkinson Jr., Chairman, April 1, 1943, A. A. Mitten to Carolyn M. Davenport, April 1, 1943, Carolyn Davenport Moore to Albert M. Greenfield, November 12, 1943, file 7, box 117, Greenfield Papers; Reginald A. Johnson to Frank McNamee, August 8, 1944, file Strikes, box 2370, Records of the War Manpower Commission, National Archives, Philadelphia, Record Group 211 (hereafter RG 211); Wolfinger, *Philadelphia Divided*, 114, chap. 5.

42. Wolfinger, *Philadelphia Divided*, 115–16, 143; Du Bois, *Black Reconstruction*, 700; *PM*, August 4, 1944, 2. On the "psychological wage" see also Roediger, *Wages of Whiteness*.

43. Kenny, *American Irish*, 191–92; Freeman, *In Transit*, v, 113–15; U.S. Congress, House Committee on Un-American Activities, *Investigation of Un-American Propaganda*, 8095, 8112; R. T. Senter to Albert M. Greenfield, March 21, 1941, file 14, box 107, Greenfield Papers. Joshua Freeman provides a fine comprehensive history of the TWU in *In Transit*.

44. PTC, *Annual Report*, 1941, 13; Wolfinger, *Philadelphia Divided*, 135; clippings in file Saul C. Waldbaum, EBM; *Transport Workers Bulletin*, March–April 1944, p. 9, July 1943, p. 3; Roberts, "History and Analysis," 182.

45. Roberts, "History and Analysis," 182; *Philadelphia Evening Bulletin*, November 2, 1943, pp. 1, 22, November 3, 1943, pp. 1, 22.

46. "P.T.C. Workers Get 4-Cent Raise," "Many Drop Check-Off," file Philadelphia Transportation Company, EBM; *Transport Workers Bulletin*, April 1943, p. 3, September 1943, p. 10; Wolfinger, *Philadelphia Divided*, 135; "The PRT 'Union' Proves It Is a Useless Parasite," undated, file January–February 1944, box 81, TWU Papers.

47. "Items 'Off Peak' Forgot," undated, file August–December 1943, box 81, TWU Papers.

48. *Transport Workers Bulletin*, May 1943, p. 3; Wolfinger, *Philadelphia Divided*, 135; untitled TWU speeches, 1943, file August–December 1943, "CIO Supports PTC Workers," undated, file March–May 1944, box 81, TWU Papers; "Murray Addresses P.T.C. Workers," *Evening Bulletin* Transit Collection, Urban Archives, Temple University, Philadelphia; *Philadelphia Evening Bulletin*, November 2, 1943, p. 22.

49. Wolfinger, *Philadelphia Divided*, 137; Michael J. Quill to Philip Murray, March 4, 1944, press release, undated, file March–May 1944, box 81, TWU Papers.

50. "James L. McDevitt Says Communism Is Issue at Labor Election," "Murray Addresses P.T.C. Workers," *Evening Bulletin* Transit Collection; "An Open Letter to the People of Philadelphia," undated, "'Philadelphians' Attention Please!" undated, file March–May 1944, box 81, TWU Papers; "Warning!" undated, file 7, box 123, Greenfield Papers.

51. Wolfinger, *Philadelphia Divided*, 136–38; "Behind the Smoke-Screen," undated, file January–February 1944, "Leaflet #2," undated, file August–December 1943, box 81, TWU Papers; *Philadelphia Tribune*, March 11, 1944, p. 4.

52. "P.T.C. Election Is Peaceful," *Evening Bulletin* Transit Collection; "Election Returns," March 14, 1944, file March–May 1944, box 81, TWU Papers; Wolfinger, *Philadelphia Divided*, 139.

53. Irving Potash to Michael J. Quill, March 15, 1944, President to Michael J. Quill, March 15, 1944, Morris Yanoff to Michael J. Quill, March 17, 1944, Joseph Curran to Michael J. Quill, March 20, 1944, Max Yergan to Michael Quill, March 17, 1944, file March–May 1944, box 81, TWU Papers.

54. "The TWU Contract Demands," undated, file January–February 1944, "To All Members of the Transport Workers Union of Phila. and PTC Employees!," file June–August 1944, box 81, TWU Papers.

55. *PM*, August 6, 1944, p. 3, August 4, 1944, p. 4; Meier and Rudwick, "Communist Unions," 191; Wolfinger, *Philadelphia Divided*, 139–40.

56. Multiple clippings in *Evening Bulletin* Transit Collection; *Transport Workers Bulletin*, July 1944, p. 9; *Prospectus*, November 1945, file 1, box 137, Greenfield Papers; Wolfinger, *Philadelphia Divided*, 140. See Wolfinger, *Philadelphia Divided*, chap. 6 for a detailed history of the strike.

57. Wolfinger, *Philadelphia Divided*, 143–44, 152–58.

58. *PM*, August 6, 1944, p. 4; Wolfinger, *Philadelphia Divided*, 147; Max Yergan to Michael Quill, August 8, 1944, Julius Thomas to Michael Quill, August 10, 1944, "Resolution on the Philadelphia Transit Outbreak," August 18, 1944, F. B. Taylor to Michael Quill, August 24, 1944, Alan Shaw to Transport Workers Union, August 8, 1944, file June–August 1944, box 81, TWU Papers; Ernest Votaw to War Manpower Commission, August 3, 1944, file Strikes, box 2370, RG 211.

59. Wolfinger, *Philadelphia Divided*, 149–52; White, *Rising Wind*, 153–55.

60. "The Negro Problem," August 3, 1944, file 1944, box 87, "Campbell's Role in Philly Strike Points to GOP String-Pulling," August 16, 1944, file June–August 1944, Joseph E. Weckler, "The Philadelphia Traction Company Strike," November 19, 1944, file September–December 1944, box 81, TWU Papers; Philleo Nash to Jonathan Daniels, August 25, 1944, file OF 4245g, box 8, Roosevelt Papers; Wolfinger, *Philadelphia Divided*, 144.

61. Douglas L. MacMahon, radio address, August 21, 1944, file June–August 1944, box 81, TWU Papers; Wolfinger, *Philadelphia Divided*, 148–49; *Pennsylvania Labor Record*, August 18, 1944, p. 7.

62. "Charges Firm Spurred Phila. Strike," August 8, 1944, file 1944, box 87, "Orville Bullitt Quits WLB Job under Fire for Strike Role," undated, file September–December 1944, box 81, TWU Papers; Wolfinger, *Philadelphia Divided*, 146; Roberts, "History and Analysis," 191–92.

63. Joseph E. Weckler, "The Philadelphia Traction Company Strike," November 19, 1944, file September–December 1944, "Philly Transit Tieup Union-Busting Move by Company and Stooge Outfit," August 5, 1944, James J. Fitzsimon to Attorney General F. J. Biddle, August 1, 1944, CIO news release, undated, file June–August 1944, "Company Role Ignored as Philly Transit Case Is Closed," March 13, 1945, file M. J. Quill TWU of A, Local 234, Philadelphia, PA, box 81, "Philadelphia Putsch," undated article from unnamed magazine, file 1944, box 87, TWU Papers; Winkler, "Philadelphia Transit Strike," 86; "Investigation of the Philadelphia Transportation Co. Strike," undated, file Grand Jury Report, box 25, RG 228; Roberts, "History and Analysis," 193.

64. PTC, *Annual Report*, 1944, 12; PTC Press Release, January 24, 1948, file 16, box 169, Greenfield Papers; "Orville Bullitt Quits WLB Job under Fire for Strike Role," undated, file September–December 1944, box 81, TWU Papers; "Ebert Takes Post as P.T.C. Head," in file Charles Ebert, EBM; PTC, *Annual Report*, 1948, 11.

65. Roberts, "History and Analysis," 195–96; *Prospectus*, November 1945, file 1, box 137, "Digest of Agreement," July 14, 1944, file 7, box 123, Greenfield Papers; *Transport Workers Bulletin*, July 1944, p. 9, September 1944, p. 9; "Don't Let the P.T.C. Sneak in One-Man Operation!," undated, "The Truth about the Check-Off," undated, "T.W.U. Starts Negotiations for New P.T.C. Pension Plan," file September–December 1944, "Good Wages—Your Fight Against Depression," undated, file undated, box 81, TWU Papers.

66. "P.T.C. Row Looms on Resignations," *Evening Bulletin* Transit Collection; "Amalgamated Association of Street, Electric Railway and Motor Coach Employes of America, A.F. of L.," undated, "The Wreckers Are at It Again," undated, TWU press release, October 12, 1944, file September–December 1944, "Congratulations!," undated, file January–February 1944, box 81, TWU Papers; Wolfinger, *Philadelphia Divided*, 157–58; Roberts, "History and Analysis," 196.

67. "P.T.C. Union Sign New Contract," *Evening Bulletin* Transit Collection; "PTC Maintenance Employees: This Is It!," undated, "Philadelphia Transit Workers Take a Second Look," October 16, 1944, file September–December 1944, box 81, "Statement by Michael J. Quill," undated, file Philadelphia Strike, box 18-A, TWU Papers; *Transport Workers Bulletin*, July 1944, p. 9.

68. Maxwell Rosenfelt to Albert Greenfield, August 8, 1945, file 40, box 131, Greenfield Papers; photos in file Phila SEPTA-Strikes-1946, box 549, *Evening Bulletin* Transit Collection; *Philadelphia Tribune*, February 16, 1946, p. 18, April 26, 1955, p. 24; Northrup et al., *Negro Employment*, 56–57.

6. LABOR RELATIONS AND PUBLIC RELATIONS

1. "Radio Report to the People by Mayor Joseph S. Clark, Jr.," May 6, 1953, file P.T.C. January to June, box A-454, Joseph Clark Papers, City Archives, Philadelphia.

2. Lichtenstein, *Walter Reuther*; Simon, *Philadelphia*, 113; Wolfinger, *Philadelphia Divided*, 202–3; Wolfinger, "Philadelphia, PA, 1941–1952," 504; PTC, *Annual Report*, 1948, 1; statement by Charles E. Ebert, undated, file 16, box 169, Greenfield Papers; "Another Wage Increase Would Mean Another *Fare* Increase," undated, file January–June 1950, box 82, TWU Papers.

3. TWU to Dear Mr. and Mrs. Citizen, undated, file M. J. Quill, TWU of A, Local 234, Philadelphia, PA, box 81, TWU Papers.

4. The literature on the labor movement and labor relations in postwar America, especially the influence of the Communist Party and the impact of anticommunism, is vast. As a starting point see Zieger, *CIO*, chaps. 8–13; Rosswurm, *CIO's Left-Led Unions*; Fones-Wolf, *Selling Free Enterprise*. For labor relations in public service in Philadelphia see Ryan, *AFSCME's Philadelphia Story*.

5. Simon, *Philadelphia*, 134–35; Wolfinger, *Philadelphia Divided*, 189. For the negative impact of lower population density on public transportation across the United States see Pushkarev and Zupan, *Urban Rail in America*; Cudahy, *Cash, Tokens, and Transfers*, chap. 16.

6. Wells, *Economic Characteristics*; Dewess, "Decline of the American Street Railways," 563–81; Cudahy, *Cash, Tokens, and Transfers*, 191; untitled report, April 17, 1949, file Dept. of City Transit, box A-414, Bernard Samuel Papers, City Archives, Philadelphia; PTC, *Annual Report*, 1952, 10; Sechler, *Speed Lines*, 65; Pushkarev and Zupan, *Urban Rail*, 286–87; Bauer and Costello, *Transit Modernization*, 38–39; Post, *Urban Mass Transit*, 94–103.

7. Cox, *Surface Cars of Philadelphia*, 94; Skelsey, "Streetcar Named Endure, Pt. 1," 12–13; Bauer and Costello, *Transit Modernization*, 267. For a list of the fate of all PTC streetcar lines see Borgnis, *Inside Story*, 168–69.

8. Bauer and Costello, *Transit Modernization*, 2, 52, 142; Miller, "History of the Transit System," 88; Dewess, "Decline of the American Street Railways," 579; PTC, *Annual Report*, 1955, 3; PTC, *Annual Report*, 1945, 1.

9. Wells, *Economic Characteristics*, 1–4, 1–5; Goddard, *Getting There*, 216; PTC, *Annual Report*, 1955, 8. For more on the racialized nature of Philadelphia's suburbanization see Wolfinger, *Philadelphia Divided*, chap. 7.

10. Bauer and Costello, *Transit Modernization*, 38; Borgnis, *Inside Story*, 60; PTC, *Annual Report*, 1962, 3; PTC, *Annual Report*, 1963, 2. Scholarly studies similarly found buses to be inferior. See St. Clair, "Motorization and Decline," 579–600.

11. PTC, *Annual Report*, 1952, unpaginated; PTC, *Annual Report*, 1954, unpaginated; PTC, *Annual Report*, 1956, 1; H. E. Ehlers, "Letter of Transmittal and Summarization of Report No. 3516–41 on Existing Shop Operations, Facilities and Controls and Suggested Changes and Potential Economies to the Philadelphia Transportation Company," pp. 24, 29, 35, 40–41, file 3, box 127, Greenfield Papers.

12. Pushkarev and Zupan, *Urban Rail*, 288–89; Cudahy, *Cash, Tokens, and Transfers*, 250; Goddard, *Getting There*, 215–28; Wells, *Economic Characteristics*, 8–5, 8–6; PTC, *Annual Report*, 1951, unpaginated; PTC, *Annual Report*, 1953, unpaginated; PTC, *Annual Report*, 1956, 1.

13. Amalgamated Transit Union Staff, *ATU*, 97, 112–13; PTC, *Annual Report*, 1952, 6; PTC, *Annual Report*, 1951, 7.

14. PTC, *Annual Report*, 1955, 4; "President's Report to Directors on Post-Strike Efforts to Increase Riding," undated, file 5, box 2, Tucker Papers; PTC, *Annual Report*, 1952, 4; *Philadelphia Evening Bulletin*, December 20, 1952, p. 3; UTTB, *Plan and Program 1955*, x; multiple articles in Mounted Clippings, Urban Archives, Temple University, Philadelphia.

15. "Draft of Report of Fact Finding Board," 1950, file 17, box 181, Greenfield Papers; UTTB, "Preview of Conclusions and Recommendations," undated, file Urban Traffic and Transp. Board, 1955, box A-483, Clark Papers.

16. Goddard, *Getting There*, chap. 11; UTTB, *Plan and Program 1955*, ix–x; Clow, "House Divided." For more on Edmund Bacon and Philadelphia's urban development in the mid-twentieth century see Heller, *Ed Bacon*; Knowles, *Imagining Philadelphia*.

17. Untitled report, April 17, 1949, file Dept. of City Transit, box A-414, Samuel Papers; UTTB, *Plan and Program 1955*, xi–xiv; Bureau of Municipal Research, *Proposed Organization*, 8.

18. PTC, *Annual Report*, 1953, 1; Wolfinger, "Philadelphia, PA, 1941–1952," 501–3.

19. "The Problem of What to Do about Phila's Public Transit," September 20, 1956, held at Free Library of Philadelphia; "PTC Has Paid $1 Million Dividends in Four Years," file David Phillips, EBM; "Radio Report to the People by Mayor Joseph S. Clark, Jr.," May 6, 1953, file P.T.C. January to June, box A-454, Clark Papers; PTC, *Annual Report*, 1956, 6; "Draft of Report of Fact Finding Board," 1950, file 17, box 181, Greenfield Papers; UTTB, *Plan and Program 1955*, xix.

20. "City Ownership of P.T. Co. Is Not the Answer," *Philadelphia Evening Bulletin*, February 10, 1950, Kern Dodge to Joseph S. Clark, March 10, 1950, file P.T.C. Co. (January to April), box A-428, Samuel Papers; "Statement by Charles E. Ebert, PTC President," April 10, 1953, file 38, box 212, Greenfield Papers.

21. "The Problem of What to Do about Phila's Public Transit," September 20, 1956, held at Free Library of Philadelphia; UTTB, "Preview of Conclusions and Recommendations," undated, file Urban Traffic and Transp. Board, 1955, box A-483, Clark Papers. On the growth of public authorities in the twentieth-century United States see Radford, *Rise of the Public Authority*; Adams, *From the Outside In*.

22. UTTB, *Plan and Program 1955*, xv, 90, 93–94, appendix C; Sechler, *Speed Lines*, 68; UTTB, "Preview of Conclusions and Recommendations," undated, file Urban Traffic and Transp. Board, 1955, box A-483, Clark Papers.

23. *New York Times*, January 21, 2011, http://www.nytimes.com/2011/01/22/business/22union.html?_r=0; "Transit Union Holds Big Party on Anniversary," "TWU Opening Labor School," file Transport Workers Union, EBM.

24. For more on public relations as corporate strategy in dealing with labor see Fones-Wolf, *Selling Free Enterprise*; Harris, *Right to Manage*, 158, 192.

25. Bureau of Labor Statistics quoted in Brecher, *Strike!*, 228; Zieger, *CIO*, 241–52; Harris, *Right to Manage*, 84, 130; Lichtenstein, *State of the Union*, chap. 3; Lichtenstein, *Walter Reuther*, chap. 11; McColloch, "Glory Days," 222–23, 242–44; Metzgar, *Striking Steel*, chap. 2; Pappalardo, "Philip Murray," 259–60; Licht, *Getting Work*, 212–16; Ryan, *AFSCME's Philadelphia Story*. For more on the strike wave see Fones-Wolf, *Selling Free Enterprise*, chap. 1; Metzgar, *Striking Steel*, chap. 2; Lichtenstein, *State of the Union*, chap. 3; Zieger, *CIO*, chap. 8.

26. "Work-Win Together," March 22, 1946, "Vacancies? No! New Homes? No!," 1946, "A Thousand to One," September 12, 1946, "Phila. Needs 75,000 Homes," September 10, 1946, "Your Dues Go a Long Way," undated, file 1946, box 81, TWU Papers.

27. "You Should Be There," undated, "The Democratic Slate for Unity," undated, file M. J. Quill, TWU of A, Local 234, Philadelphia, PA, "Brother Veteran," undated, "Think," undated, "Look at the Record," undated, "For Your Information," April 10, 1946, unsigned letter to Joseph B. Dougherty, March 6, 1946, file 1946, box 81, TWU Papers.

28. "A Thousand to One," September 12, 1946, "Political Action Committee," July 24, 1946, "What Is PAC?" June 13, 1946, "Republican Failure! Failure! Failure!" undated, "Brother Veteran," undated, "Pointers for Election," September 30, 1946, "Remember This," undated, file 1946, box 81, TWU Papers.

29. Harris, *Right to Manage*, 101, 105–9, 196; Lichtenstein, *State of the Union*, 98–99; Holtzman and Hughes, "Where the Rubber Meets," http://newpol.org/content/where-rubber-meets-road-indefinite-future-interview-jefferson-cowie.

30. Lichtenstein, *State of the Union*, 103; Harris, *Right to Manage*, 62, 127, 143, 145, 150; Brecher, *Strike!*, 230; McColloch, "Glory Days," 224–25; Rosswurm, *CIO's Left-Led Unions*.

31. *Philadelphia Inquirer*, January 15, 1953, p. 1.

32. "Draft of Report of Fact Finding Board," 1950, file 17, box 181, Greenfield Papers; "Union Wage Rates," 1–7.

33. Roberts, "History and Analysis," 197; "PTC Workers Ask $2 Boost," Sept. 21, 1945, file M. J. Quill, TWU of A, Local 234, Philadelphia, PA, "Address of Michael J. Quill," Jan. 29, 1946, "A Strike Will Be Our Weapon," undated, file 1946, box 81, TWU Papers.

34. "Address of Michael J. Quill," January 29, 1946, "It's All Too Clear!" undated, file 1946, TWU to Dear Mr. and Mrs. Citizen, undated, file M. J. Quill, TWU of A, Local 234, Philadelphia, PA, box 81, TWU Papers; photos in *Philadelphia Record* Photos; *Philadelphia Evening Bulletin*, January 14, 1953, p. 7, January 6, 1955, p. 3, December 9, 1952, pp. 1, 3; PTC, *Annual Report*, 1950, 1.

35. *Philadelphia Evening Bulletin*, January 14, 1955, p. 21; "A Message from International President Michael J. Quill," undated, Joseph S. Clark Jr. to Paul O'Rourke, January 23, 1953, file 1953, box 82, "Yessirree We Mean Business!!" undated, "Hurry! Hurry! Hurry!" undated, file 1946, box 81, TWU Papers; W. J. MacReynolds to Andrew J. Kaelin, November 21, 1949, file 2, box 181, Greenfield Papers; Borgnis, *Inside Story*, 45–50; PTC, *Annual Report*, 1949, 11; "P.T.C. and Union Still at Odds," August 1, 1946, "Union to Observe P.T.C. Safety Code," July 27, 1946, file Joseph B. Doughtery, EBM.

36. "Section Set-Up for Strike," undated, "Do You Ride?" undated, file 1946, box 81, TWU Papers.

37. Borgnis, *Inside Story*, 112; Post, *Urban Mass Transit*, 94–95; "One Man Operation—PTC System," April 17, 1951, file 7, box 193, Greenfield Papers; *Philadelphia Inquirer*, January 17, 1953, p. 3; Roberts, "History and Analysis," 229; *Philadelphia Evening Bulletin*, January 7, 1955, p. 3. More comments from senior drivers at "PTC Men Speak Out against One-Man Cars," May 28, 1951, file P.T.C., box A-442, Samuel Papers.

38. Roberts, "History and Analysis," 212; "'Wildcat Action' Hinted at Three PTC Depots," June 20, 1952, file 10, box 205, Greenfield Papers; "Protest One-Man Operation," April 21, 1952, file January–June 1952, box 82, TWU Papers; "PTC Assails TWU 'Threats' over One-Man Trolley Plans," April 18, 1950, in file William J. MacReynolds, EBM; Borgnis, *Inside Story*, 95; "City Council Hears Union Story," May 23, 1951, file P.T.C., box A-442, Samuel Papers.

39. "PTC Goes on the Offensive and Wins Back Public Opinion," *Bus Transportation*, December 1951, p. 30, file 8, box 193, "PTC's Merchandising and Advertising Activities," undated, file 9, box 205, Greenfield Papers; PTC, *Annual Report*, 1948, 9; *Philadelphia Evening Bulletin*, December 11, 1952, p. 52.

40. "PTC Must Be a Good Company to Work For," undated, file July–December 1948, box 81, TWU Papers; "PTC Plan for Returned Veterans," undated, file 10, box 127, Greenfield Papers.

41. Roberts, "History and Analysis," 316–18, 347.

42. W. J. MacReynolds to Paul O'Rourke, April 28, 1952, "To All Transportation and Maintenance Employes," May 29, 1952, file 18, box 205, "An Open Letter to PTC Operating Employees," undated, "Statement by William J. MacReynolds," December 30, 1949, file 23, box 167, Albert Greenfield to Louis Ganz, March 15, 1950, file 27, box 176, Greenfield Papers; "Excerpts from a Talk by Charles E. Ebert," January 6, 1950, file P.T.C. Co. (January to April), box A-428, Samuel Papers; *Philadelphia Evening Bulletin*, December 12, 1952, p. 1.

43. Roberts, "History and Analysis," 205, 298.

44. Whittemore, *Man Who Ran the Subways*, 147; Roberts, "History and Analysis," 238, 308, 312; "Quill and Kaelin Are Assailed by PTC Union Board," February 22, 1949, file Clips Bulletin Strikes—Transit—1949 Phila February 20 to End of Strike, box 227A, Mounted Clippings; "Quill Urges TWU to Oust Rioters," February 17, 1953, file Transport Workers Union, EBM; letter to Mayor Samuel, undated, file P.T.C. Co. (January to April), box A-428, Samuel Papers; unsigned letter, January 5, 1950, file January–June 1950, box 82, TWU Papers.

45. Roberts, "History and Analysis," 202–3; "Quill, Back in N.Y., Says He Beat Reds Again in PTC Fight," February 21, 1949, file Clips Bulletin Strikes—Transit—1949 Phila February 20 to End of Strike, box 227A, Mounted Clippings; Freeman, *In Transit*, chap. 13.

46. "Open Letter to the Membership of Local 234," undated, Philip Murray to Andrew J. Kaelin, undated, "Answer These Questions Tonight!" undated, Resolution, September 14, 1948, file July–December 1948, box 81, "To members of Local 234 TWU," undated, file January–July 1951, box 82, TWU Papers.

47. "Right Wing Man Heads PTC Union," September 30, 1948, file Joseph Dougherty, "O'Rourke Wins PTC Union Vote; Attacks Contract," October 13, 1951, file William MacReynolds, articles in file Amalgamated Association of Street and Electric Railway Employes, "Quill Erases Last of His Big TWU 'Enemies,'" file Transport Workers Union, articles in file Paul O'Rourke, EBM; *Philadelphia Inquirer*, October 20, 1957, p. 24; *Report of Proceedings*, 33; Roberts, "History and Analysis," 209–10; "Our Answer to Red Mike!" October 21, 1950, Michael Quill to John Elliott, October 12, 1950, file July–December 1950, box 82, TWU Papers.

48. Correspondence especially in letters to the editor sections of the *Philadelphia Evening Bulletin* and *Philadelphia Inquirer*, and in multiple files of the TWU, the mayor's office, and other political officials. Specific letters and files are cited in these endnotes but only suggest the number of letters and telegrams extant.

49. "PTC Executive Blasts Kaelin," April 23, 1950, "Quill Is Called Irresponsible," May 4, 1950, "PTC Assails TWU 'Threats' over One-Man Trolley Plans," April 18, 1950, file William MacReynolds, EBM; W. J. MacReynolds to Mr. Quill, May 5, 1950, Statement by R. F. Tyson, PTC Executive Vice President before City Council's Special Committee on Transportation, April 28, 1950, Charles Ebert to All Surface Trainmen, May 2, 1950, W. J. Mack to All Transportation Department Employees, July 29, 1946, file 13, box 142, statement by William J. MacReynolds, April 23, 1950, file 10, box 181, W. J. MacReynolds to Paul O'Rourke, April 28, 1952, file 18, box 205, "PTC Union Votes Not to Strike on 1-Man Car Issue," June 2, 1951, file 4, "One-Man Operation—PTC System," undated, file 7, box 193, Greenfield Papers; PTC, *Annual Report*, 1952, 7.

50. W. J. MacReynolds to Andrew Kaelin, November 21, 1949, file 2, box 181, Greenfield Papers; PTC, *Annual Report*, 1952, 7; "A Record to Be Proud of!" undated, file 1946, box 81, TWU Papers; Roberts, "History and Analysis," 196–99.

51. *Philadelphia Inquirer*, January 20, 1953, p. 12; PTC, *Annual Report*, 1947, 2–3; *Philadelphia Evening Bulletin*, December 21, 1950, pp. 1, 18; PTC, *Annual Report*, 1950, 1, 3–4; Miller, "History of the Transit System," 68, 70, 73, 79–80; PTC, *Annual Report*, 1951, 4; PTC, *Annual Report*, 1953, 4.

52. "Statement by Charles E. Ebert," February 3, 1949, news release, February 14, 1949, file 16, box 169, news release, March 2, 1951, file 7, box 193, Greenfield Papers; PTC, *Annual Report*, 1953, 1; *Philadelphia Evening Bulletin*, December 16, 1952, p. 43, January 5, 1953, p. 12.

53. *Philadelphia Evening Bulletin*, December 9, 1952, p. 3; PTC, *Annual Report*, 1949, 1; Miller, "History of the Transit System," 70, 80; PTC, *Annual Report*, 1945, 1; Roberts, "History and Analysis," 218; "Another Wage Increase Would Mean Another Fare Increase," file January–June 1950, box 82, TWU Papers; "TWU Labor Bosses Threaten Strike to Prevent Emergency Economies Required by City Opposition to Adequate Fares," file 14, box 205, Greenfield Papers.

54. "Union Says PTC Foments Strike with Layoffs," August 7, 1952, file Paul O'Rourke, EBM; *Philadelphia Evening Bulletin*, January 4, 1955, p. 51, January 6, 1953 (night extra), p. 3, January 6, 1953, p. 3; PTC, *Annual Report*, 1945, 5; PTC, *Annual Report*, 1947, 5; "Transit for Saturday," undated, file P.T.C. January to June, box A-454, Clark Papers.

55. *Philadelphia Evening Bulletin*, December 16, 1952, p. 1; PTC, *Annual Report*, 1947, 4; PTC, *Annual Report*, 1949, 4; PTC, *Annual Report*, 1950, 5; PTC, *Annual Report*, 1952, 2; "Transit for Saturday," undated, file P.T.C. January to June, box A-454, Clark Papers.

56. "Radio Address of Mayor Joseph S. Clark, Jr.," undated, "New from PTC," April 2, 1953, file P.T.C. January to June, box A-454, Clark Papers; *Philadelphia Inquirer*, January 16, 1953, p. 2; Roberts, "History and Analysis," 286.

57. Roberts, "History and Analysis," 220–26; *Philadelphia Inquirer*, January 14, 1953, pp. 1, 2, January 18, 1953, p. 2; *Philadelphia Evening Bulletin*, December 12, 1952, p. 1, December 13, 1952, p. 1, December 11, 1952, p. 52; "Radio Address of Mayor Joseph S. Clark, Jr.," January 16, 1953, Michael Quill to Joseph Clark Jr., February 16, 1953, Joseph Clark Jr. to Michael Quill, February 19, 1953, file P.T.C. January to June, box A-454, Clark Papers; "Vote and Win! Democratic Ticket" undated, file January–June 1950, box 82, TWU Papers.

58. *Philadelphia Inquirer*, January 11, 1953, p. 6, January 17, 1953, p. 6, January 16, 1953, p. 12; *Philadelphia Evening Bulletin*, January 15, 1955, p. 1, December 13, 1952, p. 6; "The Silent Partner," *Philadelphia Evening Bulletin*, February 7, 1946, file 14, box 142, "Let's Have the Facts for Everybody in Transit Dispute," *Philadelphia Daily News*, January 11, 1950, file 28, box 176, Greenfield Papers.

59. *Philadelphia Evening Bulletin*, December 14, 1952, p. 1.

60. Ibid.; *United States v. Bryan*, 339 U.S. 323 (1950); Barnum, "From Private to Public," 105–8; Amalgamated Transit Union Staff, *ATU*, 77–78; http://www.legis.state.pa.us/WU01/LI/LI/US/PDF/1947/0/0492..PDF; http://articles.philly.com/2008-01-18/news/25253458_1_philadelphia-suburban-water-harold-stassen-pennsylvania-state-senator; *Philadelphia Inquirer*, January 16, 1953, p. 12; "Transit Unionists against Even Cooling Off Period," *Philadelphia Inquirer*, March 9, 1949, file 26, box 167, Greenfield Papers; Charles and Edith States to Sir, undated, file 13, box 66, Scranton Papers; "Anti-Strike Legislation No Coincidence," file P.T.C., box A-442, Samuel Papers.

61. Unsigned letter to Bernard Samuel, September 22, 1947, file Philadelphia Transportation, box A-408, Just a Taxpayer Trying to Own a Home to Barnard Samuel, January 6, 1949, Joseph Schatz to Mayor Samuels, February 14, 1949, Jos W. King to Mayor Samuels, February 13, 1949, J. Lithgow to Bernard Samuels, February 14, 1949, Alice Liveright to Mayor Samuel, February 15, 1949, file P.T.C. Co. (January–April), box A-414, Samuel Papers.

62. Alice Stratton to Mr. Samuel, February 21, 1949, file P.T.C. Co. (January to April), box A-414, Samuel Papers; James Waide to Jos. S. Clark Jr., December 12, 1954, file 34, box 223, Greenfield Papers; Happy Devotion to James F. Duff, February 15, 1949, file Philadelphia Transit Strike, 1949–50, box 27, James H. Duff Papers, Pennsylvania State Archives, Harrisburg.

63. *Philadelphia Evening Bulletin*, January 14, 1955, p. 12; A Good Republican to Governor Duff, February 19, 1949, undated letter, file Philadelphia Transit Strike, 1949–50, box 27, Duff Papers; A Citizen to Dear Mayor, undated, file P.T.C. Strike February 11–20, 1949, box A-414, Samuel Papers.

64. Sam Mathews, "The Scandalous Strike Racket in Our United States of America," file January–February 17th 1949, box 81, TWU Papers.

65. J. H. Rummelman to Bernard Samuel, October 10, 1949, Barney Gallagher to Honorable Sir, undated, file P.T.C. Co. (September–December), box A-414, Samuel Papers; Phila. Voter and Taxpayer to James Duff, February 15, 1949, file Philadelphia Transit Strike, 1949–50, box 27, Duff Papers; A Citizen of Phila to Mr. Quill, February 16, 1949, file January–February 17th 1949, box 81, TWU Papers.

66. Ross Mateer to James Duff, undated, telegram to James Duff, February 15, 1949, A Phila. Voter and Taxpayer to James Duff, February 15, 1949, Edith Thompson to James Duff, February 10, 1949, file Philadelphia Transit Strike, 1949–50, box 27, Duff Papers; Ed Chamberlain to B. Samuel, January 5, 1951, file P.T.C., box A-442, Samuel Papers; unsigned letter, February 15, 1949, multiple telegrams and cards, file January–February 17, 1949, box 81, One Shot Charlie to Mike Quill-inski, undated, file February 18–February 29, 1949, box 82, TWU Papers; *Philadelphia Inquirer*, January 15, 1953, p. 2.

67. Stanley Root to Charles Ebert, January 17, 1949, file January–February 17th 1949, box 81, TWU Papers; unsigned to James Duff, telegram to James Duff, February 15, 1949, A Phila. Voter and Taxpayer to James Duff, February 15, 1949, file Philadelphia Transit Strike, 1949–50, box 27, Duff Papers; A Voter and Taxpayer to Mayor Samuel, February 14, 1949, Jacob Richman to *Philadelphia Evening Bulletin*, file February 11–20, 1949, box A-414, Samuel Papers.

7. NATIONAL CITY LINES AND THE IMPERATIVES OF POSTWAR CAPITALISM

1. *Philadelphia Evening Bulletin*, March 1, 1955, pp. 1, 68. For more on NCL's Baltimore career see Glazer, "Fade to Gas."

2. *Philadelphia Evening Bulletin*, March 1, 1955, p. 68, March 2, 1955, p. 3.

3. Snell, *American Ground Transport*; Goddard, *Getting There*, 126–35; Post, *Urban Mass Transit*, 153–56; Cudahy, *Cash, Tokens, and Transfers*, 188–91; McShane, Review of *Getting There*, 230; Bottles, *Los Angeles and the Automobile*; McShane, *Down the Asphalt Path*, chap. 7.

4. Pennsylvania Congress, *Report of the House Select Committee to Investigate SEPTA*, 90.

5. Roberts, "History and Analysis," 241.

6. "History of National City Lines," 4–11, 18; PTC, *Annual Report*, 1956, 4, 7; PTC, *Annual Report*, 1957, 15; NCL, *Annual Report*, 1957, 6–7; Lipson, "General Motors," 39.

7. NCL, *Annual Report*, 1954, 6; NCL, *Annual Report*, 1955, 10; "History of National City Lines," 17; PTC, *Annual Report*, 1955, 7; *Philadelphia Evening Bulletin*, May 5, 1957, p. 19, March 11, 1955, p. 1, March 1, 1955, pp. 1, 68; "Statement by Charles E. Ebert," March 1, 1955, "Joint Statement," March 1, 1955, file 3, box 238, Greenfield Papers.

8. *Philadelphia Evening Bulletin*, February 26, 1955, p. 11, March 2, 1955, p. 22; Roberts, "History and Analysis," 318, 348.

9. "PTC to Eliminate City's Sole Voice on Policy Board," April 22, 1955, "Schwartz Raps PTC Owners as 'Selfish, Carpetbaggers,'" May 6, 1955, "PTC Policy: Keep 'City Snoopers'

Out," April 23, 1955, "PTC to Name City Man to Board—with Gimmick," April 26, 1955, "Council Fights PTC Move to Oust Representative on Executive Board," April 22, 1955, file 1, box 238, Greenfield Papers; *Philadelphia Evening Bulletin*, March 8, 1955, p. 3; PTC, *Annual Report*, 1956, 2; *James H. J. Tate, Thomas McIntosh, George X. Schwartz and City of Philadelphia v. Philadelphia Transportation Company*, December 1962, Court of Common Pleas, file 1, box 15, SEPTA Collection.

10. Fitzherbert, "50 Years," 12; *Philadelphia Inquirer*, February 24, 1956, pp. 1, 10; "National City Lines Reports Consolidated Net Earning of $5,158,028 Last Year," *Passenger Transport*, May 26, 1962, file 4, box 308, Greenfield Papers; "PTC Shows First Profit in 3 Years," file Albert Lyons, EBM; PTC, *Annual Report*, 1962, 4; PTC, *Annual Report*, 1963, 2; PTC, *Annual Report*, 1958, 2; PTC, *Annual Report*, 1966, unpaginated; PTC, *Annual Report*, 1956, 4; PTC, *Annual Report*, 1964, 4.

11. "Transit Progress Is Linked with City-PTC Cooperation," April 30, 1962, file 5, box 317, "Mayor Says City Must Run PTC," *Philadelphia Daily News*, January 7, 1957, file 12, box 249, "New Prescription for Transit Ills," November 3, 1956, file 17, box 249, Greenfield Papers; PTC, *Annual Report*, 1958, 3; *Philadelphia Inquirer*, July 16, 1958, p. 11; "PTC Declares 15c Dividend," "D'Ortona Calls PTC 'Greedy' for Fare Boost," file Robert Stier, "PTC Denies Fare Rise Will Pay Dividend," file Frederick Benton, EBM; *Philadelphia Evening Bulletin*, June 27, 1957, p. 56, June 17, 1958, p. 1, July 2, 1955, p. 3; *James H. J. Tate, Thomas McIntosh, George X. Schwartz and City of Philadelphia v. Philadelphia Transportation Company*, December 1962, Court of Common Pleas, file 1, box 15, SEPTA Collection; untitled document, undated, file 2, box 2, Charles H. Frazier Jr. Papers, Urban Archives, Temple University, Philadelphia.

12. Roberts, "History and Analysis," 287; Wells, *Economic Characteristics*, 6–90, 6–92; PTC, *Annual Report*, 1965, 3; Siddiqi, "Functional Analysis," 512; PTC, *Annual Report*, 1956, 1; PTC, *Annual Report*, 1959, 1; PTC, *Annual Report*, 1962, 1; PTC, *Annual Report*, 1967, 1; *Philadelphia Evening Bulletin*, March 16, 1955, p. 33.

13. *Philadelphia Evening Bulletin*, March 6, 1955, p. 3, July 5, 1955, p. 2; Roberts, "History and Analysis," 233–34; "Ebert to Be Dismissed in Shakeup of PTC," February 26, 1955, "PTC's 'New Look' Reflected in Stockholders' Questions," April 19, 1955, file 1, box 238, Greenfield Papers; "'Jungle Law' Bargaining Best with PTC, Says TWU," February 17, 1959, file Charles Robinson, EBM.

14. Joe Marks to Michael Quill, January 20, 1956, A Need for Action to Sir, February 21, 1956, Dear Member, undated, file January–June 1956, box 82, TWU Papers; *Philadelphia Evening Bulletin*, July 21, 1955, p. 3, July 26, 1955, p. 3.

15. Michael Quill to D. M. Pratt, November 2, 1960, file September–December 1960, box 83, TWU Papers; "TWU Men Seen Resenting Pact," March 4, 1960, file Transport Workers Union, "Union Assails Threat by PTC of 30-Cent Fare," September 1, 1964, file Albert G. Lyons, EBM.

16. *Philadelphia Evening Bulletin*, June 24, 1957, p. 14; Manuel Newmark to Mayor Tate, January 13, 1964, file Transportation Board, box A-4426, James H. J. Tate Papers, City Archives, Philadelphia; Rose Levin to William Scranton, February 3, 1963, file 13, box 66, William Scranton Papers, Pennsylvania State Archives, Harrisburg; PTC, *Annual Report*, from 1958 to 1965.

17. Pushkarev and Zupan, *Urban Rail*, 11, 287; Lipson, "General Motors," 1.

18. *Philadelphia Evening Bulletin*, March 1, 1955, p. 68; Cox, *Surface Cars*, 94; "Buses or Trackless Trolleys?" file 10, box 249, Greenfield Papers; Borgnis, *Inside Story*, 168–69; Skelsey, "Streetcar named Endure," 13–14.

19. PTC, *Annual Report*, 1961, 3; "PTC to Reduce Runs by 2 Pct. on Jan. 19; Union Assails Move," January 12, 1964, file 1964, box 88, TWU Papers; Cox, *Surface Cars*, 94; *Philadelphia Evening Bulletin*, March 9, 1955, p. 3, February 21, 1956, p. 13; PTC, *Annual*

Report, 1956, 8; PTC, *Annual Report*, 1957, 5, 7; PTC, *Annual Report*, 1966, unpaginated; PTC, *Annual Report*, 1967, 8; Kramer and James, *PTC Rails*, 22–23; Hackbridge, "When National City Lines Came to Town," http://thethirdrail.net/0005/phil1.htm; PTC, *Annual Report*, 1959, 9.

20. Meyers and Spivak, *Philadelphia Trolleys*, 83; *Philadelphia Evening Bulletin*, March 9, 1955, p. 1, March 16, 1955, p. 33, July 1, 1955, p. 1; "PTC Subway Cleanup Is Stopgap, TWU Says," undated, file 1962, box 88, TWU Papers; "Tate Hits PTC for 'Dirty Cars, Lack of Help,'" file 8, box 308, "Face-Lifting Is Urged to Bolster PTC," March 2, 1962, "Inexcusable Subway Neglect," March 26, 1962, file 10, box 317, Greenfield Papers.

21. *Philadelphia Evening Bulletin*, July 1, 1955, pp. 1, 5, July 2, 1955, p. 3; resolution, undated, file 1966–1967, box 84, "Union Charges PTC Policies Perils Riders and Operators," January 13, 1964, "Union Assails PTC Service," undated, file 1964, box 88, TWU Papers.

22. Roberts, "History and Analysis," xxii; "PTC Strike Record: Eighth Disruption in Last 17 Months," September 27, 1956, file 1956, box 87, TWU Papers.

23. Roberts, "History and Analysis," 241–63; *Philadelphia Evening Bulletin*, October 2, 1958, p. 3.

24. Hood, *722 Miles*, 219; "Corridors of Crime" article series, *Philadelphia Inquirer*, February 24 and 25, 1956. The literature on crime on urban transportation systems is insufficient, but see Patty, "Crime on the Bus."

25. *Philadelphia Evening Bulletin*, March 4, 1955, p. 1, February 22, 1956, p. 3; "PTC Operators Demand Halt in N. Phila. Pupil Rowdyism," November 18, 1953, "2d Man Sought as Kidnaper," January 20, 1952, "Union, PTC Split on Protection of Trolley Women," January 26, 1952, "Union, PTC Split on Protection of Trolley Women," January 26, 1952, file Transport Workers Union, "Union Offers Help to Make Subways Safe," April 7, 1965, "PTC Plan 'Exact Fare' to Cut Down Holdups," August 9, 1968, file Dominic DiClerico, EBM; "PTC Gets Help with Rowdies," June 15, 1961, file 5, box 308, Statement by William J. MacReynolds, January 25, 1952, file 14, box 205, Greenfield Papers; Northrup, *Negro Employment*, 56–57; "Bias Charged in TWU Hiring of Local Staff," September 7, 1963, file 1963, box 88, Robert McGlotten to Michael Quill, February 25, 1964, file 1964, box 84, TWU Papers; Albert Lyons to Fred T. Corleto, September 13, 1968, file Philadelphia Transportation Company, box A-4607, Tate Papers; Amalgamated Transit Union Staff, *ATU*, 94–96; Booz, Allen & Hamilton Inc., "SEPTA Management Study," ix.

26. "PTC Violates Pact, TWU Says," January 3, 1956, file Paul O'Rourke, "TWU Is Sued for $100,000 in Walkout," November 2, 1956, file Charles Robinson, "Advance 'Fee' for Walkouts Asked by PTC," October 17, 1956, file Douglass Pratt, EBM; Roberts, "History and Analysis," 243–44, 253, 261–62, 361; *Philadelphia Evening Bulletin*, October 2, 1958, p. 3, October 23, 1958, p. 3.

27. *Philadelphia Evening Bulletin*, October 16, 1958, p. 30, October 8, 1958, p. 10, October 3, 1958, p. 16; "Transport Workers Union International Council Condemns Wild-Cat Strikes in Philadelphia," October 15, 1958, file October–December 1958, box 83, "Big Party: All Maintenance Men Are Invited," undated, "What Do You Know," undated, "Who's Kidding Who, or the Big Laugh," undated, Dues paying members of Local 234 to Mr. Quill, November 16, 1956, file July–December 1956, box 82, TWU Papers; Roberts, "History and Analysis," 307.

28. *Philadelphia Evening Bulletin*, October 16, 1958, p. 30, October 30, 1956, p. 36.

29. *Philadelphia Evening Bulletin*, August 8, 1957, pp. 1, 4; *Philadelphia Inquirer*, March 20, 1958, p. 3; Eli Rock to Mayor Dilworth, December 4, 1956, file Labor—Eli Rock, box A-493, Clark Papers; Roberts, "History and Analysis," 265, 358.

30. "PTC Out to Milk City, Tate Says," April 18, 1962, "Tate Says City Will Sue If PTC Cuts Payments," June 12, 1962, file Douglas Pratt, EBM; "TWU Predicts Cuts by PTC," March 3, 1955, "PTC Policy: Keep 'City Snoopers' Out," April 23, 1955, file 1, box 238,

Greenfield Papers; speech by Michael J. Quill, undated, file May–June 1961, box 84, Ernest Mozer to Michael Quill, undated, file September–December 1960, box 83, TWU Papers; *Philadelphia Evening Bulletin*, July 5, 1955, p. 32; Hague, "Outline of Transit Service," file 1, box 4, Tucker Papers.

31. "Market St. El Pillars Sink but PTC Sees No Danger," November 14, 1961, "Engineers Find 671 Faults in Check of Work on El," October 17, 1961, "PTC Accuses Union of Creating El Scare," November 16, 1961, file 5, box 308, "Ex-PTC Man Blames Bill Padding on Pratt," May 24, 1961, file 6, box 308, Phil Crockett to Albert M. Greenfield, February 19, 1962, file 6, box 317, Greenfield Papers; "Pratt's Fate with PTC Hinges on Board Probe," May 24, 1961, "Ex-Aide Accuses Pratt in 700Gs El Repair Padding," May 23, 1961, "PTC Will Hold Open Investigation of Its Finances, Greenfield Says," May 24, 1961, "City Readies $2 Million Suit against PTC," May 19, 1961, file Philadelphia, Frankford El Scandal, box 53, TWU Papers; PTC, *Annual Report*, 1961, 8; PTC, *Annual Report*, 1962, 11.

32. "Stier, Overruled by His Bosses, Calls Resignation Unavoidable," January 26, 1963, file Albert Lyons, EBM; Robert Stier to Men and Women of PTC, July 12, 1962, Robert Stier to PTC Management, July 12, 1962, file 5, box 317, statement by Robert H. Stier, November 26, 1962, file 8, box 317, Greenfield Papers.

33. "Stier, Overruled by His Bosses, Calls Resignation Unavoidable," January 26, 1963, file Albert Lyons, EBM; "Day-to-Day Account of PTC Strike," February 2, 1963, file SEPTA History, box 215, Mounted Clippings; "This Is One Pratt Fall Our Town Can't Laugh Off," January 26, 1963, file 1963, box 88, TWU Papers; "Transit: Is Public Ownership the Answer?" February 17, 1963, file 5, box 326, Greenfield Papers.

34. *1968 Conference on Mass Transportation*, 417; "Mass Transportation in Philadelphia," undated, file Mass Transit 1963, "Council of the City of Philadelphia," undated, "Remarks of the President at the Signing of S. 6 Urban Mass Transportation Act of 1964 in the Cabinet Room," July 9, 1964, "Section-by-Section Summary of the Urban Mass Transportation Act of 1964," July 16, 1964, file Mass Transit, box A-4416, Tate Papers; Smerk, *Urban Mass Transportation*, 23–39. For a fine study of urban transportation in the context of the Great Society see Schrag, *Great Society Subway*.

35. Barnum, "From Private to Public," 99, 103; Cudahy, *Cash, Tokens, and Transfers*, 180–83, 201, 209–10; Smerk, *Urban Mass Transportation*, 58–59; Wells, *Economic Characteristics*, 2–3, 6–10–6–18, 6–56; SEPTA, *First Annual Report*, 1966, 8–9; Abrams, "Story of Rapid Transit," 22–23; SEPTA, *1969 Annual Report*, 1; "Subways Pushed Everywhere but Here," May 4, 1957, file Broad Street Subway—History, box 30, Mounted Clippings; Sechler, *Speed Lines*, 78.

36. Bureau of Municipal Research and Pennsylvania Economy League, *Improved Transportation*, 104–11; Mayor's Transit Study Task Force, *Public Transit Authority*, 5, 8, 49.

37. Mayor's Transit Study Task Force, *Public Transit Authority*, 1, 5, 22.

38. Louis Johanson to James H. J. Tate, President, City Council, April 4, 1961, file unlabeled, box A-6368, Tate Papers.

39. "Urban Transportation," undated, file unlabeled, box A-6368, Tate Papers; Siddiqi, "Functional Analysis," 340–41.

40. SEPTA, *Second Annual Report*, 4; Skelsey, "Streetcar Named Endure," 54; James H. J. Tate to William Scranton, August 15, 1963, Resolution No. 381, undated, James H. J. Tate to William Scranton, July 25, 1963, file SEPACT, box A-6368, Tate Papers.

41. Wolfinger, *Philadelphia Divided*, 189; Soyode, "Performance Measurement," 148; Skelsey, "Streetcar Named Endure," 96; Hague, Outline of Transit Service, unpaginated, file 1, box 4, Tucker Papers; Simon, *Philadelphia*, 115, 118; William W. Scranton to James H. J. Tate, August 12, 1963, file SEPACT, box A-6368, Tate Papers. For more on the racial politics of Philadelphia and its suburbs see Wolfinger, *Philadelphia Divided*, chaps. 7 and 8; Wolfinger, "American Dream."

42. SEPTA, *1972 Annual Report*, unpaginated; SEPTA, *First Annual Report*, 1966, 2; Bureau of Municipal Research and Pennsylvania Economy League, *Improved Transportation*, 116–19, 123–24; Sechler, *Speed Lines*, 85–96, 100.

43. UTTB, History of Public Transportation, XII-2; Siddiqi, "Functional Analysis," 470–71; Soyode, "Performance Measurement," 144; PSTC, *Annual Report*, 1959, 1966, 1962, 1967, 1968, 1963, 1965, 1969, unpaginated; "Red Arrow Comes Aboard," August 2, 1969, file SEPTA Editorials 1969, box 213, Mounted Clippings; SEPTA, *1969 Annual Report*, 11.

44. *Philadelphia Evening Bulletin*, October 17, 1958, p. 3; Middleton, Smerk, and Diehl, *Encyclopedia of North American Railroads*, 931; Transportation Act of 1958, 56 Stat. 284 (1958); PTC, *Annual Report*, 1958, 9; Sechler, *Speed Lines*, 78; Abrams, "Story of Rapid Transit," 18–20; *Philadelphia Inquirer*, February 15, 1957, p. 12; Albert Greenfield to Editor, *Philadelphia Inquirer*, January 15, 1962, file 6, box 317, Greenfield Papers; SEPTA, *1972 Annual Report*, unpaginated; Siddiqi, "Functional Analysis," 478–79. On the decline of passenger service more broadly and the creation of Amtrak see Goddard, *Getting There*, chap. 14.

45. SEPTA news release, November 15, 1966, file SEPTA, box A-4530, Tate Papers; "Pratt Rehires Aid He Retired," June 3, 1961, "Transit Authority Bill Called Unfair by PTC President," July 26, 1953, file Albert Lyons, "Pratt and Reavis Lose Key Posts, O'Brien Resigns in Shakeup at PTC," April 27, 1965, file Douglas Pratt, EBM; "Ownership by Public Could Improve Service, PTC President Concedes," March 1, 1964, file 1964, box 88, TWU Papers.

46. Kramer and James, *PTC Rails*, 23; PTC, *Annual Report*, 1966, 2; "Chronology—SEPTA Attempts to Acquire PTC," undated, file 5, box 6, Tucker Papers; PTC, *Annual Report*, 1967, 10–11; PTC, *Annual Report*, 1968, 7.

47. Booz, Allen & Hamilton Inc., "SEPTA Management Study," I-2; SEPTA, *1971 Annual Report*, 9; SEPTA, *1972 Annual Report*, unpaginated; Wells, *Economic Characteristics*, 6–60–6–61; Skelsey, "Street Car Named Endure," 97; Pennsylvania Congress, *Report of the House Select Committee to Investigate SEPTA*, 42–44; "SEPTA's Money Woes," February 2, 1970, file SEPTA Editorials 1970, box 213, Mounted Clippings.

48. "GE Executive Is Chosen to Run SEPTA," July 22, 1970, file David Phillips, EBM; Carol Conaway to Peter Rothberg, March 8, 1972, file Urban Transportation Coord, box A-3543, Frank Rizzo Papers, City Archives, Philadelphia; Pennsylvania Congress, *Report of the House Select Committee to Investigate SEPTA*, 45–49, 91; Ellsworth, "Transit," 42; SEPTA, *1971 Annual Report*, unpaginated.

49. Pennsylvania Congress, *Report of the House Select Committee to Investigate SEPTA*, 89–93, 134; "Pulling Together on Transit," June 26, 1970, "SEPTA and the City," August 18, 1970, file SEPTA Editorials 1970, "Mass Transit Planning," August 28, 1969, file SEPTA Editorials 1969, box 213, Mounted Clippings; Urban Mass Transportation Administration, *Managing SEPTA*; Booz, Allen & Hamilton Inc., "SEPTA Management Study," iii; Siddiqi, "Functional Analysis," 605; Cudahy, *Cash, Tokens, and Transfers*, 202; SEPTA, *1972 Annual Report*, unpaginated; SEPTA, *1974 Annual Report*, unpaginated; "Joseph T. Mack, 75, SEPTA Administrator," http://articles.philly.com/2000-08-31/news/25595848_1_septa-general-manager-joseph-t-mack-william-g-stead. For a list of routes that SEPTA closed see Skelsey, "Street Car Named Endure," 54.

50. Siddiqi, "Functional Analysis," 606; Soyode, "Performance Measurement," 140; SEPTA, *1969 Annual Report*, 1, 12; SEPTA, *1970 Annual Report*, unpaginated; Pennsylvania Congress, *Report of the House Select Committee to Investigate SEPTA*, 14, 91, 98; SEPTA, *1968 Annual Report*, 2; SEPTA, *1971 Annual Report*, 6; Booz, Allen & Hamilton Inc., "SEPTA Management Study," unpaginated, 106.

51. "To All Members of Local 234," June 24, 1960, "TWU Demands Pact with City," undated, file April–August 1960, box 83, TWU Papers; "TWU Local Hails PTC Sale Accord," December 8, 1965, "From World War II to Strike Stand, Jailed TWU Leaders Stick Together," April 15, 1971, file Dominic DiClerico, EBM; Urban Mass Transportation Administration, *Managing SEPTA*, 49–51.

52. Jerome Shestack to Casimir Sienkiewicz, January 10, 1966, file SEPTA, box A-4530, Tate Papers; Simpson & Curtin, *Appraisal of Physical and Intangible Assets*, 24–25; Pennsylvania Congress, *Report of the House Select Committee to Investigate SEPTA*, 114–15, 133.

53. Barnum, "From Private to Public," 96–98, 113; SEPTA, *This Is SEPTA*, 11. The literature on the problems facing the working class in the 1970s is vast, but a worthy starting point is Cowie, *Stayin' Alive*.

54. Barnum, "From Private to Public," 103–15; Kerrigan, "Why Public-Sector Strikes Are So Rare," http://www.governing.com/topics/public-workforce/col-why-public-sector-strikes-are-rare.html; Pennsylvania Public Employe Relations Act, http://www.portal.state.pa.us/portal/server.pt?open=514&objID=552991&mode=2#X.

55. "Can't Afford Raises, SEPTA Says," February 24, 1975, "Philadelphia: Will SEPTA, Union Avert a Strike?" March 2, 1975, file SEPTA Employees 1975 January–March 17, box 214, "SEPTA 'Short' to Meet Payroll, McConnon Says," June 14, 1973, file SEPTA Employees April 1974, "SEPTA Approves $6-Million Loan to Pay Employes," June 26, 1975, file SEPTA Employees 1975 March 18–December, "Talks Resume as TWU Strike Deadline Nears," March 12, 1973, file SEPTA Employees 1973 March, box 214, "SEPTA Keeps Rolling," March 21, 1973, file SEPTA Editorials 1972, "Perils of Fact-Finding," March 24, 1971, "SEPTA Turns to the State," April 1, 1971, file SEPTA Editorials 1971–72, "The Price of Transit Peace," January 14, 1969, file SEPTA Editorials, box 213, Mounted Clippings.

56. SEPTA, *1968 Annual Report*, unpaginated; SEPTA, *1971 Annual Report*, 4; "The Price of Transit Peace," January 14, 1969, file SEPTA Editorials, "SEPTA's Workers Should Measure Their Gains and Avoid a Strike," March 13, 1975, file SEPTA Editorials 1975, box 213, multiple articles in SEPTA files in box 213 and box 214, Mounted Clippings; "SEPTA Depends on City and Suburb," March 21, 1986, http://articles.philly.com/1986-03-21/news/26081821_1_septa-board-septa-services-transit-strike; "SEPTA Strikes," undated, http://philadelphia.about.com/od/transportation/a/SEPTA_strikes.htm.

57. "As the Strike Ends," April 21, 1971, file SEPTA Editorials 1971–72, box 213, "TWU Head 'Overjoyed' at Layoff Change," June 20, 1973, file SEPTA Employees April 1974, "Talks Resume as TWU Strike Deadline Nears," March 12, 1973, file SEPTA Employees 1973 March, box 214, Mounted Clippings; "TWU Will Ask Court to Bar SEPTA Layoffs," June 14, 1973, "Transit Bodies Unfair to TWU, Tate Charges," December 18, 1968, file Transport Workers Union, "Tate Accuses SEPTA of Antilabor Acts in Dealing with TWU," December 10, 1968, file Dominic DiClerico, EBM; Amalgamated Transit Union Staff, *ATU*, 116; Matthew Guinan to Frank Rizzo, April 8, 1975, file Labor Unions, box A-3875, Rizzo Papers; Booz, Allen & Hamilton Inc., "SEPTA Management Study," IV-5.

58. "Bitter Strikers Shout Their Defiance at SEPTA Depots around the City," April 17, 1971, file Transport Workers Union, EBM.

Bibliography

Archival Materials in Repositories

City Archives, Philadelphia
   Bernard Samuel Papers
   Frank Rizzo Papers
   James H. J. Tate Papers
   Joseph Clark Papers
Franklin D. Roosevelt Library, Hyde Park, N.Y.
   Franklin D. Roosevelt Papers
Hagley Museum and Library, Wilmington, Del.
   John F. Tucker Papers
   SEPTA Collection
Historical Society of Pennsylvania, Philadelphia
   Albert Greenfield Papers
   George Forman Scrapbook
   Harold Cox Papers
   Philadelphia Record Photos
Library of Congress, Washington, D.C.
   National Association for the Advancement of Colored People Papers
National Archives, Philadelphia Branch, Philadelphia
   Records of District Courts of the United States, Record Group 21
   Records of the Committee on Fair Employment Practice, Record
      Group 228
   Records of the National Labor Relations Board, Record Group 25
   Records of the War Manpower Commission, Record Group 211
Pennsylvania State Archives, Harrisburg
   James H. Duff Papers
   William Scranton Papers
Tamiment Library, New York University, New York City
   Transport Workers Union Papers
Urban Archives, Temple University, Philadelphia
   Charles H. Frazier Jr. Papers
   Evening Bulletin Morgue
   Evening Bulletin Transit Collection

John MacKay Shaw Papers
Mounted Clippings

Newspapers, Magazines, and Other Periodicals

American Federationist
American Magazine
Catholic Times
Co-Operative Bulletin
Crisis
Current Literature
McClure's Magazine
Motorman and Conductor
New York Evening Call
New York Times
The Outlook
Pennsylvania Labor Record
Philadelphia Afro American
Philadelphia Evening Bulletin
Philadelphia Inquirer
Philadelphia North American
Philadelphia Public Ledger
Philadelphia Tribune
PM
Rapid Transit Talks
Service Talks
Trade Union News
Transport Workers Bulletin

MICROFILM

AFL-CIO, *Philadelphia Council (Pa.) Records*. (Held at Urban Archives.)
Albert, Peter J., and Harold L. Miller, eds. *American Federation of Labor Records: The Samuel Gompers Era*. Madison: State Historical Society of Wisconsin, 1981.

GOVERNMENT DOCUMENTS

Agreement of July 1, 1907, Authorized by Ordinance of Councils, Approved July 1, 1907. Philadelphia: Press of Allen, Lane & Scott, 1907.
Annual Report of the Department of City Transit. Philadelphia: n.p., multiple years.
Bureau of Municipal Research. *A Proposed Organization to Recommend an Urban Transportation and Traffic Policy and Program for the City of Philadelphia*. Philadelphia: Bureau of Municipal Research, 1953.
Bureau of Municipal Research and Pennsylvania Economy League. *Improved Transportation for Southeastern Pennsylvania*. Philadelphia: n.p., 1960.

Department of Commerce and Labor. *Bulletin of the Bureau of Labor.* Vol. 10. Washington, D.C.: GPO, 1905.

Facts Respecting Street Railways; The Substance of a Series of Official Reports from the Cities of New York, Brooklyn, Boston, Philadelphia, Baltimore, Providence, Newark, Chicago, Quebec, Montreal, and Toronto. London: P. S. King, 34, Parliament Street, 1866.

Feustel, Robert M. *Report on Behalf of the City of Philadelphia on the Valuation of the Property of the Philadelphia Rapid Transit Co.* N.p.: n.p., 1922.

Ford, Bacon & Davis. "Pennsylvania State Railroad Commission in the Matter of the Complaints against the Philadelphia Rapid Transit Company." N.p.: n.p., 1911.

Mayor's Transit Study Task Force. *The Public Transit Authority: A Study of Five Cities.* N.p.: n.p., n.d.

McChord, C. C. *Report of C. C. McChord to the Public Service Commission of the Commonwealth of Pennsylvania in Re Philadelphia Transit Situation.* N.p.: n.p., 1927.

Pennsylvania Congress, House. *Report of the House Select Committee to Investigate SEPTA.* Harrisburg, Pa., 1980.

Report of Transit Advisory Committee to General Conference on Transit Situation in Philadelphia. Philadelphia: n.p., 1930.

"Review of the Proposal for Condemnation of Purchase by the City of Philadelphia of the Property of the Companies Underlying the Philadelphia Rapid Transit Company." N.p.: n.p., n.d.

SEPTA (Southeastern Pennsylvania Transportation Authority). *SEPTA History.* Philadelphia: n.p., 1982.

——. *This Is SEPTA.* Philadelphia: n.p., n.d.

Snell, Bradford C. *American Ground Transport: A Proposal for Restructuring the Automobile, Truck, Bus, and Rail Industries.* U.S. Congress, Senate, Committee on the Judiciary, 93rd Cong., 2 sess., February 26, 1974.

Supreme Court of Pennsylvania. *Pennsylvania State Reports.* West Publishing.

Transportation Act of 1958. 56 Stat. 284, 1958.

Twining, William S. *A Report to the City Council of Philadelphia on the Frankford Elevated Railway.* N.p.: n.p., 1920.

United States v. Bryan 339. U.S. 323 (1950).

Urban Mass Transportation Administration. *Managing SEPTA (Southeastern Pennsylvania Transportation Authority) Strategically.* Washington, D.C.: GPO, 1979.

U.S. Commission on Industrial Relations. *Industrial Relations: Final Report and Testimony.* Washington, D.C.: GPO, 1916.

U.S. Congress. House. Committee on Un-American Activities (HUAC). *Investigation of Un-American Propaganda Activities in the United States.* 76th Cong., 3rd sess., 1940. Washington, D.C.: GPO, 1938.

UTTB (Urban Traffic and Transportation Board). *History of Public Transportation in Philadelphia.* Philadelphia: Urban Traffic and Transportation Board, 1955.

——. *Plan and Program 1955.* Philadelphia: Urban Traffic and Transportation Board, 1956.

——. *A Survey of Investigations, Reports, and Proposals for Public Transit in Philadelphia.* Philadelphia: Urban Traffic and Transportation Board, 1956.

SECONDARY SOURCES

Abrams, Robert L. "The Story of Rapid Transit: Philadelphia." *Bulletin of the National Railway Historical Society* 26 (First Quarter 1961): 4–24.

Adamic, Louis. *Dynamite: The Story of Class Violence in America.* Gloucester, Mass.: Peter Smith, 1963.

Adams, Carolyn T. *From the Outside In: Suburban Elites, Third-Sector Organizations, and the Reshaping of Philadelphia*. Ithaca, N.Y.: Cornell University Press, 2014.

Amalgamated Transit Union Staff. *ATU 100 Years: A History of the Amalgamated Transit Union*. Washington, D.C.: Amalgamated Transit Union, 1992.

Anderson, Karen. *Wartime Women: Sex Roles, Family Relations, and the Status of Women during World War II*. Westport, Conn.: Greenwood Press, 1981.

Andrews, J. H. M. *A Short History of the Development of Street Railway Transportation in Philadelphia*. Philadelphia: n.p., 1945.

Arnesen, Eric. *Waterfront Workers of New Orleans: Race, Class, and Politics, 1863–1923*. New York: Oxford University Press, 1991.

Arnold, Stanley. "Building the Beloved Community: Philadelphia's Interracial Civil Rights Organizations and Race Relations, 1930–1970." PhD diss., Temple University, 1999.

———. *Building the Beloved Community: Philadelphia's Interracial Civil Rights Organizations and Race Relations, 1930–1970*. Jackson: University Press of Mississippi, 2014.

Barger, Harold. *The Transportation Industries, 1889–1946: A Study of Output, Employment, and Productivity*. New York: National Bureau of Economic Research, 1951.

Barnum, Darold T. "From Private to Public: Labor Relations in Urban Transit." *Industrial and Labor Relations Review* 25 (October 1971): 95–115.

Barrett, James. *William Z. Foster and the Tragedy of American Radicalism*. Urbana: University of Illinois Press, 1999.

Barrett, Paul. *The Automobile and Urban Transit: The Formation of Public Policy in Chicago, 1900–1930*. Philadelphia: Temple University Press, 1983.

Bauer, John, and Peter Costello. *Transit Modernization and Street Traffic Control: A Program of Municipal Responsibility and Administration*. Chicago: Public Administration Service, 1950.

Bayor, Ronald, and Timothy Meagher. *The New York Irish*. Baltimore: Johns Hopkins University Press, 1996.

Beckert, Sven. "History of American Capitalism." In *American History Now*, edited by Eric Foner and Lisa McGirr, 314–35. Philadelphia: Temple University Press, 2011.

Bernstein, Irving. *Turbulent Years: A History of the American Worker, 1933–1941*. Boston: Houghton Mifflin, 1971.

Blatz, Perry K. "Titanic Struggles, 1873–1916." In *Keystone of Democracy: A History of Pennsylvania Workers*, edited by Howard Harris and Perry K. Blatz, 83–160. Harrisburg: Pennsylvania Historical and Museum Commission, 1999.

Bonsall, Ward, comp. *Handbook of Social Laws of Pennsylvania*. Pittsburgh: Associated Charities of Pittsburgh and the Philadelphia Society for Organizing Charity, 1914.

Booz, Allen & Hamilton Inc. *SEPTA Management Study*. N.p.: n.p., 1978.

Borgnis, Mervin E. *An Inside Story of PRT and PTC*. N.p.: Mervin E. Borgnis, 1995.

———. *The Legacy of Thomas E. Mitten*. N.p.: n.p. 1983.

———. *The Near Side Car and the Legacy of Thomas E. Mitten*. Winchester, Va.: Winchester Printers, 1994.

Bottles, Scott L. *Los Angeles and the Automobile: The Making of the Modern City*. Berkeley: University of California Press, 1987.

Brandes, Stuart D. *American Welfare Capitalism, 1880–1940*. Chicago: University of Chicago Press, 1976.

Brecher, Jeremy. *Strike!* San Francisco: Straight Arrow Books, 1972.

Brenner, Aaron, Benjamin Day, and Immanuel Ness, eds. *The Encyclopedia of Strikes in American History*. Armonk, N.Y.: M. E. Sharpe, 2009.

Bruce, Robert V. *1877: Year of Violence*. Chicago: Ivan R. Dee, 1989.

Bulletin Almanac. Philadelphia: *Evening Bulletin*, multiple years.

Burt, Nathaniel, and Wallace E. Davies. "The Iron Age, 1876–1905." In *Philadelphia: A 300-Year History*, edited by Russell F. Weigley, 471–523. New York: Norton, 1982.

Cameron, Ardis. *Radicals of the Worst Sort: Laboring Women in Lawrence, Massachusetts, 1860–1912*. Urbana: University of Illinois Press, 1993.

Caskie, John J. Kerr. "The Philadelphia Rapid Transit Plan." *Annals of the American Academy of Political and Social Science* 85 (September 1919): 189–204.

Chafe, William. *The American Woman: Her Changing Social, Economic, and Political Roles, 1920–1970*. New York: Oxford University Press, 1972.

Chandler Brothers & Co., comp. *The Philadelphia Rapid Transit Company: A Descriptive and Statistical Analysis*. Philadelphia: n.p., 1904.

Cheape, Charles. *Moving the Masses: Urban Public Transit in New York, Boston, and Philadelphia, 1880–1912*. Cambridge, Mass.: Harvard University Press, 1980.

Clark, Dennis. *The Irish Relations: Trials of an Immigrant Tradition*. East Brunswick, N.J.: Associated University Presses, 1982.

Clow, David. *House Divided: Philadelphia's Controversial Crosstown Expressway*. Hilliard, Ohio: Society for American City and Regional Planning History, 1989.

Cohen, Lizabeth. *Making a New Deal: Industrial Workers in Chicago, 1919–1939*. New York: Cambridge University Press, 1990.

Cole, Peter. *Wobblies on the Waterfront: Interracial Unionism in Progressive-Era Philadelphia*. Urbana: University of Illinois Press, 2007.

Connelly, James, ed. *The History of the Archdiocese of Philadelphia*. Philadelphia: Archdiocese of Philadelphia, 1976.

Conway, Thomas, Jr. "The Decreasing Financial Returns upon Urban Street Railway Properties." *Annals of the American Academy of Political and Social Science* 37 (January 1911): 14–30.

——. "Street Railways in Philadelphia since 1900." *Annals of the American Academy of Political and Social Science* 24 (September 1904): 70–76.

Co-Operative Plan. Philadelphia: n.p., 1911.

Countryman, Matthew. *Up South: Civil Rights and Black Power in Philadelphia*. Philadelphia: University of Pennsylvania Press, 2006.

Cowie, Jefferson. *Capital Moves: RCA's Seventy-Year Quest for Cheap Labor*. Ithaca, N.Y.: Cornell University Press, 1999.

——. *Stayin' Alive: The 1970s and the Last Days of the Working Class*. New York: New Press, 2010.

Cox, Harold. *Early Electric Cars of Philadelphia, 1885–1911*. Forty Fort, Pa.: Harold E. Cox, 1969.

——. *The Road from Upper Darby: The Story of the Market Street Subway-Elevated*. N.p.: Electric Railroaders' Association, 1967.

——. *Surface Cars of Philadelphia, 1911–1965*. Forty Fort, Pa.: n.p., 1965.

——. "The Tram Subways of Philadelphia—a History and a Forward Look." *Modern Tramway* 26 (June 1963): 205–14.

——. *Utility Cars of Philadelphia, 1892–1971*. Forty Fort, Pa.: Harold E. Cox, 1971.

Cox, Harold E., and John F. Meyers. "The Philadelphia Traction Monopoly and the Pennsylvania Constitution of 1874: The Prostitution of an Ideal." *Pennsylvania History* 35 (October 1968): 406–23.

Cudahy, Brian J. *Cash, Tokens, and Transfers: A History of Urban Mass Transit in North America*. New York: Fordham University Press, 1990.

Cunningham, Wallace McCook. "Electric Railway Stocks." *Annals of the American Academy of Political and Social Science* 35 (May 1910): 175–91.

Currarino, Rosanne. *The Labor Question in America: Economic Democracy in the Gilded Age.* Urbana: University of Illinois Press, 2011.

Cutler, William W., III, and Howard Gillette Jr. *The Divided Metropolis: Social and Spatial Dimensions of Philadelphia, 1800–1975.* Westport, Conn.: Greenwood Press, 1980.

Delmont, Matthew. *The Nicest Kids in Town: American Bandstand, Rock 'n' Roll, and the Struggle for Civil Rights in 1950s Philadelphia.* Berkeley: University of California Press, 2012.

Dennison, Henry S. "The Employee Investor." *Proceedings of the Academy of Political Science in the City of New York* 11 (April 1925): 29–31.

Dewess, Donald. "The Decline of the American Street Railways." *Traffic Quarterly* 24 (October 1970): 563–81.

Douglas, Paul H. "An Analysis of Strike Statistics, 1881–1921." *Journal of the American Statistical Association* 18 (September 1923): 866–77.

——. "Shop Committees: Substitute for, or Supplement to, Trades-Unions?" *Journal of Political Economy* 29 (February 1921): 89–107.

Drayer, Robert E. "J. Hampton Moore: An Old Fashioned Republican." PhD diss., University of Pennsylvania, 1961.

Du Bois, W. E. B. *Black Reconstruction in America, 1860–1880.* New York: Free Press, 1992 (1935).

Dudden, Arthur P. "The City Embraces 'Normalcy,' 1919–1929." In *Philadelphia: A 300-Year History,* edited by Russell F. Weigley, 566–600. New York: Norton, 1982.

Easton, Alexander. *A Practical Treatise on Street or Horse-Power Railways: Their Location, Construction and Management; With General Plans and Rules for Their Organization and Operation.* Philadelphia: Crissy & Markley, 1859.

Ellsworth, Kenneth G. "Transit: SEPTA Shows the Way." *Railway Age* 176 (March 1975): 40–46.

Emmons, C. D. "The Relations of the Electric Railway Company with Its Employees." *Annals of the American Academy of Political and Social Science* 37 (January 1911): 88–92.

Fairchild, C. B., Jr. *Training for the Electric Railway Business.* Philadelphia: J. B. Lippincott, 1919.

Federal Writers' Project. *Philadelphia: A Guide to the Nation's Birthplace.* Philadelphia: William Penn Association of Philadelphia, 1937.

Fink, Leon. *Sweatshops at Sea: Merchant Seamen in the World's First Globalized Industry, from 1812 to the Present.* Chapel Hill: University of North Carolina Press, 2011.

——. *Workingmen's Democracy: The Knights of Labor and American Politics.* Urbana: University of Illinois Press, 1983.

Fitzherbert, Tony. "50 Years of the Broad Street Subway." *Headlights* 41 (January–March 1979): 2–14.

Fogelson, Robert M. *America's Armories: Architecture, Society, and Public Order.* Cambridge, Mass.: Harvard University Press, 1989.

Foner, Philip S. *History of the Labor Movement.* Vol. 5. New York: International Publishers, 1987.

Fones-Wolf, Elizabeth A. *Selling Free Enterprise: The Business Assault on Labor and Liberalism, 1945–60.* Urbana: University of Illinois Press, 1994.

Fones-Wolf, Ken. "An Industrial Giant Takes Shape, 1800–1872." In *Keystone of Democracy: A History of Pennsylvania Workers,* edited by Howard Harris and

Perry K. Blatz, 37–79. Harrisburg: Pennsylvania Historical and Museum Commission, 1999.

——. "Notes and Documents: Mass Strikes, Corporate Strategies: The Baldwin Locomotive Works and the Philadelphia General Strike of 1910." *Pennsylvania Magazine of History and Biography* 110 (July 1986): 447–57.

——. *Trade Union Gospel: Christianity and Labor in Industrial Philadelphia, 1865–1915*. Philadelphia: Temple University Press, 1989.

Frankford Elevated. N.p.: n.p., 1919.

Freeman, Joshua. *In Transit: The Transport Workers Union in New York City, 1933–1966*. New York: Oxford University Press, 1989.

Fundamentals of Industrial Prosperity. N.p.: n.p., 1926.

Gabin, Nancy. "The Hand That Rocks the Cradle Can Build Tractors, Too." *Michigan History Magazine* 76 (March/April 1992): 12–21.

Gemmill, Paul F. "The Literature of Employee Representation." *Quarterly Journal of Economics* 42 (May 1928): 479–94.

Gilje, Paul A. *Rioting in America*. Bloomington: Indiana University Press, 1996.

Glazer, Aaron Michael. "Fade to Gas: The Conversion of Baltimore's Mass Transit System from Streetcars to Diesel-Powered Buses." *Maryland Historical Magazine* 97 (Fall 2002): 337–57.

Goddard, Stephen B. *Getting There: The Epic Struggle between Road and Rail in the American Century*. New York: Basic Books, 1994.

Golab, Caroline. "The Immigrant and the City: Poles, Italians, and Jews in Philadelphia, 1870–1920." In *The Peoples of Philadelphia: A History of Groups and Lower-Class Life, 1790–1940*, edited by Allen F. Davis and Mark H. Haller, 203–30. Philadelphia: University of Pennsylvania Press, 1973.

——. *Immigrant Destinations*. Philadelphia: Temple University Press, 1977.

Gottlieb, Peter. "Shaping a New Labor Movement, 1917–1941." In *Keystone of Democracy: A History of Pennsylvania Workers*, edited by Howard Harris and Perry K. Blatz, 161–212. Harrisburg: Pennsylvania Historical and Museum Commission, 1999.

Green, James. *Death in the Haymarket: A Story of Chicago, the First Labor Movement and the Bombing That Divided Gilded Age America*. New York: Random House, 2006.

Greene, Julie. *Pure and Simple Politics: The American Federation of Labor and Political Activism, 1881–1917*. New York: Cambridge University Press, 1999.

Hague, George T. *An Outline of Transit Service Development in Philadelphia, 1837–1970*. N.p.: n.p., 1980.

Hallowell, Guernsey A., and Thomas Creighton, comp. *Frankford: Direction of a Greater Philadelphia*. N.p.: n.p., 1922.

Harring, Sidney L. "Car Wars: Strikes, Arbitration, and Class Struggle in the Making of Labor Law." *NYU Review of Law and Social Change* 14 (1986): 849–72.

Harris, Howard, and Perry K. Blatz, eds. *Keystone of Democracy: A History of Pennsylvania Workers*. Harrisburg: Pennsylvania Historical and Museum Commission, 1999.

Harris, Howell John. *Bloodless Victories: The Rise and Fall of the Open Shop in the Philadelphia Metal Trades, 1890–1940*. New York: Cambridge University Press, 2000.

——. *The Right to Manage: Industrial Relations Policies of American Business in the 1940s*. Madison: University of Wisconsin Press, 1982.

Hartmann, Susan. *The Home Front and Beyond: American Women in the 1940s*. Boston: Twayne, 1982.

Heller, Gregory L. *Ed Bacon: Planning, Politics and the Building of Modern Philadelphia*. Philadelphia: University of Pennsylvania Press, 2013.

Hepp, John, IV. *The Middle-Class City: Transforming Space and Time in Philadelphia, 1876–1926*. Philadelphia: University of Pennsylvania Press, 2003.

Herod, Andrew. "From a Geography of Labor to a Labor Geography: Labor's Spatial Fix and the Geography of Capitalism." *Antipode* 29 (1997): 1–31.

Hershberg, Theodore, Harold E. Cox, Dale B. Light Jr., and Richard R. Greenfield. "The 'Journey-to-Work': An Empirical Investigation of Work, Residence and Transportation, Philadelphia, 1850 and 1880." In *Philadelphia: Work, Space, Family, and Group Experience in the Nineteenth Century*, edited by Theodore Hershberg, 128–73. New York: Oxford University Press, 1981.

Higbie, Frank Tobias. *Indispensable Outcasts: Hobo Workers and Community in the American Midwest, 1880–1930*. Urbana: University of Illinois Press, 2003.

Higgins, Edward E. *Street Railway Investments: A Study in Values*. New York: Street Railway Publishing Co., 1895.

"History of National City Lines." *Motor Coach Age* 26 (February 1974): 4–20.

Hood, Clifton. *722 Miles: The Building of the Subways and How They Transformed New York*. Baltimore: Johns Hopkins University Press, 1993.

Jacoby, Sanford M. *Employing Bureaucracy: Managers, Unions, and the Transformation of Work in the 20th Century*. Mahwah, N.J.: Lawrence Erlbaum, 2004.

Johanningsmeier, Edward. "Philadelphia 'Skittereen' and William Z. Foster: The Childhood of an American Communist." *Pennsylvania Magazine of History and Biography* 117 (October 1993): 287–308.

Johnson, Emory R. "Public Regulation of Street Railway Transportation." *Annals of the American Academy of Political and Social Science* 29 (January 1907): 31–47.

Johnson, F. W. *The Case for Co-operation*. Philadelphia: n.p., 1922.

Kahn, Albert. *High Treason: The Plot against the People*. New York: Lear, 1950.

Kaplan, Temma. *Red City, Blue Period: Social Movements in Picasso's Barcelona*. Berkeley: University of California Press, 1992.

Kaufman, Bruce E. *The Origin and Evolution of the Field of Industrial Relations in the United States*. Ithaca, N.Y.: ILR Press, 1993.

Keller, Richard. *Pennsylvania's Little New Deal*. New York: Garland, 1982.

Kelley, Blair M. *Right to Ride: Streetcar Boycotts and African American Citizenship in the Era of* Plessy v. Ferguson. Chapel Hill: University of North Carolina Press, 2010.

Kelley, Robin D. G. *Race Rebels: Culture, Politics, and the Black Working Class*. New York: Free Press, 1994.

Kennedy, David M. *Freedom from Fear: The American People in Depression and War, 1929–1945*. New York: Oxford University Press, 2005.

Kenny, Kevin. *The American Irish: A History*. Edinburgh Gate, UK: Longman, 2000.

Kessler-Harris, Alice. *Gendering Labor History*. Urbana: University of Illinois Press, 2007.

Knowles, Scott Gabriel, ed. *Imagining Philadelphia: Edmund Bacon and the Future of the City*. Philadelphia: University of Pennsylvania Press, 2009.

Kramer, Frederick A., and Samuel L. James Jr. *PTC Rails: A Historical Review and Scenes of the Trolley Years*. Flanders, N.J.: RAE Publishing, 1996.

Lane, Roger. *Roots of Violence in Black Philadelphia, 1860–1900*. Cambridge, Mass.: Harvard University Press, 1986.

Lauck, W. Jett. *Political and Industrial Democracy, 1776–1926*. New York: Funk & Wagnalls, 1926.

Lefebvre, Henri. *Writings on Cities*. Cambridge, Mass.: Blackwell, 1996.

Leidenberger, Georg. *Chicago's Progressive Alliance: Labor and the Bid for Public Streetcars.* DeKalb: Northern Illinois University Press, 2006.

Levinson, Edward. *I Break Strikes! The Technique of Pearl L. Bergoff.* New York: R. M. McBride & Co., 1935.

Lewis, Edwin O. "Philadelphia's Relation to Rapid Transit Company." *Annals of the American Academy of Political and Social Science* 31 (May 1908): 66–77.

——. *The Street Railway Situation in Philadelphia.* Philadelphia: n.p., 1907.

Licht, Walter. *Getting Work: Philadelphia, 1840–1950.* Cambridge, Mass.: Harvard University Press, 1992.

Lichtenstein, Nelson. *State of the Union: A Century of American Labor.* Princeton, N.J.: Princeton University Press, 2002.

——. *Walter Reuther: The Most Dangerous Man in Detroit.* New York: Basic Books, 1995.

Lipson, David. "General Motors, National City Lines and the Motor Bus: The Motor Bus' Role in the Decline of Mass Transit in the United States." BA thesis, Harvard University, 1987.

Lovitt, James L. *Educating for Industry.* Philadelphia: n.p., n.d.

Marsh, Margaret S. "The Impact of the Market Street 'El' on Northern West Philadelphia: Environmental Change and Social Transformation, 1900–1930." In *The Divided Metropolis: Social and Spatial Dimensions of Philadelphia, 1800–1975,* edited by William W. Cutler III and Howard Gillette Jr., 169–92. Westport, Conn.: Greenwood Press, 1980.

McCaffery, Peter. *When Bosses Ruled Philadelphia: The Emergence of the Republican Machine, 1867–1933.* University Park: Pennsylvania State University Press, 1993.

McCartin, Joseph A. *Labor's Great War: The Struggle for Industrial Democracy and the Origins of Modern American Labor Relations, 1912–1921.* Chapel Hill: University of North Carolina Press, 1997.

McColloch, Mark. "Glory Days: 1941–1969." In *Keystone of Democracy: A History of Pennsylvania Workers,* edited by Howard Harris and Perry K. Blatz, 213–74. Harrisburg: Pennsylvania Historical and Museum Commission, 1999.

McKee, Guian. *The Problem of Jobs: Liberalism, Race, and Deindustrialization in Philadelphia.* Chicago: University of Chicago Press, 2008.

McLain, Frank D. "The Street Railways of Philadelphia." *Quarterly Journal of Economics* 22 (February 1908): 233–60.

McShane, Clay. *Down the Asphalt Path: The Automobile and the American City.* New York: Columbia University Press, 1994.

——. Review of *Getting There: The Epic Struggle between Road and Rail in the American Century* by Stephen B. Goddard. *Wisconsin Magazine of History* 78 (Spring 1995): 230.

——. *Technology and Reform: Street Railways and the Growth of Milwaukee, 1887–1900.* Madison: University of Wisconsin Press, 1974.

McShane, Clay, and Joel A. Tarr. *The Horse in the City: Living Machines in the Nineteenth Century.* Baltimore: Johns Hopkins University Press, 2007.

Meier, August, and Elliott Rudwick. "Communist Unions and the Black Community: The Case of the Transport Workers Union, 1934–1944." *Labor History* 23 (Spring 1982): 165–97.

Metropolitan Rail Road Company. *Rapid Transit for the City of Philadelphia by Subway Railroads.* Philadelphia: Metropolitan Rail Road Co., 1886.

Metzgar, Jack. *Striking Steel: Solidarity Remembered.* Philadelphia: Temple University Press, 2000.

Meyers, Allen, and Joel Spivak. *Philadelphia Trolleys*. Charleston, S.C.: Arcadia
    Publishing, 2003.
Middleton, William D., George M. Smerk, and Roberta L. Diehl, eds. *Encyclopedia of
    North American Railroads*. Bloomington: Indiana University Press, 2007.
Milkman, Ruth. *Gender at Work: The Dynamics of Job Segregation during World
    War II*. Urbana: University of Illinois Press, 1987.
——, ed. *Women, Work and Protest: A Century of US Women's Labor History*. Boston:
    Routledge & Kegan Paul, 1985.
Miller, Fredric M., Morris J. Vogel, and Allen F. Davis. *Philadelphia Stories: A
    Photographic History, 1920–1960*. Philadelphia: Temple University Press, 1988.
——. *Still Philadelphia: A Photographic History, 1890–1940*. Philadelphia: Temple
    University Press, 1988.
Miller, Harry A. "History of the Transit System in Philadelphia." MA thesis,
    University of Pennsylvania, 1951.
Mitten, A. A. "Maintenance of Contact with Employes of the Philadelphia Rapid
    Transit Company." *Annals of the American Academy of Political and Social
    Science* 119 (May 1925): 108–14.
——. "Results of Collective Bargaining in the Street Railway Industry of
    Philadelphia." *Annals of the American Academy of Political and Social Science*
    90 (July 1920): 57–60.
Mitten, T. E. *Philadelphia's Answer to the Traction Question*. N.p.: n.p., 1919.
Moffat, Bruce G. *The "L": The Development of Chicago's Rapid Transit System,
    1888–1932*. Chicago: Central Electric Railfans' Association, 1995.
Molloy, Scott. "Trolley Wars." In *The Encyclopedia of Strikes in American
    History*, edited by Aaron Brenner, Benjamin Day, and Immanuel Ness,
    519–33. Armonk, N.Y.: M. E. Sharpe, 2009.
——. *Trolley Wars: Streetcar Workers on the Line*. Durham: University of New
    Hampshire Press, 1996.
Montgomery, David. *Citizen Worker: The Experience of Workers in the United States
    with Democracy and the Free Market during the Nineteenth Century*. New York:
    Cambridge University Press, 1993.
——. *Workers' Control in America: Studies in the History of Work, Technology, and
    Labor Struggles*. New York: Cambridge University Press, 1979.
Morley, Christopher. *Travels in Philadelphia*. Philadelphia: David McKay Co., 1920.
Murray, Robert K. *Red Scare: A Study in National Hysteria, 1919–1920*. Minneapolis:
    University of Minnesota Press, 1955.
NCL (National City Lines). *Annual Report*. N.p.: n.p., multiple years.
1968 Conference on Mass Transportation. United Transportation Union: n.p., 1968.
Northrup, Herbert, et al. *Negro Employment in Land and Air Transport: A Study of
    Racial Policies in the Railroad, Airline, Trucking, and Urban Transit Industries*.
    Philadelphia: University of Pennsylvania Press, 1971.
Norwood, Stephen. *Strikebreaking and Intimidation: Mercenaries and Masculinity in
    Twentieth-Century America*. Chapel Hill: University of North Carolina Press, 2002.
The Overhead Electric Trolley Ordinances. Philadelphia: n.p., 1892.
Painter, Nell Irvin. *Standing at Armageddon: The United States, 1877–1919*. New York:
    Norton, 2008.
Palmer, Gladys. *The Manpower Outlook in Philadelphia in 1943*. Philadelphia:
    Wharton School, 1943.
——. *The Philadelphia Labor Market in 1944*. Philadelphia: Wharton School, 1944.
——. *War Labor Supply Problems in Philadelphia and Its Environs*. Philadelphia:
    Wharton School, 1942.

Pappalardo, Louis. "Philip Murray: Pennsylvania's Mid-Century Labor Giant." In *Keystone of Democracy: A History of Pennsylvania Workers*, edited by Howard Harris and Perry K. Blatz, 259–60. Harrisburg: Pennsylvania Historical and Museum Commission, 1999.

Parry, D. E., comp. *History of Rapid Transit Development: City of Philadelphia, 1886 to 1928*. Philadelphia: Bureau of Engineering, 1928.

Patty, William Jordan. "Crime on the Bus: Bus Driver Safety in Postwar Washington, D.C." *Washington History* 25 (Summer 2013): 36–51.

Perkiss, Abigail. *Making Good Neighbors: Civil Rights, Liberalism, and Integration in Postwar Philadelphia*. Ithaca, N.Y.: Cornell University Press, 2014.

Perlman, Selig, and Philip Taft. *History of Labor in the United States, 1896–1932*. Vol. 4, *Labor Movements*. New York: Macmillan, 1935.

Persion, Vera E. "The Mitten Plan of the Philadelphia Rapid Transit Company." MA thesis, Columbia University, 1930.

Philadelphia Rapid Transit: Being an Account of the Construction and Equipment of the Market-Street Subway Elevated and Its Place in the Great System and Service of the Philadelphia Rapid Transit Company. Philadelphia: Arnold & Dyer, 1908.

Philadelphia Rapid Transit: Stotesbury-Mitten Agreement, 1911–1920. N.p.: n.p., 1921.

The Philadelphia Transit System: A Short Study in Social Justice. N.p.: n.p., 1934.

Pierce, Daniel T. "The Strike Problem upon American Railways." *Annals of the American Academy of Political and Social Science* 37 (January 1911): 93–103.

Post, Robert C. *Urban Mass Transit: The Life Story of a Technology*. Westport, Conn.: Greenwood Press, 2007.

Pratt, C. O. "The Sympathetic Strike." *Annals of the American Academy of Political and Social Science* 36 (September 1910): 137–42.

President Mitten's Talk. N.p.: n.p., 1923.

PRT (Philadelphia Rapid Transit Company). *Annual Report*. Philadelphia: n.p., multiple years.

——. *Mitten Men and Management*. N.p.: n.p., 1922.

——. *A Proposed Garage Corporation to Serve Philadelphia Central Business District*. N.p.: n.p., 1925.

PSTC (Philadelphia Suburban Transportation Company). *Annual Report*. N.p.: n.p., multiple years.

PTC (Philadelphia Transportation Company). *Annual Report*. Philadelphia: n.p., multiple years.

Pushkarev, Boris, and Jeffrey Zupan. *Urban Rail in America: An Exploration of Criteria for Fixed-Guideway Transit*. Washington, D.C.: Urban Mass Transit Administration, 1980.

Radford, Gail. *The Rise of the Public Authority: Statebuilding and Economic Development in Twentieth-Century America*. Chicago: University of Chicago Press, 2013.

"Recent Cases." *University of Pennsylvania Law Review Register* 84 (January 1936): 421–23.

Report of Proceedings, 13th Constitutional Convention. New York: Transport Workers Union of America, 1969.

Righter, Robert W. *The Battle over Hetch Hetchy: America's Most Controversial Dam and the Birth of Modern Environmentalism*. New York: Oxford University Press, 2005.

Roberts, Thomas. "A History and Analysis of Labor-Management Relations in the Philadelphia Transit Industry." PhD diss., University of Pennsylvania, 1959.

Roediger, David R. *The Wages of Whiteness: Race and the Making of the American Working Class.* New York: Verso Books, 1991.

Roediger, David R., and Elizabeth D. Esch. *The Production of Difference: Race and the Management of Labor in U.S. History.* New York: Oxford University Press, 2012.

Roediger, David R., and Philip S. Foner. *Our Own Time: A History of American Labor and the Working Day.* New York: Greenwood Press, 1987.

Rose, James D. *Duquesne and the Rise of Steel Unionism.* Urbana: University of Illinois Press, 2001.

Rosenthal, Anton. "The Arrival of the Electric Streetcar and the Conflict over Progress in Early Twentieth-Century Montevideo." *Journal of Latin American Studies* 27 (May 1995): 319–41.

——. "Streetcar Workers and the Transformation of Montevideo: The General Strike of May 1911." *Americas* 51 (April 1995): 471–94.

Rosenzweig, Roy. *Eight Hours for What We Will: Workers and Leisure in an Industrial City, 1870–1920.* New York: Cambridge University Press, 1983.

Rosswurm, Steven, ed. *The CIO's Left-Led Unions.* New Brunswick, N.J.: Rutgers University Press, 1992.

Ryan, Francis. *AFSCME's Philadelphia Story: Municipal Workers and Urban Power in the Twentieth Century.* Philadelphia: Temple University Press, 2011.

Ryan, Mary. *Women in Public: Between Banners and Ballots, 1825–1880.* Baltimore: Johns Hopkins University Press, 1990.

Schmidt, Emerson P. *Industrial Relations in Urban Transportation.* Minneapolis: University of Minnesota Press, 1937.

Schneirov, Richard. *Labor and Urban Politics: Class Conflict and the Origins of Modern Liberalism in Chicago, 1864–97.* Urbana: University of Illinois Press, 1998.

Schrag, Zachary. *The Great Society Subway: A History of the Washington Metro.* Baltimore: Johns Hopkins University Press, 2006.

Scranton, Philip. "Large Firms and Industrial Restructuring: The Philadelphia Region, 1900–1980." *Pennsylvania Magazine of History and Biography* 116 (October 1992): 419–65.

Scranton, Philip, and Walter Licht. *Work Sights: Industrial Philadelphia, 1890–1950.* Philadelphia: Temple University Press, 1986.

Sechler, Robert P. *Speed Lines to City and Suburbs: A Summary of Rapid Transit Development in Metropolitan Philadelphia from 1879 to 1974.* N.p.: n.p., n.d.

SEPTA (Southeastern Pennsylvania Transportation Authority). *Annual Report.* N.p.: n.p., multiple years.

Shaw, John M. *Mitten: The Story of a Life.* Unpublished ms., 1930.

Siddiqi, M. Kamil. "Functional Analysis and Long Term Performance Evaluation of the Evolution of a Public Transit Agency: A Case Study of SEPTA (1968–1988)." PhD diss., University of Pennsylvania, 1995.

Sidorick, Daniel. *Condensed Capitalism: Campbell Soup and the Pursuit of Cheap Production in the Twentieth Century.* Ithaca, N.Y.: ILR Press, 2009.

——. "The 'Girl Army': The Philadelphia Shirtwaist Strike of 1909–1910." *Pennsylvania History* 71 (Summer 2004): 323–69.

Simon, Roger. *Philadelphia: A Brief History.* University Park: Pennsylvania Historical Association, 2003.

Simpson & Curtin, Transportation Engineers. *Appraisal of Physical and Intangible Assets of Philadelphia Transportation Company.* Philadelphia: n.p., 1966.

Skelsey, Geoffrey. "A Streetcar Named Endure: Rail Transit in Philadelphia." *Tramways and Urban Transit* 69 (January 2006): 10–15.

——. "A Streetcar Named Endure: Rail Transit in Philadelphia (II)." *Tramways and Urban Transit* 69 (2006): 53–55.

Sklansky, Jeffrey. "The Elusive Sovereign: New Intellectual and Social Histories of Capitalism." *Modern Intellectual History* 9 (2012): 233–48.

——. "Labor, Money, and the Financial Turn in the History of Capitalism." *Labor: Studies in Working-Class History of the Americas* 11 (Spring 2014): 23–46.

Smerk, George M. *Urban Mass Transportation: A Dozen Years of Federal Policy.* Bloomington: Indiana University Press, 1974.

Smith, Robert Michael. *From Blackjacks to Briefcases: A History of Commercialized Strikebreaking and Unionbusting in the United States.* Athens: Ohio University Press, 2003.

——. "King of the Strikebreakers: The Notorious Career of James A. Farley." *Labor's Heritage* 11 (Spring/Summer 2000): 20–37.

Soyode, Afolabi. "Performance Measurement and Control in Public Transit Authorities." PhD diss., University of Pennsylvania, 1973.

Speirs, Frederic. *The Street Railway System of Philadelphia, Its History and Present Condition.* New York: Johnson Reprint Corporation, 1973. First published in 1897 by Johns Hopkins University Press.

The Spirit of PRT. N.p.: n.p., 1924.

St. Clair, David J. "The Motorization and Decline of Urban Public Transit, 1935–1950." *Journal of Economic History* 41 (September 1981): 579–600.

Steffens, Lincoln. *The Shame of the Cities.* New York: McClure, Phillips & Co., 1904.

Stowell, David O. *Streets, Railroads, and the Great Strike of 1877.* Chicago: University of Chicago Press, 1999.

Stromquist, Shelton. *A Generation of Boomers: The Pattern of Railroad Labor Conflict in Nineteenth-Century America.* Urbana: University of Illinois Press, 1987.

——. *Reinventing "The People": The Progressive Movement, the Class Problem, and the Origins of Modern Liberalism.* Champaign: University of Illinois Press, 2006.

Taylor, A. Merritt. "Philadelphia's Transit Problem." *Annals of the American Academy of Political and Social Sciences* 57 (January 1915): 28–32.

——. *The Rapid Transit Problem.* Philadelphia: n.p., 1913.

Tinkcom, Margaret B. "Depression and War, 1929–1946." In *Philadelphia: A 300-Year History,* edited by Russell F. Weigley, 601–48. New York: Norton, 1982.

Tomlins, Christopher. *The State and the Unions: Labor Relations, Law, and the Organized Labor Movement in America, 1880–1960.* New York: Cambridge University Press, 1985.

Toynbee Society of Philadelphia. *The Philadelphia Trolley Companies and Their Employees.* Philadelphia: Toynbee Society, 1895.

Trachtenberg, Alan. *The Incorporation of America: Culture and Society in the Gilded Age.* New York: Hill & Wang, 1982.

"Union Wage Rates and Hours of Local Transit Operating Employees, Oct. 1947." *Union Wages and Hours: Local Transit Operating Employees* (1958): 1–7.

Vance, James E., Jr. *Capturing the Horizon: The Historical Geography of Transportation since the Transportation Revolution of the Sixteenth Century.* New York: Harper & Row, 1986.

Warner, Sam Bass, Jr. *The Private City: Philadelphia in Three Periods of Its Growth.* Philadelphia: University of Pennsylvania Press, 1968.

——. *Streetcar Suburbs: The Process of Growth in Boston, 1870–1900.* Cambridge, Mass.: Harvard University Press, 1962.

Weckler, Joseph E., and Robert C. Weaver. *Negro Platform Workers.* Chicago: American Council on Race Relations, 1945.

Welch, Emmett H. *Employment Trends in Philadelphia*. N.p.: Commonwealth of Pennsylvania, 1933.

Welke, Barbara Young. *Recasting American Liberty: Gender, Race, Law, and the Railroad Revolution, 1865–1920*. New York: Cambridge University Press, 2001.

Wells, John D., et al. *Economic Characteristics of the Urban Public Transportation Industry*. Washington, D.C.: GPO, 1972.

White, Richard. *Railroaded: The Transcontinentals and the Making of Modern America*. New York: Norton, 2011.

White, Walter. *A Rising Wind*. Garden City, N.Y.: Doubleday, Doran and Co., 1945.

Whittemore, L. H. *The Man Who Ran the Subways: The Story of Mike Quill*. New York: Holt, Rinehart and Winston, 1968.

Wiebe, Robert H. *Businessmen and Reform: A Study of the Progressive Movement*. Chicago: Ivan R. Dee, 1989. First published in 1962 by Harvard University Press.

Wike, Jessee Roffe. "The Pennsylvania Manufacturers' Association: A Study of a Political Interest Group in the Governmental Process." PhD diss., University of Pennsylvania, 1955.

Winkler, Allan M. "The Philadelphia Transit Strike of 1944." *Journal of American History* 59 (June 1972): 73–89.

Wolfinger, James. "'The American Dream—for All Americans': Race, Politics, and the Campaign to Desegregate Levittown." *Journal of Urban History* 38 (May 2012): 430–51.

———. "'An Equal Opportunity to Make a Living—and a Life': The FEPC and Postwar Black Politics." *Labor: Studies in Working-Class History of the Americas* 4 (Summer 2007): 65–94.

———. *Philadelphia Divided: Race and Politics in the City of Brotherly Love*. Chapel Hill: University of North Carolina Press, 2007.

———. "Philadelphia, PA, 1929–1941." In *Cities in American Political History*, edited by Richardson Dilworth, 436–42. Thousand Oaks, Calif.: CQ Press, 2011.

———. "Philadelphia, PA, 1941–1952." In *Cities in American Political History*, edited by Richardson Dilworth, 501–6. Thousand Oaks, Calif.: CQ Press, 2011.

———. "'We Are in the Front Lines in the Battle for Democracy': Carolyn Moore and Black Activism in World War II Philadelphia." *Pennsylvania History* 72 (January 2005): 1–23.

Woodruff, Clinton Rogers. "Philadelphia Street-Railway Franchises." *American Journal of Sociology* 7 (September 1901): 216–33.

Yearbook of the Amalgamated Association of Street and Electric Railway Employees of America. N.p.: n.p., 1910.

Young, Dina M. "The St. Louis Streetcar Strike of 1900: Pivotal Politics at the Century's Dawn." *Gateway Heritage* 12 (Summer 1991): 4–17.

Zieger, Robert H. *The CIO: 1935–1955*. Chapel Hill: University of North Carolina Press, 1995.

Index